Precision Medicine and Artificial Intelligence

Precision Medicine and Artificial Intelligence

The Perfect Fit for Autoimmunity

First Edition

Edited by

Michael Mahler
Inova Diagnostics, Inc. (a Werfen company),
San Diego, CA, United States

Academic Press is an imprint of Elsevier
125 London Wall, London EC2Y 5AS, United Kingdom
525 B Street, Suite 1650, San Diego, CA 92101, United States
50 Hampshire Street, 5th Floor, Cambridge, MA 02139, United States
The Boulevard, Langford Lane, Kidlington, Oxford OX5 1GB, United Kingdom

Library of Congress Cataloging-in-Publication Data
A catalog record for this book is available from the Library of Congress

British Library Cataloguing-in-Publication Data
A catalogue record for this book is available from the British Library

ISBN 978-0-12-820239-5

Publisher: Andre Gerhard Wolff
Acquisitions Editor: Linda Versteeg-Buschman
Editorial Project Manager: Pat Gonzalez
Production Project Manager: Swapna Srinivasan
Cover Designer: Miles Hitchen

Typeset by SPi Global, India

Contents

6. The evolving potential of precision medicine in the management of autoimmune liver disease

Gary L. Norman, Nicola Bizzaro, Danilo Villalta, Diego Vergani, Giorgina Mieli-Vergani, Gideon M. Hirschfield, and Michael Mahler

7. Precision medicine in autoimmune disease

Kevin D. Deane

8. Development of multi-omics approach in autoimmune diseases

May Y. Choi, Marvin J. Fritzler, and Michael Mahler

9. N-of-1 trials: Implications for clinical practice and personalized clinical trials

Joanne Bradbury and Michael Mahler

10. Health risk assessment and family history: Toward disease prevention

Lily W. Martin, Lauren C. Prisco, Laura Martinez-Prat, Michael Mahler, and Jeffrey A. Sparks

Contributors

Numbers in parenthesis indicate the pages on which the authors' contributions begin.

Roger Albesa (237), Inova Diagnostics, Inc., San Diego, CA, United States

Mary Ann Aure (65), Inova Diagnostics, Inc., San Diego, CA, United States

Carolina Auza (237), Inova Diagnostics, Inc., San Diego, CA, United States

Nicola Bizzaro (135), Laboratory of Clinical Pathology, San Antonio Hospital, Tolmezzo, Italy

Joanne Bradbury (203), Southern Cross University, Gold Coast, QLD, Australia

May Y. Choi (189), Cumming School of Medicine, University of Calgary, Calgary, AB, Canada

Kevin D. Deane (169), Division of Rheumatology, University of Colorado Anschutz Medical Campus, Aurora, CO, United States

Marvin J. Fritzler (39, 109, 189), Cumming School of Medicine, University of Calgary, Calgary, AB, Canada

Gideon M. Hirschfield (135), Division of Gastroenterology and Hepatology, Toronto Centre for Liver Disease, University of Toronto, Toronto, ON, Canada

Marie Hudson (109), Department of Medicine, Division of Rheumatology, McGill University, Jewish General Hospital, and Lady Davis Institute for Medical Research, Montreal, QC, Canada

Jan Trøst Jørgensen (97), Dx-Rx Institute, Fredensborg, Denmark

Olga Kubassova (1), Image Analysis Group, London, United Kingdom; Image Analysis Group, Philadelphia, PA, United States

Kim MacMartin-Moglia (267), Inova Diagnostics, Inc., San Diego, CA, United States

Michael Mahler (1, 39, 65, 109, 135, 189, 203, 215, 237, 267), Inova Diagnostics, Inc., San Diego, CA, United States

Lily W. Martin (215), Division of Rheumatology, Inflammation, and Immunity, Brigham and Women's Hospital, Boston, MA, United States

Laura Martinez-Prat (215), Inova Diagnostics, Inc., San Diego, CA, United States

Carlos Melus (1, 65, 237), Inova Diagnostics, Inc., San Diego, CA, United States

Giorgina Mieli-Vergani (135), Paediatric Liver, GI & Nutrition Centre, MowatLab, King's College London Faculty of Life Sciences & Medicine at King's College Hospital, London, United Kingdom

Gary L. Norman (135), Inova Diagnostics, Inc., San Diego, CA, United States

Lauren C. Prisco (215), Division of Rheumatology, Inflammation, and Immunity, Brigham and Women's Hospital, Boston, MA, United States

Brenden Rossin (65), Inova Diagnostics, Inc., San Diego, CA, United States

Minoru Satoh (109), Department of Clinical Nursing, University of Occupational and Environmental Health, Kitakyushu, Japan

Savino Sciascia (109), Department of Clinical and Biological Sciences, and SCDU Nephrology and Dialysis, Center for Research of Immunopathology and Rare Diseases-Coordinating Center of Piemonte and Valle d'Aosta Network for Rare Diseases, S. Giovanni Bosco Hospital, Turin, Italy

Faiq Shaikh (1), Image Analysis Group, London, United Kingdom; Image Analysis Group, Philadelphia, PA, United States

Jeffrey A. Sparks (215), Division of Rheumatology, Inflammation, and Immunity, Brigham and Women's Hospital; Inova Diagnostics, Inc., San Diego, CA; Harvard Medical School, Boston, MA, United States

Diego Vergani (135), Institute of Liver Studies, MowatLab, King's College London Faculty of Life Sciences & Medicine at King's College Hospital, London, United Kingdom

Danilo Villalta (135), Immunology and Allergy Unit, Santa Maria degli Angeli Hospital, Pordenone, Italy

Jungen Andrew Wu (237), Rook Quality Systems Taiwan Branch, New Taipei, Taiwan

Preface

Artificial intelligence (AI) and precision medicine are poised to disrupt the healthcare industry which is reflected by significant investments from large companies and by the genesis of various startups. Key drivers for such strong commitments are the increasing healthcare expenditures, the associated demand for value-based medicine, as well as the increasing prevalence of chronic conditions. Especially autoimmune conditions which are usually multifactorial and heterogeneous can benefit from a precision medicine approach. However, most advances in precision medicine occurred in oncology and were mainly based on genetic and molecular information.

This book entitled "**Precision Medicine and Artificial Intelligence: The perfect fit for autoimmunity**" was created to increase awareness and to facilitate collaborations mainly targeting the scientific community at the interface between medicine, biotechnology, diagnostics, health economics (e.g., insurance companies), and computer science (i.e., artificial intelligence). The balanced content covering the background on AI and autoimmunity, regulatory aspects of AI in healthcare, an overview of biomarkers (ranging from traditional or conventional to digital biomarkers and their companion diagnostic applications), data visualization tools, as well as specific examples, aims to provide valuable insights for a broad scientific audience. Since AI in medical devices is controlled by regulatory agencies such as the FDA, knowledge about the process for regulatory approval is important for every company embarking on AI as medical devices. In addition, the content around trends might generate ideas and concepts for business development and company strategy. Lastly, patient advocacy groups can gain insights on how to leverage new trends and help to educate patients at all education who will play a central role in PM. All chapters include references, especially for readers that desire to go into depth on certain topics related to precision medicine in autoimmunity.

Currently, there are very few books that cover the scientific side of precision medicine and at the same time discuss commercial opportunities and challenges. The existing books also do not elaborate much on the scalability of concepts as well as their regulatory aspects.

In summary, I truly hope the readers will benefit from the content to advance collaborations in precision medicine with the ultimate goal to improve patient care of autoimmune disease patients. Finally, I would like to thank all my colleagues and collaborators for fruitful discussions and especially all the authors for their excellent contributions.

Chapter 1

History, current status, and future directions of artificial intelligence

Olga Kubassova[a,b], Faiq Shaikh[a,b], Carlos Melus[c], and Michael Mahler[c]
[a]*Image Analysis Group, London, United Kingdom,* [b]*Image Analysis Group, Philadelphia, PA, United States,* [c]*Inova Diagnostics, Inc., San Diego, CA, United States*

Abbreviations

AI	artificial intelligence
ANN	artificial neural network
ASL	ambient assisted living
BCT	blockchain technology
CAD	computer-aided diagnosis
CAP	community acquired pneumonia
CDS	clinical decision support
COVID-19	coronavirus disease 2019
CT	computed tomography
DL	deep learning
DSRPAI	Dartmouth Summer Research Project on Artificial Intelligence
ECG	electrocardiography
EHR	electronic health record
EMR	electronic medical record
GPU	graphic processing unit
IBD	inflammatory bowel disease
IoT	Internet of things
ML	machine learning
MRI	magnetic resonance imaging
NAFLD	nonalcoholic fatty liver disease
NGS	next-generation sequencing
NLP	natural language processing
NN	neural networks
OCR	optical character recognition
OESS	OMERACT-EULAR Synovitis Scoring

Precision Medicine and Artificial Intelligence. https://doi.org/10.1016/B978-0-12-820239-5.00002-4

RA	rheumatoid arthritis
ROI	return of investment
RPM	remote patient monitoring
SDA	stacked denoising autoencoder
SLE	systemic lupus erythematosus
SVM	support vector machine
US	ultrasound

1 Introduction: History of artificial intelligence

Although it is broadly acknowledged that the genesis of artificial intelligence (AI) dates back many decades, it is more challenging to define the exact birth date of AI. However, there are several important events that triggered the evolution of what we today define as AI (Fig. 1). As with many other technologies, AI has developed in phases, through hypes and disappointments. The history of AI started with the inception of computers with "Electronic Numerical Integrator and Computer" (ENIAC) being one of the first which was presented in 1946. Just four years later, in "Computing Machinery and Intelligence," Alan Turing proposed a method to classify a machine as "intelligent." Many experts today believe that the Turing test is a very poor test, because it only looks at external behavior. Consequently, a number of scientists now plan to develop an updated version of the Turing test. In addition, the field of AI has become much broader than just the pursuit of true, humanlike intelligence.

The Dartmouth Conference of 1956, a summer workshop, is considered the inception of AI as a scientific field. John McCarthy, an Assistant Professor of Mathematics at Dartmouth College, in collaboration with Marvin Minsky, Nathaniel Rochester and Claude Shannon, started the project to clarify the emerging field of thinking machines, cybernetics and information analysis. This group introduced the concept of AI leading to the first boom in the 50s and 60s characterized by many AI prototype developments.

From 1957 to 1974, as computer hardware quickly evolved and new programming languages were created, AI flourished. Computers could store more information and became faster, cheaper, and more accessible. Machine Learning (ML) algorithms also improved, and people got better at knowing which algorithm to apply to solve a given problem. Several examples include

- Newell and Simon's General Problem Solver.
- Joseph Weizenbaum's ELIZA showed promise toward the goals of problem solving and the interpretation of spoken language, respectively.
- Dartmouth Summer Research Project on Artificial Intelligence (DSRPAI) convinced government agencies such as the Defense Advanced Research Projects Agency (DARPA) to fund several AI research initiatives.

One of the earliest commercial applications of AI was the use of Optical Character Recognition (OCR) to automate and speed up mail processing in the 1970s by the US Postal Service. However, after several reports criticizing the

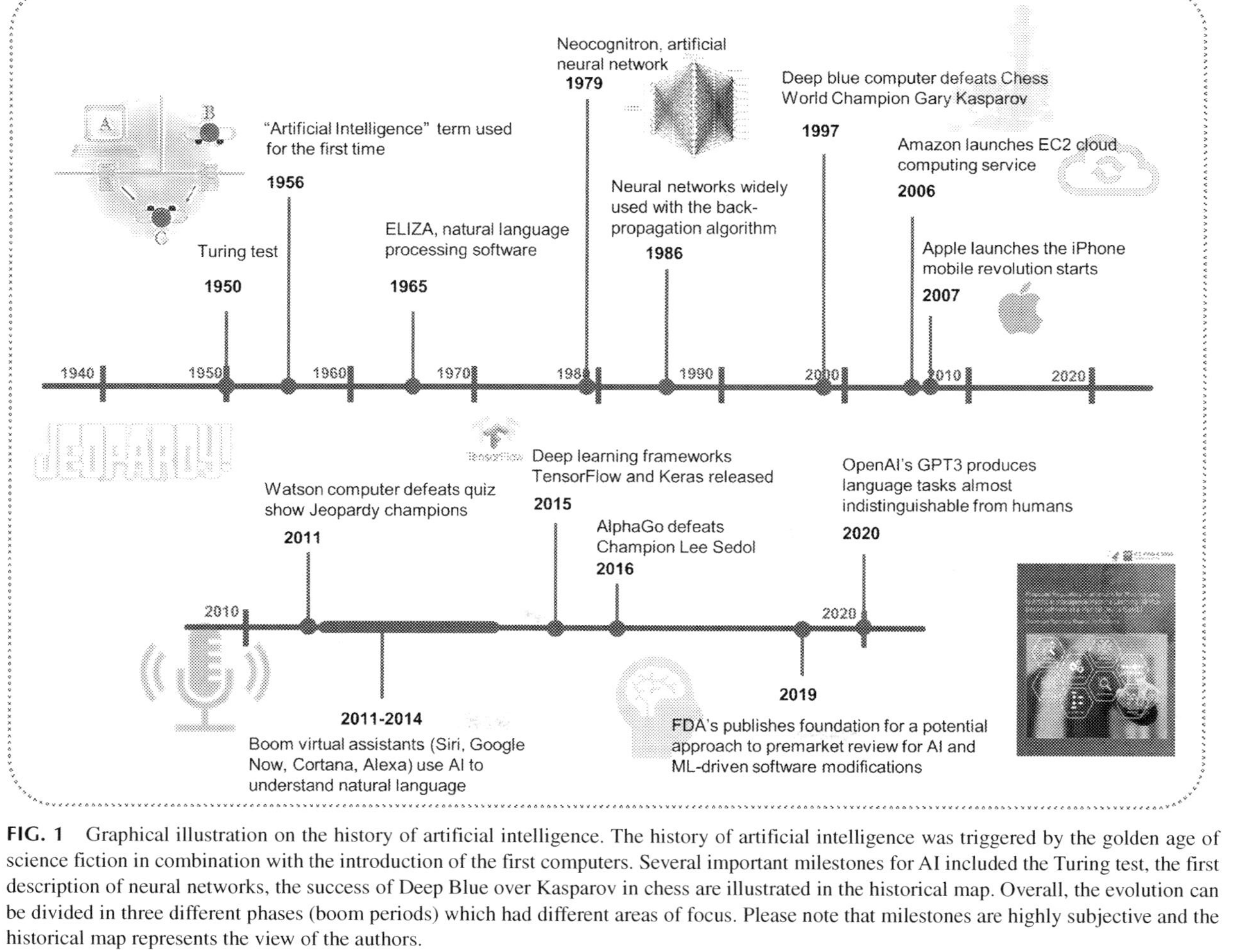

FIG. 1 Graphical illustration on the history of artificial intelligence. The history of artificial intelligence was triggered by the golden age of science fiction in combination with the introduction of the first computers. Several important milestones for AI included the Turing test, the first description of neural networks, the success of Deep Blue over Kasparov in chess are illustrated in the historical map. Overall, the evolution can be divided in three different phases (boom periods) which had different areas of focus. Please note that milestones are highly subjective and the historical map represents the view of the authors.

progress in AI, government funding and interest in the field dropped. A period from 1974 to 1980 is known as the "AI winter." The field reignited in the 1980s, when the British government started funding again, in part to compete with efforts by the Japanese leadership [Fifth Generation Computer Project (FGCP)]. In 1979, Kunihiko Fukushima introduced the concept of Neural Networks (NN) which paved the way for more advanced and modern ML approaches. This triggered the second AI boom in the 80s and 90s (age of knowledge representation) resulting in development of systems that were able to reproduce human decision-making. The last and still ongoing boom in AI started in 2010 and is heavily focused on Deep Learning (DL) approaches to solve data-based problems. The different technologies and disciplines that contribute to AI are illustrated in Figs. 2–4.

1.1 Artificial intelligence success in games and competitions

Deep Blue was able to defeat Gary Kasparov in chess in 1997. Later, in 2011, IBM's question-answering system Watson won the quiz show "Jeopardy!", beating reigning champions Brad Rutter and Ken Jennings [1]. In March 2016, Google's Alpha Go was able to defeat Chinese Go champion, Ke Jie. All these events are considered significant milestones in the history and evolution of AI.

1.2 Ever expanding use of artificial intelligence

As of today, AI became fruitful in several industries such as technology, banking, marketing, and entertainment (see Table 1). Although it is difficult to define

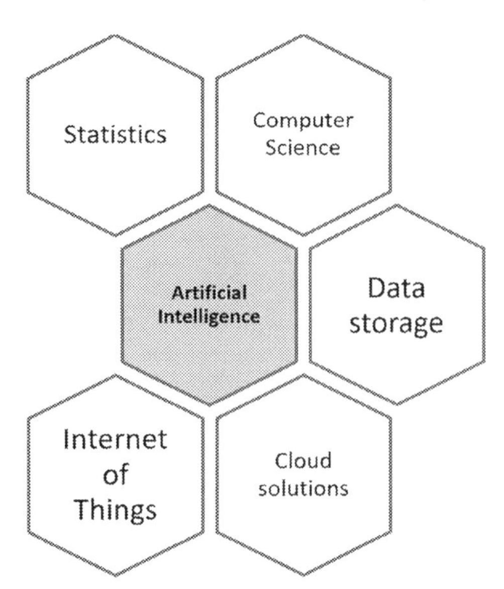

FIG. 2 The interplay of technologies and disciplines that influence and shape artificial intelligence (AI). AI requires significant interplay with various technologies and industries which are instrumental in enabling AI. Several of them have experienced significant progress during the past years. The figure illustrates some of the major technologies that are required for AI or are highly beneficial.

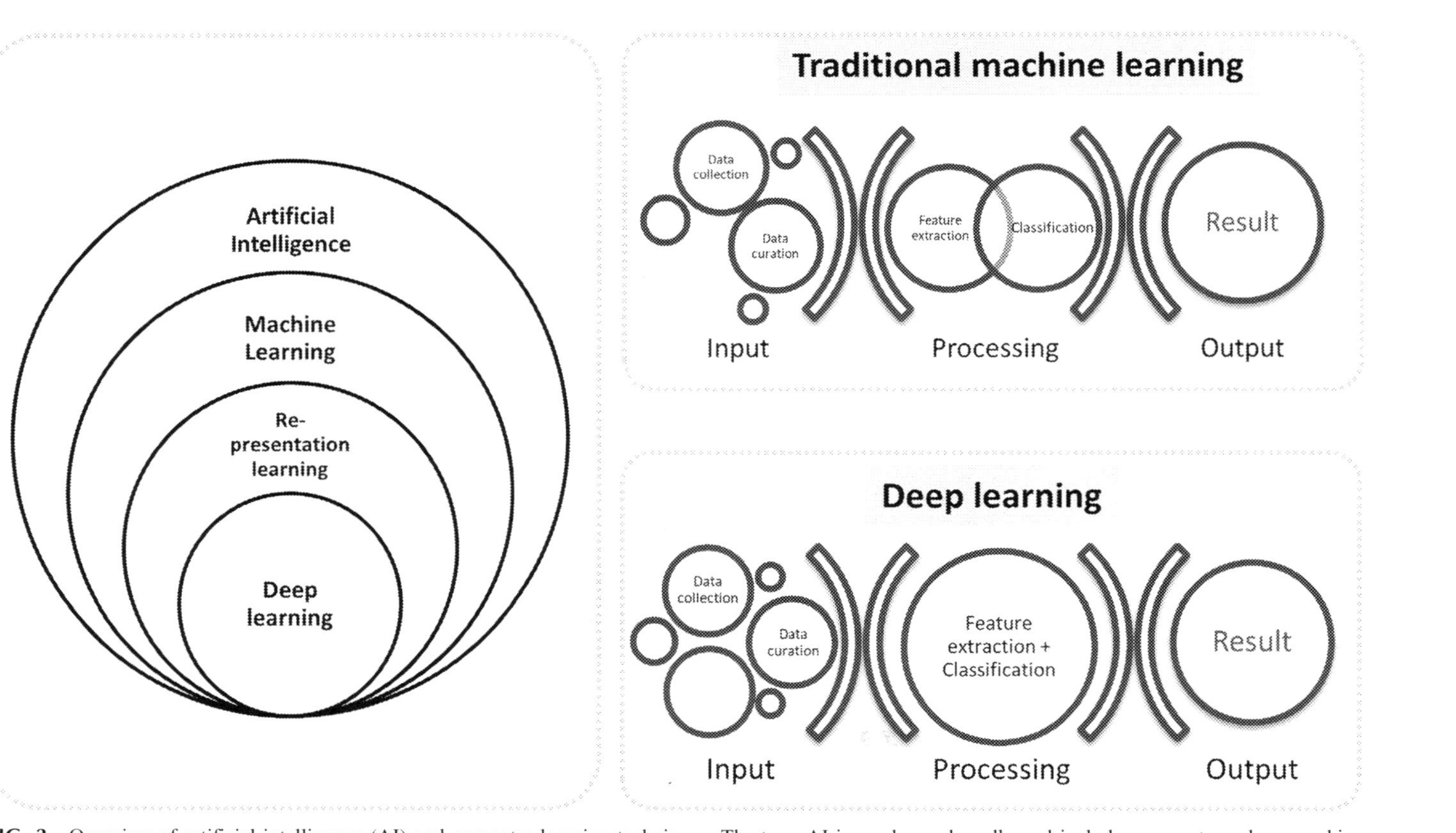

FIG. 3 Overview of artificial intelligence (AI) and computer learning techniques. The term AI is used very broadly and includes concepts such as machine learning, representation learning as well as deep learning. The right panels illustrate the difference between traditional ML and deep learning which differ mostly during the data processing phase.

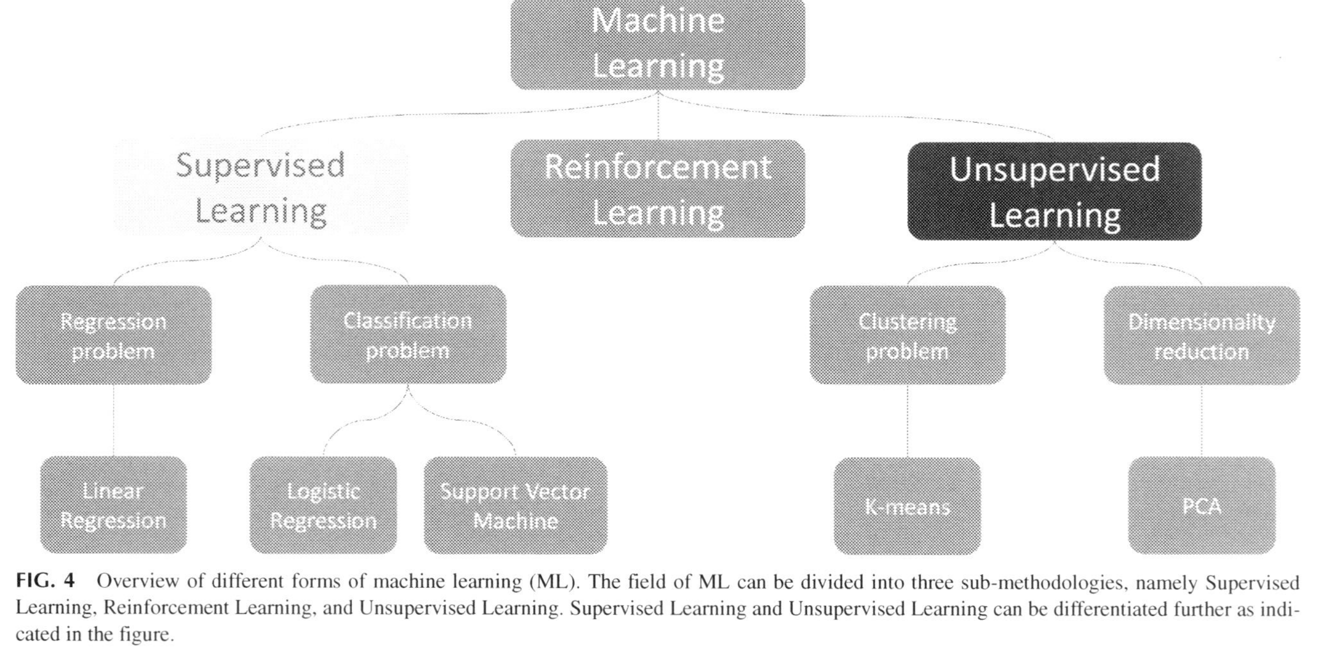

FIG. 4 Overview of different forms of machine learning (ML). The field of ML can be divided into three sub-methodologies, namely Supervised Learning, Reinforcement Learning, and Unsupervised Learning. Supervised Learning and Unsupervised Learning can be differentiated further as indicated in the figure.

the most significant successes, it is well acknowledged that AI revolutionized personalized marketing [2] and fraud detection [3, 4]. In the future, AI will be superior to human intelligence in areas where large and curated datasets are available and support from massive computing is available. Navigation systems, like Google Maps, have become extremely accurate thanks to AI [5]. These systems provide directions and useful real-time traffic details, analyzing information from millions of users to create powerful predictive capabilities to inform users ahead of time if their means of transportation is going to be disrupted. Facebook users might experience more and more custom-tailored advertisement in their profiles not only considering products of interest, but also timing of the posting which significantly increases the chance of a purchase. Those are only few of many examples of the expanding use and successes of AI (Tables 2–4).

1.3 Commercial aspects

The global AI and DL market is projected to grow significantly over the next years. Depending on the data source, the projected growth is as high as an annual compound growth rate of 46%. Consequently, many start-ups are being generated based on the vision to utilize AI and DL to evolve or disrupt different industries including computer vision, business intelligence, sales, customer relation management, voice interface, robotic/auto, security and last but not least healthcare. Examples of AI are summarized in Table 1.

TABLE 1 **Examples of artificial intelligence in today's life and potential future evolutions.**

Example	Industry
Investments	Finance
Customer service	Many
Fraud protection	Finance, healthcare
Voice recognition	Technology, communication
Driverless cars	Car, semi-trucks
Price prediction	Travel and transportation
Smart home devices	Technology, electronics
Online adds	Many
Face recognition	Technology, social media
Spam filters	Communication
Sales and business forecasting	Many
Remote analytics	Many

TABLE 2 Overview of artificial intelligence technologies.

	Advantages	Disadvantages	Additional comments
Logistic regression	Nice probabilistic interpretation; can be regularized to avoid overfitting; can be updated through stochastic gradient effect	Underperforms when there are multiple or non-linear decision boundaries; not flexible enough to naturally capture more complex relationships	Not a true AI method, but often considered as AI
Supervised learning	Predefined before giving the data for training allowing for very specific classification; only decision boundary (mathematical formula) needs to be retained instead of the training set; good for predicting a target numerical value from the dataset	Cannot handle complex tasks; cannot provide unknown information from the training set; training sets are large and need high computation time	
Unsupervised learning	Useful for large data set annotations; useful for classification and data mining	More subjective than supervised learning	
Reinforcement learning	Can solve very complex problems that cannot be solved by other techniques; can correct errors that occur during training; it can learn from experience without a training set	Overdoing it can diminish results; high computational need	It is much like a human learning a new skill. Robots are taught how to walk by Reinforcement learning
Deep learning	Performs well on image, audio, and text data; easily updated with new data using batch propagation; can be adapted to many types of problems; hidden layers architecture reduces the need for feature engineering	Require a very large amount of data; outperformed by tree ensembles for classical machine learning problems; they are computationally intensive to train; require much more expertise to tune	

TABLE 2 Overview of artificial intelligence technologies—cont'd

	Advantages	Disadvantages	Additional comments
Natural language processing	Natural language understanding, machine translation, sentiment analysis	Designed for one specific task only, can't adapt to new domains	
Expert systems	Provide solutions for repetitive processes and tasks Manage big amounts of information Centralize decision making process Reduce time to solve problems Reduce human error Combine knowledge from multiple human experts	Difficult to automate complex processes No flexibility to adapt to changing problems	
Computer vision and hearing	Speech recognition Handwriting recognition Optical character recognition Image and video recognition Facial recognition	Privacy Biased models (race, accent)	
Generation	Speech synthesis Natural language generation	Deepfake: ability to generate convincing material that can be used to deceive audiences	
Robotics	Cost effective Increased productivity Can work in any environment (even hazardous ones) Repeatability leading to increase quality	Initial investment Potential job loses Limited scope of work	

TABLE 3 Examples of artificial intelligence in health care by technology or process.

Example	Comment	References
Virtual assistants for staff	AI-based bots for telemedicine—triaging, screening patients	
Automated image diagnosis with AI/ML	Radiology assistance software, CAD/CDASS	[6–10]
AI-powered chatbots	Main application for telemedicine	[11, 12]
Robots for explaining lab results	Automated result interpretation	[13–15]
Robot-assisted surgery	In clinical practice (e.g., DaVinci systems)	[16–24]
Pathology	Pathology assistants, computer-aided diagnosis/clinical decision support systems (CDSS), pathomics	[25, 26]
Rare disease detection	Genomics applications, digital pathology, pathomics	[27]
Cybersecurity applications of AI in healthcare	Blockchain-based electronic health record (EHR), billing, other back-end business solutions	[28]
Medication management	CDSS within EHR (such as medicine reconciliation, and other clinical pharmacology support)	[29, 30]
Health monitoring with wearables	AI for pattern detection (fitbits, Apple watch, etc.)	[31–39]
Clinical trial design and participation	AI-based patient recruitment, go/no-go decisions	[40–44]
Fraud detection	Forensic accounting, malware, cybersecurity considerations	[45]
Workflow optimization	AI-augmented informatics solutions	[46]
Natural language processing	Clinical context generation, radiologic-pathologic correlation	[41, 47–52]
Electric health record	AI-based CDSS	[46, 53, 54]
Antibiotic resistance	AI used for target discovery, drug development	[55]
Phones, selfies etc.	Dermatology and ophthalmology	[31, 56–64]
Decision support	CDSS as above	[65–69]

TABLE 4 Examples of artificial intelligence in health care by disease area.

Example	Comments	References
Dermatology	Teleconsultation Skin cancer screening Diagnose skin lesions, particularly melanoma	[70–73]
Gastroenterology/ hepatology	Liver biopsy specimens for histologic features of NAFLD Endoscopic detection and differentiation of esophageal lesions Risk scoring systems for upper gastrointestinal bleeding	[74–81]
Neurology	Management of Parkinson's disease	[37, 41, 47, 82–85]
Rheumatology	Image analysis of joints	[36, 61, 86–89]
Ophthalmology	Detection of cataracts	[8, 19, 90–98]
Oncology	Early detection of breast cancer	[80, 99–109]
Infectious disease	Antibiotic resistance management Chest CT AI/DL model to detect COVID-19 related lung disease	[55, 110–114]
Intensive care	Sepsis management in urgent care	[21, 67, 115–118]
Dentistry	Computer vision algorithms to identify oral anatomy and diseases from dental X-rays and other patient records	[119]

AI, artificial intelligence; *CT*, computed tomography; *DL*, deep learning; *NAFLD*, nonalcoholic fatty liver disease.

1.4 Setting and managing proper expectations

It is well-known and as a matter of fact accepted that humans make mistakes. In contrast, it is expected that AI-based systems and computers have to be perfect which is a miss-perception and will slow down progress in AI. The most important aspect for AI application is to differentiate between random and systematic mistakes [120–124]. Humans tend to make random mistakes which contrasts with AI-based systems that can be prone to systemic mistakes if not trained and validated correctly. Setting the appropriate expectations with end-users is key for a smooth adoption of systems using these techniques.

1.5 Ethic and privacy consideration

Handling large amounts of patient data is a prerequisite for AI and ML which poses a significant ethic concern mostly regarding security of data [31, 86, 125–140].

Although this is not restricted to healthcare applications, it becomes a very sensitive and important topic when managing or utilizing patient data. The Health Insurance Portability and Accountability Act of 1996 (HIPAA) required the Secretary of the US Department of Health and Human Services (HHS) to develop regulations protecting the privacy and security of certain health information. Following the HIPAA regulations is important but also provides additional obstacles toward large datasets and AI application. Future solutions that both protect sensitive information and provides access to data will be instrumental for the success in this field.

2 Enabling technologies

2.1 Computing power

Computer hardware has evolved quickly in the past decades, both increasing the performance and energy efficiency, while decreasing the size and lowering costs. This improvement is partially explained by the ability to increase the number of transistors in an integrated circuit, observation known as Moore's law, but also, by the technical limitations to increase the number of transistors. Those limitations pushed engineers in different directions, for example developing multi-core processing units with the ability to run several program threads simultaneously, reducing the need to keep producing smaller, faster chips, that generate increased heat.

One of the key hardware developments occurred in Graphic Processing Units (GPUs). Starting in the late 1990s, 3D-capable graphics cards were initially used for gaming. These devices were capable of processing millions of polygons per second to render realistic images onto computer monitors. In the mid-2000s, general purpose GPUs began production including dedicated programming models for GPU computing. The increased performance per watt and software frameworks enabled dedicating these devices to fields other than graphics, like scientific image processing, linear algebra, 3D reconstruction and, of course, ML. Most of the AI/ML development frameworks available today can run on GPUs to speed up computing times. This is especially critical when training large models with large datasets.

2.2 Sensors

The development of mobile technology used in cell phones and tablets has brought a great number of signal capturing devices to everyone's pockets:

- Camera
- Global Positioning System (GPS)
- Proximity sensor
- Compass
- Accelerometer

- Gyroscope
- Fingerprint scanner
- Ambient light sensor
- Microphone
- Touch screen sensors
- Pedometer
- Barometer
- Heart rate sensor
- Thermometer
- Etc.

Those sensors can be used to capture characteristics of the user environment and behavior to provide very rich data input to algorithms running on mobile devices. Many of them having potential utility for health care.

2.3 Electronic health record systems

The adoption of electronic health record (EHR) systems in hospitals and laboratories has been driven by economics of health care [141]. Some beneficial side effects for medical professionals are increased legibility of the medical record and easier access to information while reducing errors. Having patient information available also helps researchers to identify subjects and track quality of care. The EHR deployment process involves interfacing with other information devices (instruments and other data recording systems), which has pushed industry into developing communications protocols and standards to make those interactions and deployments less cumbersome. The availability of data together with standard communication protocols have enabled data analysis and development of ML algorithms [142].

2.4 Blockchain technologies

Blockchain technology (BCT) is, in the simplest of terms, a time-stamped series of immutable records of data that is managed by a cluster of computers not owned by any single entity [143–146]. Each of these blocks of data (i.e., block) is secured and bound to each other using cryptographic principles (i.e., chain). Since it is a shared and immutable ledger, the contained information is by its very nature transparent and everyone involved is accountable for their actions. BCT allows patients to assign access rules for their medical data, for example, permitting specific researchers to access parts of their data for a fixed period of time. With BCT, patients can connect to other hospitals and collect their medical data automatically. To further described the concept of BCT, a blockchain is a distributed, public ledger, recording transaction and tracking assets, and of which immutability is guaranteed by a peer-to-peer network of computers, not by any centralized authority. Assets can be tangible, such as homes or cash, or they can be intangible, such as patents or copyrights. A blockchain consists of ordered records arranged in a block structure. Each data block contains a hash (digital fingerprint or unique identifier), timestamped batches of recent transactions, and a hash of the previous block. With this design, each block is connected in chronological order and the connected blocks are called a BC. It is practically impossible to modify one of the blocks in the middle of the chain because all of the blocks after the modified block must be modified at the same time.

A smart contract is a computer protocol that runs automatically when the prerequisites are met, and it is an entity separate from the original BCT. Many aspects of BCT, such as the immutability of the data stored in a blockchain, are drawing the attention of the healthcare sector, and rosy prospects for many available cases are being discussed. BCT is expected to improve EHR management and the insurance claim process, accelerate clinical and biomedical research, and advance biomedical and healthcare data ledger [143–150]. These expectations are based on the key aspects of BCT, such as decentralized management, immutable audit trail, data provenance, robustness, and improved security and privacy. BCT can facilitate the transition from institution-driven interoperability to patient-centered interoperability [3]. BCT allows patients to assign access rules for their medical data. With BCT, patients can connect to other hospitals and collect their medical data automatically.

2.5 Internet of things

Another field that is relevant for AI in the context of data collection is the Internet of things (IoT) [32, 47, 56, 143, 151–162]; namely physical devices embedded with internet connectivity that can communicate and interact with humans and other devices. These devices, usually in the form of wearables or smart home accessories [31, 40, 57–60, 70, 156, 163], are becoming key to collecting big data from end-users to enable the creation of new AI models or improving existing ones. The confluence of sensors, real-time analytics and AI is enabling applications like elder care and health monitoring and emergency notification systems [162–165]. While the IoT field is not free of criticism, mostly around privacy and security, the number of devices was expected to reach over 50 billion by 2020 (https://www.mesh-net.co.uk/what-is-the-internet-of-things-iot/), 35 times the number of websites on the world wide web in 2019 (https://news.netcraft.com/archives/2019/07/26/july-2019-web-server-survey.html).

2.6 Frameworks and libraries

Currently the field of AI is flooded with frameworks, libraries and tools to create new AI-based tools. Multiple companies are trying to set the standards for the field through their tools, like Google with TensorFlow [166–170], and Facebook with PyTorch [82, 171–173] and Prophet. In the future, many of these libraries will likely disappear or merge and the crowded space will be dominated by fewer and more standard options.

The programming language landscape has been shaped by the recent AI surge, with programming languages like Python getting increased attention thanks to many of the new libraries mentioned above running on Python, enabling fast research and prototyping cycles.

3 Introduction of artificial intelligence in life science and healthcare

One of the newer and more exciting developments in the field of AI has been its application in the field of medicine [1, 6, 16, 17, 28, 31, 33, 41, 45, 48, 55, 57–59, 61, 65, 66, 71, 72, 90–92, 99, 121, 123, 133, 134, 152–154, 174–244], with the ultimate goal to improve patient care and outcomes as well as to reduce costs while increasing quality of healthcare delivery. Some of the most fertile areas for the growth of AI in medicine have been radiology [7, 28, 100, 189, 245–254], pathology [25, 74, 157, 255], EMR [46, 53, 54, 238, 256], and more recently, specific applications in clinical management, such as the clinical decision support systems (CDSS) [55, 66, 67, 257].

The advent of AI in healthcare may be traced back to the research done in the 1960s and 1970s that led to the first problem-solving program, known as Dendral (Dendritic Algorithm), an organic chemistry application which paved the way for MYCIN [258] which is considered one of the most significant early uses of AI in medicine. Some AI-powered technologies [259–261] were developed but failed to find clinical acceptance [262]. The evolution of AI has taken a unique path in each of its fields of study: clinical medicine, clinical imaging [31, 56, 93, 100, 251, 253, 263–266], pathology/laboratory medicine [25, 132, 267, 268], healthcare management, etc. In each of these areas, AI has found major areas of application and continues to evolve into an indelible mark in the practice and management of medicine.

In the field of clinical medicine, ML has been routinely applied to capture clinical data, which is getting increasingly complex, to generate potentially new insights to guide clinicians. This application of AI/ML in clinical decision making is coined as Clinical Decision Support (CDS) [68] and is being increasingly applied in all important areas of clinical management. But another important application of AI is in automated/semi-automated monitoring of clinical metrics. One example of the latter is seen with electrocardiography (ECG), whereby ML-driven applications can extract useful information which could be missed by the clinician who cannot continuously observe all ongoing ECG patterns. The spectrum and the sophistication of such AI tools in clinical medicine is increasing by leaps and bounds. Having said that, many of these applications are not quite ready for prime time yet. There are gaps in their statistical validity and more work is needed to improve their accuracy, reproducibility, and predictive power to provide the foundation for regulatory pathways (see also Chapter 11).

There have been several advances in the realm of AI-based clinical management, such as detecting and treating new-onset atrial fibrillation [18, 31, 34, 161], but there has been some concern of having very minimal impact on clinical outcomes and how such an intervention could become yet another example of "overuse" of healthcare resources [18]. As AI tools continue to get introduced in other areas of clinical management, it is very important to monitor overdiagnosis as well as the increased sensitivity of detection methods [18].

3.1 Three-tiered application approach of artificial intelligence in healthcare

In general, AI has impacted healthcare through overhauling three main areas:

(1) Business intelligence: Automation and efficiency of business processes.
(2) Clinical decision support: Utilization of large volumes of data for nuanced clinical guidance.
(3) AI-based data analytics.

3.1.1 Business intelligence in healthcare

This type of AI relates to automation of what we can broadly call "back-office" function and it is deployed through more efficient handling of digital and physical tasks using technology. Such tasks may include but are not limited to

- Transferring data from a site to a storage or transferring the data from a call or email into a record storing system.
- Reconciling failures in systems or data by combining and automatically checking information in multiple systems and multiple document types.
- Reading clinical reports through natural language processing (NLP) techniques to extract provisions, data points, and ultimately making conclusions.

This AI application is the simplest and least complex to implement. Without modern trail infrastructure, in the clinical research industry, data transfer, data quality control, and data management processes may feel somewhat archaic. Traditionally, they would inherit a lot of inefficiencies founded in utilization of manual labor, outdated, disconnected technologies and hardware. The efficiencies and quality of the outcomes is being improved with the use of cloud enterprise solutions that allow companies to create an efficient infrastructure between the data suppliers (sites or hospitals), data users (biotechnology or pharmaceutical companies) and data analytics function of the business. AI-powered infrastructure allows oversight of the performance of each data supplier, ensures full transparency of the overall delivery, assists each stakeholder in communicating their needs and requests but most importantly allows biotechnology or pharmaceutical companies to make fast decisions based on real data.

3.1.2 AI-based clinical decision support

AI-based CDS is often feared by specialists and referred to as better decision making or replacing a human in the decision-making process [22,26,72,122,155,218,269]. Strictly speaking, this represents a misperception [16,18]. The AI-driven methodologies in clinical research and practice are generally designed to support expert decisions and not to replace them. We can broadly separate AI methods into semi- or fully-automated. An example of fully automated AI-driven method would be if an MRI scan gets processed by a software leading to a medical decision (e.g., diagnosis of a medical condition). However, as of today, most of the methods are semi-automated and will require either the initial input or the final decision made by an expert, with the AI algorithm handling everything in between in a fully automated manner [6,8,10,270,271].

Clinical decisions made with the use of AI will be made faster, be potentially more accurate, precise and objective. Use of AI also allows to look deeper into the data, extracting valuable information or pattern which might not be noticed by a human eye [27,43,94,272,273].

At the very beginning of clinical research, all data were analyzed by experts, an approach that is currently referred to as the "state-of-the-art." The industry has moved to bringing automation in pathology, assay analysis and of course medical image interpretation, which perhaps is the most apparent example of them all. It is important to note that an expert and the machine will most likely make the same decision (at least in most situations). This implies that we are speaking about the actual expert and a well-designed robust algorithm.

This type of AI is easy to implement alongside with more traditional state-of-the-art methods. In drug development, the use of AI or automated analysis in decision making in late phase clinical trials as a methodology for assessment of efficacy of novel treatments will require consensus of regulatory bodies. In many therapeutic areas, including serious conditions such as osteoarthritis, glioblastoma, and many rare diseases, use of quantitative methodologies is paramount to the success of novel drug development; several academic and commercial groups are working toward making this possible. During early phases of clinical trials, when go/no-go decisions are critical AI can bring high return of investment (ROI) by providing robust rich data and thus ensuring less failures during later development stages (which are most costly).

3.1.3 Artificial intelligence and insight into the data/focus on business decisions

This type of AI is used to detect patterns in large volumes of data and to interpret their meaning using a blend of ML techniques and true AI. Examples of such applications include the prediction of a particular cancer type based on gene combinations, automation of personalized therapy selection, choosing the right dosing based on the treated patient statistics. In related markets, this type of AI will inform insurance companies with more accurate and detailed actuarial modeling, provide drug discovering organizations with the most likely combinations to succeed or fail, intelligently influence the design of next clinical trials based on the competitive intelligence, and utilize endpoints and patient populations.

An example of big data analytics could encompass the assessment of the www.clinicaltrials.gov database and all relevant publications by a machine for information on a particular treatment to provide insights into which sub-populations of patients would most likely benefit.

AI-driven insights are very different from the traditional analytics as they are much more data intensive and detailed and they improve as more data becomes available.

Such AI is typically used to improve performance or decisions only a machine can do.

Other examples of such AI type include

- Intelligent agents that offer 24/7 customer service or technical support in the referral of customers to the right specialist.
- Treatment recommendation systems that help providers create customized care plans considering individual patient history, previous treatments and patient analytics.
- A biotechnology or pharmaceutical company combining all their early stage data to determine the next indication for which a drug is most likely to succeed.

The implementation of AI in this domain is the most expensive as it would require the first two types to be in place for added efficiencies. This is not a low hanging fruit and will require consensus and readiness to change from all levels of the organization. It is the author's personal view that this concept is very actual and critical for smaller biotechnology companies with platform technologies who are in early stages of development or just got their first success. Capitalizing on this success with the use of the best technologies will yield long term benefits and sustainable growth.

We have come a long way since the advent of AI applications in medicine, and there is tremendous potential of more significant use cases. However, there are some key challenges that hinder progress, such as the availability of rich and refined data especially for the most pressing clinical questions, which limits our ability to develop effective and robust DL models. Moreover, the available data in healthcare tends to be highly heterogeneous, ambiguous, noisy and incomplete, which severely limits proper training of ML models. Another issue is that of temporality, which means that the diseases are always progressing and changing over time in a nondeterministic way, which are incompatible with ML models designed with static vector-based inputs, which is why the future DL models will need to be designed to handle temporal health care for real time applications in medicine. Other challenges including interpretability and domain complexity also abound.

As a summary of this section, AI can provide a powerful mechanism of improving business performance and decision power. However, before embarking on any AI initiative, the user (company/institution) must understand which technologies to leverage, what type of tasks to improve and the strength and limitations of each method (Fig. 5). In order not to waste valuable time and resources, it is important to set the right expectations from the very beginning. For instance, rule-based systems can make decisions but are not going to learn; ML driven analysis can outperform a human when given the right data but might fail if the variation is too significant. One of the fundamental obstacles of AI products is the "black box" concept, meaning that outputs from a model are difficult to explain. When using black box AI, extensive validation and certification of results quality should be provided, in order to produce trustworthy models and to drive adoption of those systems. Institutions and companies without data science AI/ML capabilities in house, might seek to build an ecosystem to enable seamless data flow to ensure success of the entire operation.

3.2 Medical areas

In the following section, AI applications are reviewed and discussed based on the disease state or the medical area. Although the section does not contain a comprehensive overview, the section will provide some key areas and examples.

3.2.1 Electronic health record

In the domain of healthcare informatics, AI/ML has made a big impact on EHR's [122]. More specificaly, ML has been applied to process aggregated EHRs, with both structured and unstructured data entering from various clinical silos. Traditionally, only supervised ML approaches have been used for EHRs, but recently unsupervised DL approaches have shown better results vs. conventional supervised ML models with respect to certain metrics, such as Area Under the Receiver Operating Characteristic Curve, accuracy and F-score (see also Chapter 3). One specific example of this approach is DeepCare, an end-to-end deep dynamic network that infers current illness states and predicts future medical outcomes [225]. More recently, a three-layer Stacked Denoising Autoencoder (SDA) has been used for disease risk prediction models using random forest as classifiers, which significantly improved the accuracy of results [225]. Natural language deep models were also applied to EHRs for analysis of medical conditions and development of prediction models [225]. One major bottleneck of using AI-driven programs on mobile devices, which are now omnipresent, is the compute/processing power traditionally required to run the models. A lot of effort has been invested to enable AI on mobile devices. DeepX is a new software accelerator capable of lowering the device resources required by DL, which can propel the mobile adoption of these programs [225]. Google's TensorFlow Lite is TensorFlow's lightweight solution for mobile and embedded

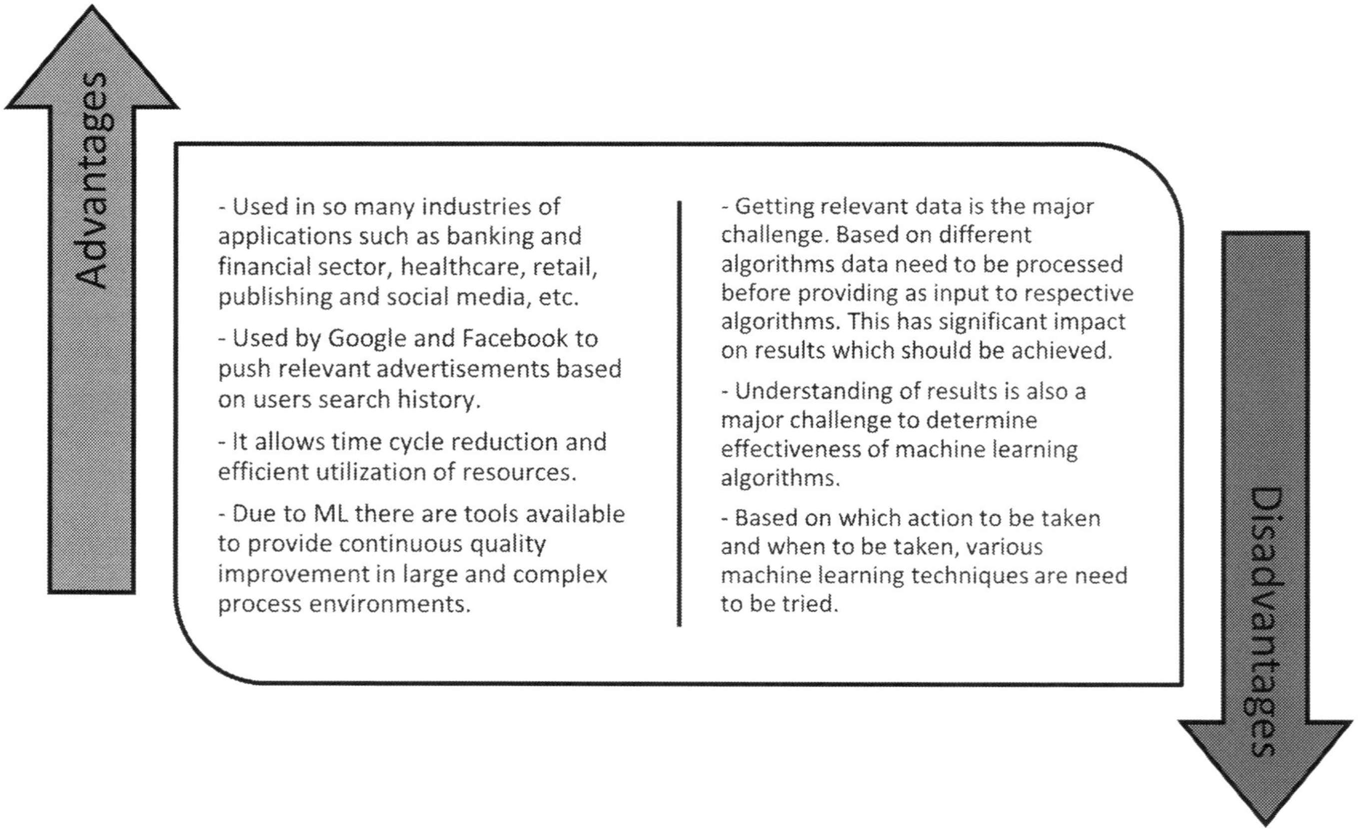

Advantages

- Used in so many industries of applications such as banking and financial sector, healthcare, retail, publishing and social media, etc.

- Used by Google and Facebook to push relevant advertisements based on users search history.

- It allows time cycle reduction and efficient utilization of resources.

- Due to ML there are tools available to provide continuous quality improvement in large and complex process environments.

Disadvantages

- Getting relevant data is the major challenge. Based on different algorithms data need to be processed before providing as input to respective algorithms. This has significant impact on results which should be achieved.

- Understanding of results is also a major challenge to determine effectiveness of machine learning algorithms.

- Based on which action to be taken and when to be taken, various machine learning techniques are need to be tried.

FIG. 5 Advantages and disadvantages of artificial intelligence (AI) and machine learning (ML). Although there are many advantages of AI and ML, there are also disadvantages or challenges that need to be considered before embarking on AI projects.

devices. It enables low-latency inference of on-device ML models with a small binary size and fast performance supporting hardware acceleration [199].

3.2.2 TeleHealth and remote patient monitoring

TeleHealth and remote patient monitoring (RPM) leveraging AI technologies are increasingly used in certain jurisdictions and have to potential to improve healthcare logistics [11, 35, 87, 243, 256, 274–279]. While telehealth/telemedicine has existed for decades, there were several challenges that stifled its growth. These varied from technological challenges to patient acceptance and physician behavior change. However, AI by way of automated bots that can triage or escort these patients and take brief history and create clinical synopsis, and providing guidance through CDSS, has facilitated the real-world utilization of this technology. This has happened at a very opportune time as telemedicine has become almost essential in the wake of COVID-19 pandemic [280–282]. Telemedicine as enabled with nuanced AI tools, will become an essential tool for preventative and primary care.

3.2.3 Radiology

Within the realm of radiology, AI/ML has made significant strides [6, 17, 28, 100, 146, 213, 247–249, 251–253, 265, 266, 283, 284]. There have been significant contributions by way of statistical analysis algorithms for autonomous predictions, in which the computer program performance improves automatically with experience. Advanced analytic methodologies, such as Radiomics [247, 285] allow extraction of features from images and applying a large knowledge base to interpret those features using ML. This approach is rapidly changing the scope of medical imaging in terms of how it can detect, classify, and prognosticate diseases based on novel imaging biomarkers [6, 31, 254, 263, 265, 266, 285, 286]. Furthermore, it allows multi-specialty (multi-omics) correlative analytics utilizing large data sets [86, 88, 101, 115] previously siloed in disparate domains, such as radiology, pathology and genetics [40, 287, 288].

There are several challenges preventing full application of AI/ML in radiology at its true potential [253, 289], which is why successful radiology ML projects have thus far focused on tightly defined and narrow scopes governed by the availability of the training datasets. In many situations, the most pressing questions in clinical diagnostics that could hugely benefit from AI application are not met with the availability of the training dataset needed to develop such AI models. However, there has been development of numerous techniques for segmentation and analysis of organs and tissues, lesion identification, computer-aided diagnosis (CAD) and CDS blackboxes, including traditional CAD and ML approaches. The most prominent success stories of AI and DL applications in radiology include early detection of breast cancer with mammography and ultrasound [102, 103, 213, 290], classification of chest radiographs, and differentiating malignant from benign nodules on chest CT images [246].

Over the past several years, there have been many efforts to apply DL methods to healthcare as an industry. Google DeepMind and other biotech enterprises, such as Enlitic, are using DL to detect lesions on medical imaging modalities, such as X-rays and CT scans [225]. ML has been leveraged in clinical medicine to derive predictive models without a need for strong assumptions about the underlying mechanisms, especially when the real-world data is lacking [225].

As the technology involving AI/ML became more sophisticated, the possibilities for its application in healthcare increased multifold. The traditional Artificial Neural Networks (ANNs) have been usually limited to three layers and trained to obtain supervised representations optimized only for the specific task. However, DL involves each layer to produce a representation of the observed patterns based on the received input from the previous layer, in an unsupervised fashion [155, 218]. This allows DL to discover intricate structures in high-dimensional data, which is especially the case in clinical medicine, such as the use of DL in the detection of diabetic retinopathy in retinal fundus photographs, melanoma classification and predicting of the sequence specificities of DNA- and RNA-binding proteins. DL has been leveraged in the field of medical imaging, such as with MRI scans to predict Alzheimer disease and its variations, to infer a hierarchical representation of low-field knee MRI scans to automatically segment cartilage and predict the risk of osteoarthritis [225].

3.2.4 Pathology

AI has played a major role in the field of digital pathology [75, 92]. This includes automated pattern recognition and laboratory data analytics. More recently, more ambitious projects leveraging accomplished AI tools in pathology (e.g., IBM's Watson Genomics [1, 41, 49, 289, 291], which is an assistive AI to improve next-generation sequencing (NGS) interpretations), have been initiated [292]. With more recent advances in AI technology, an emerging question is whether AI will fully replicate and thus replace the role of a pathologist [292]. For now, the answer is negative, because human cognition along with its biases and heuristics is still superior to AI-driven interpretations, but the gap may be narrowing fast. For now, the philosophy behind AI-based applications in pathology, and in medicine overall, should be complimenting human knowledge and productivity instead of competing with it. There is a tremendous potential in this synergy, in the form of AI-assistants, quantitative analyses, or CDSS, or predictive models, and the field is making strides in realizing that potential (*Source*: (1) https://www.cancer.gov/publications/dictionaries/cancer-terms/def/solid-tumor. (2) https://www.kinderkrebsinfo.de/patients/asking_the_doctor/what_is_a_solid_tumour/index_eng.html. (3) https://onlinelibrary.wiley.com/doi/full/10.1002/cac2.120124. https://arxiv.org/ftp/arxiv/papers/1905/1905.06871.pdf).

3.2.5 Infectious diseases

AI has also shown some promise in the field of infectious diseases [54, 55, 178, 227, 293] including hepatitis C (HCV) and human immunodeficiency virus (HIV) virus infections [227] as well as tuberculosis [293]. In addition, recent focus turned on the outbreak of the corona virus infectious disease 19 (COVID-19) [110–113] for which an AI framework based on mobile phone-based survey has been proposed [111]. Due to the potential of severe lung disease in patients with COVID-19, it is of high importance to not only diagnose COVID-19, but also to distinguish community acquired pneumonia (CAP) from COVID-19 lung disease. Li et al. [114] recently reported an AI/DL based model using chest CT. The sensitivity and specificity for detecting COVID-19 in the independent test set was 90% [95% Confidence interval CI: 83%, 94%] and 96% [95% CI: 93%, 98%], respectively, with an AUC of 0.96 (P-value < 0.001). The per-exam sensitivity and specificity for detecting CAP in the independent test set was 87% and 92%, respectively, with an AUC of 0.95 (95% CI: 0.93, 0.97). Last but not least, the large sets of data collected on COVID-19 are now being submitted to various AI-based concepts [113]. These examples illustrate how fast AI-based systems can be developed, but also show that regulatory approval and clinical application often become the critical path from implementation standpoint.

3.2.6 Autoimmunity

Although autoimmune diseases are an excellent target for AI application [242], the progress in those areas is less advanced. Nevertheless, a few early examples have been developed and implemented [36, 61, 86, 88, 89, 242, 294], some of them are discussed below.

There is ample potential for applications of AI and ML in autoimmunity, ranging from data analyses and the generation of diagnosis supporting algorithms to image classification. The multifactorial/variate nature of autoimmune disease testing [200], usually including several biomarkers in the diagnostic process or disease activity scores, makes autoimmunity an ideal candidate to apply ML techniques and to develop algorithms that can help laboratorians, physicians and specialists with disease diagnosis, prognosis and treatment. Big data is the key to a future where algorithms can help with early disease diagnosis, improving patient prognosis and saving healthcare costs.

One example of an algorithm developed in the autoimmunity space is IBD sgi Diagnostic® (Prometheus, San Diego, United States), that combines serologic and genetic testing into a diagnostic score for inflammatory bowel disease (IBD) [295]. Another example is AVISE CTD (Exagen, Vista, United States). AVISE CTD is an autoimmune rheumatic disease test specifically designed to aid physicians in the differential diagnosis of systemic lupus erythematosus (SLE), combining biomarkers to produce results with increased sensitivity compared to single biomarker testing [296–300]. The Helios automated indirect immunofluorescence (IIF) system by Aesku (Wendelsheim, Germany) is able to recognize the pattern of images from slides by using Support Vector Machine (SVM) technique. All suggested results obtained with the software must be confirmed by a trained user. At present, AI-based systems for IIF are not superior to systems using more traditional algorithms (human rule based interpretations; e.g. NOVA View). However, this will likely change with next generation system that extract more and more data from images. In the image analysis field, NN have been evaluated in the scoring of disease activity on Doppler ultrasound (US) images according to the OMERACT-EULAR Synovitis Scoring (OESS) system. The US scanning and evaluation of synovitis activity by the OESS system is used in the diagnosis and monitoring of patients with inflammatory arthritis (IA). Recent studies confirmed that NN technology can be used to score disease activity on Doppler US images according to the OESS system, with very good agreement with human experts [301]. Recent studies on the role of AI in assessment of peripheral joint MRI in IA [302] indicate that AI-based methods demonstrate feasibility, reliability, and validity when utilized to quantify MRI pathologies of peripheral joints in patients with IA. With the results from 28 studies, consisting of 1342 MRIs, the authors showed that computer-aided evaluation of IA on MRI could be considered as an alternative to conventional observer-based methods.

Because AI and big data in autoimmunity are relatively immature compared to other well established technologies, there is a wide spread of analytical methods used, with no consensus on what direction the field will evolve, and making tasks like data sharing, reporting and benchmarking challenging [88].

Hardware technology has become another powerful ally in developing AI and big data solutions in the autoimmunity field. Combining sensors and applications in smartphones might provide additional information in predicting disease flares [89] or disease onset in individuals at risk. Most current apps do not provide the expected user experience for autoimmune patients, and there is ample opportunity for optimization, from security and privacy to accessibility and usefulness. One example of these mobile apps developed to help patients with RA is MyRa, by Myriad Genetics, Inc. (Salt Lake City, Utah). Patients living with RA can track their condition, create snapshots of your data, and communicate about your RA. The app can record how the patient is feeling every day, including detailed joint pain severity and location, measure how the disease affects patient's activities using a rating scale and providing reports how joint pain, morning stiffness, fatigue, and daily functionality, compare and contrast over time. The app produces a summary report to help patients when they visit their physician. Using this report, patients can effectively convey how the disease has evolved since the last doctor's appointment. Medical experts should be active partners in the development of these data-based technologies, in order to incorporate their vast experience and understanding of the disease, of the patient journey and the healthcare system into new apps and systems [242]. Analytical methods and patient data must be approached considering ethical issues such as privacy and confidentiality and informed consent [86]. Ultimately, AI-based systems will be another tool helping medical experts in decision making around autoimmune patients. Based on the opportunities and challenges, guidelines on the use of big data have been created and published [86, 88, 294].

4 Future artificial intelligence development

Based on the broad opportunities of AI in healthcare, significant investments are being made which will positively impact the healthcare system. This includes business process driven AI as well as applications that directly impact the patient (e.g. clinical decision support).

Competing interests

Michael Mahler, Carlos Melus are employed at Inova Diagnostics selling autoantibody assays. Olga Kubassova and Faik Shaikh are employees of Image Analysis Group.

References

[1] S. Doyle-Lindrud, Watson will see you now: a supercomputer to help clinicians make informed treatment decisions, Clin. J. Oncol. Nurs. 19 (1) (2015) 31–32.

[2] S. Vázquez, et al., A classification of user-generated content into consumer decision journey stages, Neural Netw. 58 (2014) 68–81.

[3] M. Zhang, R.M. Alvarez, I. Levin, Election forensics: using machine learning and synthetic data for possible election anomaly detection, PLoS One 14 (10) (2019) e0223950.

[4] Y.J. Zheng, et al., Generative adversarial network based telecom fraud detection at the receiving bank, Neural Netw. 102 (2018) 78–86.

[5] O.E. Apolo-Apolo, et al., A cloud-based environment for generating yield estimation maps from apple orchards using UAV imagery and a deep learning technique, Front. Plant Sci. 11 (2020) 1086.

[6] M. Kim, et al., Deep learning in medical imaging, Neurospine 16 (4) (2019) 657–668.

[7] M.J. Diaz Candamio, S. Jha, J.M. Villagran, Overdiagnosis in imaging, Radiologia 60 (5) (2018) 362–367.

[8] D.S.W. Ting, et al., Artificial intelligence and deep learning in ophthalmology, Br. J. Ophthalmol. 103 (2) (2019) 167–175.

[9] F.L. Caetano Dos Santos, et al., Automatic classification of IgA endomysial antibody test for celiac disease: a new method deploying machine learning, Sci. Rep. 9 (1) (2019) 9217.

[10] M. Chumbita, et al., Can artificial intelligence improve the management of pneumonia, J. Clin. Med. 9 (1) (2020) 248, https://doi.org/10.3390/jcm9010248.

[11] V.T. Tran, C. Riveros, P. Ravaud, Patients' views of wearable devices and AI in healthcare: findings from the ComPaRe e-cohort, NPJ Digit. Med. 2 (2019) 53.

[12] T. Nadarzynski, et al., Acceptability of artificial intelligence (AI)-enabled chatbots, video consultations and live webchats as online platforms for sexual health advice, BMJ Sex. Reprod. Health 46 (3) (2020) 210–217, https://doi.org/10.1136/bmjsrh-2018-200271.

[13] H. Ashrafian, A. Darzi, T. Athanasiou, A novel modification of the Turing test for artificial intelligence and robotics in healthcare, Int. J. Med. Robot. 11 (1) (2015) 38–43.

[14] S. Price, The promise of artificial intelligence, Tex. Med. 115 (10) (2019) 32–35.

[15] Y. Yasuhara, et al., Potential legal issues when caring healthcare robot with communication in caring functions are used for older adult care, Enferm. Clin. 30 (Suppl. 1) (2020) 54–59.

[16] L.D. Jones, et al., Artificial intelligence, machine learning and the evolution of healthcare: a bright future or cause for concern? Bone Joint Res. 7 (3) (2018) 223–225.

[17] D. Ben-Israel, et al., The impact of machine learning on patient care: a systematic review, Artif. Intell. Med. 103 (2020) 101785.

[18] M. Komorowski, L.A. Celi, Will artificial intelligence contribute to overuse in healthcare? Crit. Care Med. 45 (5) (2017) 912–913.

[19] J. He, et al., The practical implementation of artificial intelligence technologies in medicine, Nat. Med. 25 (1) (2019) 30–36.

[20] P. Jayakumar, M.L.G. Moore, K.J. Bozic, Value-based healthcare: can artificial intelligence provide value in orthopaedic surgery? Clin. Orthop. Relat. Res. 477 (8) (2019) 1777–1780.

[21] M. Kasparick, et al., Enabling artificial intelligence in high acuity medical environments, Minim. Invasive Ther. Allied Technol. 28 (2) (2019) 120–126.

[22] R.S. Kerr, Surgery in the 2020s: implications of advancing technology for patients and the workforce, Future Healthc. J. 7 (1) (2020) 46–49.

[23] M.A. Brzezicki, et al., Artificial intelligence outperforms human students in conducting neurosurgical audits, Clin. Neurol. Neurosurg. 192 (2020) 105732.

[24] D. Boczar, et al., Artificial intelligent virtual assistant for plastic surgery patient's frequently asked questions: a pilot study, Ann. Plast. Surg. 84 (4) (2020) e16–e21.

[25] B. Acs, M. Rantalainen, J. Hartman, Artificial intelligence as the next step towards precision pathology, J. Intern. Med. (2020) 62–81.

[26] J. Nabi, Artificial intelligence can augment global pathology initiatives, Lancet 392 (10162) (2018) 2351–2352.

[27] D.G. Kiely, et al., Utilising artificial intelligence to determine patients at risk of a rare disease: idiopathic pulmonary arterial hypertension, Pulm. Circ. 9 (4) (2019). 2045894019890549.

[28] F. Pesapane, et al., Artificial intelligence as a medical device in radiology: ethical and regulatory issues in Europe and the United States, Insights Imaging 9 (5) (2018) 745–753.

[29] A. Loyola-Sanchez, et al., Qualitative study of treatment preferences for rheumatoid arthritis and pharmacotherapy acceptance: indigenous patient perspectives, Arthritis Care Res. (Hoboken) 72 (4) (2020) 544–552, https://doi.org/10.1002/acr.23869.

[30] S. Dentzer, Creating the future of artificial intelligence in health-system pharmacy, Am. J. Health Syst. Pharm. 76 (24) (2019) 1995–1996.

[31] G. Briganti, O. Le Moine, Artificial intelligence in medicine: today and tomorrow, Front. Med. (Lausanne) 7 (2020) 27.

[32] E. Jovanov, Wearables meet IoT: synergistic personal area networks (SPANs), Sensors (Basel) 19 (19) (2019) 4295.

[33] J. Dunn, R. Runge, M. Snyder, Wearables and the medical revolution, Perinat. Med. 15 (5) (2018) 429–448.

[34] D. Witt, et al., Windows into human health through wearables data analytics, Curr. Opin. Biomed. Eng. 9 (2019) 28–46.

[35] X. Li, et al., Digital health: tracking physiomes and activity using wearable biosensors reveals useful health-related information, PLoS Biol. 15 (1) (2017) e2001402.

[36] G.R. Burmester, Rheumatology 4.0: big data, wearables and diagnosis by computer, Ann. Rheum. Dis. 77 (7) (2018) 963–965.

[37] C. Hansen, A. Sanchez-Ferro, W. Maetzler, How mobile health technology and electronic health records will change care of patients with Parkinson's disease, J. Parkinsons Dis. 8 (s1) (2018) S41–S45.

[38] D. Solino-Fernandez, et al., Willingness to adopt wearable devices with behavioral and economic incentives by health insurance wellness programs: results of a US cross-sectional survey with multiple consumer health vignettes, BMC Public Health 19 (1) (2019) 1649.

[39] K. Paranjape, M. Schinkel, P. Nanayakkara, Short keynote paper: mainstreaming personalized healthcare-transforming healthcare through new era of artificial intelligence, IEEE J. Biomed. Health Inform. 24 (7) (2020) 1860–1863, https://doi.org/10.1109/JBHI.2020.2970807.

[40] C. Chen, et al., The times they are a-changin'—healthcare 4.0 is coming! J. Med. Syst. 44 (2) (2019) 40.
[41] F. Jiang, et al., Artificial intelligence in healthcare: past, present and future, Stroke Vasc. Neurol. 2 (4) (2017) 230–243.
[42] C. Wang, et al., Current strategies and applications for precision drug design, Front. Pharmacol. 9 (2018) 787.
[43] M. Woo, An AI boost for clinical trials, Nature 573 (7775) (2019) S100–S102.
[44] X. Yao, et al., Clinical trial design data for electrocardiogram artificial intelligence-guided screening for low ejection fraction (EAGLE), Data Brief 28 (2020) 104894.
[45] M. Househ, B. Aldosari, The hazards of data mining in healthcare, Stud. Health Technol. Inform. 238 (2017) 80–83.
[46] J.D. Symons, et al., From EHR to PHR: let's get the record straight, BMJ Open 9 (9) (2019) e029582.
[47] F. Sadoughi, A. Behmanesh, N. Sayfouri, Internet of things in medicine: a systematic mapping study, J. Biomed. Inform. 103 (2020) 103383.
[48] D. De Silva, et al., Machine learning to support social media empowered patients in cancer care and cancer treatment decisions, PLoS One 13 (10) (2018) e0205855.
[49] G. Jackson, J. Hu, Section Editors for the IMIA Yearbook Section on Artificial Intelligence in Health, Artificial intelligence in health in 2018: new opportunities, challenges, and practical implications, Yearb. Med. Inform. 28 (1) (2019) 52–54.
[50] S. Sheikhalishahi, et al., Natural language processing of clinical notes on chronic diseases: systematic review, JMIR Med. Inform. 7 (2) (2019) e12239.
[51] S. Nasiri, et al., Security requirements of internet of things-based healthcare system: a survey study, Acta Inform. Med. 27 (4) (2019) 253–258.
[52] N. Vaci, et al., Natural language processing for structuring clinical text data on depression using UK-CRIS, Evid. Based Ment. Health 23 (1) (2020) 21–26.
[53] A. Otokiti, Using informatics to improve healthcare quality, Int. J. Health Care Qual. Assur. 32 (2) (2019) 425–430.
[54] C.F. Luz, et al., Machine learning in infection management using routine electronic health records: tools, techniques, and reporting of future technologies, Clin. Microbiol. Infect. 26 (10) (2020) 1291–1299.
[55] N. Peiffer-Smadja, et al., Machine learning for clinical decision support in infectious diseases: a narrative review of current applications, Clin. Microbiol. Infect. 26 (5) (2020) 584–595, https://doi.org/10.1016/j.cmi.2019.09.009.
[56] S. Gupta, et al., Radiology, mobile devices, and internet of things (IoT), J. Digit. Imaging 33 (3) (2020) 735–746, https://doi.org/10.1007/s10278-019-00311-2.
[57] S.R. Steinhubl, E.D. Muse, E.J. Topol, The emerging field of mobile health, Sci. Transl. Med. 7 (283) (2015) 283rv3.
[58] J.M. Rosen, et al., Cybercare 2.0: meeting the challenge of the global burden of disease in 2030, Health Technol. (Berl) 6 (2016) 35–51.
[59] R.S.H. Istepanian, T. Al-Anzi, m-health 2.0: new perspectives on mobile health, machine learning and big data analytics, Methods 151 (2018) 34–40.
[60] S. Ashrafzadeh, O. Hamdy, Patient-driven diabetes care of the future in the technology era, Cell Metab. 29 (3) (2019) 564–575.
[61] D. Luo, et al., Mobile apps for individuals with rheumatoid arthritis: a systematic review, J. Clin. Rheumatol. 25 (3) (2019) 133–141.
[62] J.A. Moral-Munoz, et al., Smartphone applications to perform body balance assessment: a standardized review, J. Med. Syst. 42 (7) (2018) 119.

[63] M.H. Stanfill, D.T. Marc, Health information management: implications of artificial intelligence on healthcare data and information management, Yearb. Med. Inform. 28 (1) (2019) 56–64.

[64] T.D. Hadley, et al., Artificial intelligence in global health—a framework and strategy for adoption and sustainability, Int. J. MCH AIDS 9 (1) (2020) 121–127.

[65] T. Davenport, R. Kalakota, The potential for artificial intelligence in healthcare, Future Healthc. J. 6 (2) (2019) 94–98.

[66] F. Magrabi, et al., Artificial intelligence in clinical decision support: challenges for evaluating AI and practical implications, Yearb. Med. Inform. 28 (1) (2019) 128–134.

[67] L.A. Lynn, Artificial intelligence systems for complex decision-making in acute care medicine: a review, Patient Saf. Surg. 13 (2019) 6.

[68] V. Koutkias, J. Bouaud, Section Editors for the IMIA Yearbook Section on Decision Support, Contributions on clinical decision support from the 2018 literature, Yearb. Med. Inform. 28 (1) (2019) 135–137.

[69] W. Koller, et al., Augmenting analytics software for clinical microbiology by man-machine interaction, Stud. Health Technol. Inform. 264 (2019) 1243–1247.

[70] P. Elsner, et al., Position paper: telemedicine in occupational dermatology—current status and perspectives, J. Dtsch. Dermatol. Ges. 16 (8) (2018) 969–974.

[71] A.L. Fogel, J.C. Kvedar, Artificial intelligence powers digital medicine, NPJ Digit. Med. 1 (2018) 5.

[72] F. Talebi-Liasi, O. Markowitz, Is artificial intelligence going to replace dermatologists? Cutis 105 (1) (2020) 28–31.

[73] X. Du-Harpur, et al., What is AI? Applications of artificial intelligence to dermatology, Br. J. Dermatol. 183 (3) (2020) 423–430, https://doi.org/10.1111/bjd.18880.

[74] V. Patel, et al., Artificial intelligence applied to gastrointestinal diagnostics: a review, J. Pediatr. Gastroenterol. Nutr. 70 (1) (2020) 4–11.

[75] R. Forlano, et al., High-throughput, machine learning-based quantification of steatosis, inflammation, ballooning, and fibrosis in biopsies from patients with nonalcoholic fatty liver disease, Clin. Gastroenterol. Hepatol. 18 (9) (2019) 2081–2090.e9, https://doi.org/10.1016/j.cgh.2019.12.025.

[76] S.M. Schussler-Fiorenza Rose, et al., A longitudinal big data approach for precision health, Nat. Med. 25 (5) (2019) 792–804.

[77] J.K. Ruffle, A.D. Farmer, Q. Aziz, Artificial intelligence-assisted gastroenterology-promises and pitfalls, Am. J. Gastroenterol. 114 (3) (2019) 422–428.

[78] G.S. Hazlewood, et al., Patient preferences for maintenance therapy in Crohn's disease: a discrete-choice experiment, PLoS One 15 (1) (2020) e0227635.

[79] F. Nehme, K. Feldman, Evolving role and future directions of natural language processing in gastroenterology, Dig. Dis. Sci. (2020). https://link.springer.com/article/10.1007/s10620-020-06156-y#citeas.

[80] M. Ohmori, et al., Endoscopic detection and differentiation of esophageal lesions using a deep neural network, Gastrointest. Endosc. 91 (2) (2020) 301–309 e1.

[81] D.L. Shung, et al., Validation of a machine learning model that outperforms clinical risk scoring systems for upper gastrointestinal bleeding, Gastroenterology 158 (1) (2020) 160–167.

[82] M. Signaevsky, et al., Artificial intelligence in neuropathology: deep learning-based assessment of tauopathy, Lab. Investig. 99 (7) (2019) 1019–1029.

[83] Z. Tang, et al., Interpretable classification of Alzheimer's disease pathologies with a convolutional neural network pipeline, Nat. Commun. 10 (1) (2019) 2173.

[84] F. Al-Mufti, et al., Machine learning and artificial intelligence in neurocritical care: a specialty-wide disruptive transformation or a strategy for success, Curr. Neurol. Neurosci. Rep. 19 (11) (2019) 89.

[85] R.M. Galimova, et al., Artificial intelligence-developments in medicine in the last two years, Chronic Dis. Transl. Med. 5 (1) (2019) 64–68.

[86] A. Manrique de Lara, I. Pelaez-Ballestas, Big data and data processing in rheumatology: bioethical perspectives, Clin. Rheumatol. 39 (4) (2020) 1007–1014, https://doi.org/10.1007/s10067-020-04969-w.

[87] L. Gossec, et al., Detection of flares by decrease in physical activity, collected using wearable activity trackers in rheumatoid arthritis or axial spondyloarthritis: an application of machine learning analyses in rheumatology, Arthritis Care Res. (Hoboken) 71 (10) (2019) 1336–1343.

[88] L. Gossec, et al., EULAR points to consider for the use of big data in rheumatic and musculoskeletal diseases, Ann. Rheum. Dis. 79 (1) (2020) 69–76.

[89] T. Hugle, A. Dumusc, Arthritis 4.0: the digital cycle has begun, Rev. Med. Suisse 15 (641) (2019) 549–553.

[90] V.G. Honavar, Machine learning in clinical care: quo vadis? Indian J. Ophthalmol. 67 (7) (2019) 985–986.

[91] T. Alsuliman, D. Humaidan, L. Sliman, Machine learning and artificial intelligence in the service of medicine: necessity or potentiality? Curr. Res. Transl. Med. 68 (4) (2020) 245–251.

[92] Y. Xiang, et al., Implementation of artificial intelligence in medicine: status analysis and development suggestions, Artif. Intell. Med. 102 (2020) 101780.

[93] X. Wu, et al., Universal artificial intelligence platform for collaborative management of cataracts, Br. J. Ophthalmol. 103 (11) (2019) 1553–1560.

[94] D.S. Kermany, et al., Identifying medical diagnoses and treatable diseases by image-based deep learning, Cell 172 (5) (2018) 1122–1131.

[95] Z. Li, et al., Can artificial intelligence make screening faster, more accurate, and more accessible? Asia Pac. J. Ophthalmol. (Phila.) 7 (6) (2018) 436–441.

[96] C. Shen, et al., Accuracy of a popular online symptom checker for ophthalmic diagnoses, JAMA Ophthalmol. 137 (6) (2019) 690–692, https://doi.org/10.1001/jamaophthalmol.2019.0571.

[97] J. Bali, R. Garg, R.T. Bali, Artificial intelligence (AI) in healthcare and biomedical research: why a strong computational/AI bioethics framework is required? Indian J. Ophthalmol. 67 (1) (2019) 3–6.

[98] C. Kern, et al., Implementation of a cloud-based referral platform in ophthalmology: making telemedicine services a reality in eye care, Br. J. Ophthalmol. 104 (3) (2020) 312–317.

[99] K.L. Fessele, The rise of big data in oncology, Semin. Oncol. Nurs. 34 (2) (2018) 168–176.

[100] I. El Naqa, et al., Artificial Intelligence: reshaping the practice of radiological sciences in the 21st century, Br. J. Radiol. 93 (1106) (2020) 20190855.

[101] Y. Lin, et al., Data-driven translational prostate cancer research: from biomarker discovery to clinical decision, J. Transl. Med. 18 (1) (2020) 119.

[102] I. Sechopoulos, R.M. Mann, Stand-alone artificial intelligence—the future of breast cancer screening? Breast 49 (2020) 254–260.

[103] K.S. Hughes, et al., Natural language processing to facilitate breast cancer research and management, Breast J. 26 (1) (2020) 92–99.

[104] D. Dana, et al., Deep learning in drug discovery and medicine; scratching the surface, Molecules 23 (9) (2018) 2384, https://doi.org/10.3390/molecules23092384.

[105] I.S. Boon, T.P.T. Au Yong, C.S. Boon, Assessing the role of artificial intelligence (AI) in clinical oncology: utility of machine learning in radiotherapy target volume delineation, Medicines (Basel) 5 (4) (2018) 131, https://doi.org/10.3390/medicines5040131.
[106] C.R. Deig, A. Kanwar, R.F. Thompson, Artificial intelligence in radiation oncology, Hematol. Oncol. Clin. North Am. 33 (6) (2019) 1095–1104.
[107] N. Nagarajan, et al., Application of computational biology and artificial intelligence technologies in cancer precision drug discovery, Biomed. Res. Int. 2019 (2019) 8427042.
[108] C. Qiu, et al., Establishment and validation of an immunodiagnostic model for prediction of breast cancer, Oncoimmunology 9 (1) (2020) 1682382.
[109] K.C. Koo, et al., Long short-term memory artificial neural network model for prediction of prostate cancer survival outcomes according to initial treatment strategy: development of an online decision-making support system, World J. Urol. 38 (10) (2020) 2469–2476, https://doi.org/10.1007/s00345-020-03080-8.
[110] Z. Allam, D.S. Jones, On the coronavirus (COVID-19) outbreak and the smart city network: universal data sharing standards coupled with artificial intelligence (AI) to benefit urban health monitoring and management, Healthcare (Basel) 8 (1) (2020) 46, https://doi.org/10.3390/healthcare8010046.
[111] A. Rao, J.A. Vazquez, Identification of COVID-19 can be quicker through artificial intelligence framework using a mobile phone-based survey in the populations when cities/towns are under quarantine, Infect. Control Hosp. Epidemiol. (2020) 1–18.
[112] T. Ai, et al., Correlation of chest CT and RT-PCR testing in coronavirus disease 2019 (COVID-19) in China: a report of 1014 cases, Radiology (2020) 200642.
[113] A. Tarnok, Machine learning, COVID-19 (2019-nCoV), and multi-OMICS, Cytometry A 97 (3) (2020) 215–216.
[114] L. Li, et al., Artificial intelligence distinguishes COVID-19 from community acquired pneumonia on chest CT, Radiology (2020) 200905.
[115] A. Nunez Reiz, M.A. Armengol de la Hoz, M. Sanchez Garcia, Big data analysis and machine learning in intensive care units, Med. Intensiva 43 (7) (2019) 416–426.
[116] M. Komorowski, et al., The artificial intelligence clinician learns optimal treatment strategies for sepsis in intensive care, Nat. Med. 24 (11) (2018) 1716–1720.
[117] A. Chan, et al., Deep learning algorithms to identify documentation of serious illness conversations during intensive care unit admissions, Palliat. Med. 33 (2) (2019) 187–196.
[118] N.M. Murray, et al., Artificial intelligence to diagnose ischemic stroke and identify large vessel occlusions: a systematic review, J. Neurointerv. Surg. 12 (2) (2020) 156–164.
[119] D. Tandon, J. Rajawat, Present and future of artificial intelligence in dentistry, J. Oral Biol. Craniofac. Res. 10 (4) (2020) 391–396.
[120] S. Saria, A. Butte, A. Sheikh, Better medicine through machine learning: what's real, and what's artificial? PLoS Med. 15 (12) (2018) e1002721.
[121] M. Mazzanti, et al., Imaging, health record, and artificial intelligence: hype or hope? Curr. Cardiol. Rep. 20 (6) (2018) 48.
[122] S. Bhattacharya, et al., Artificial intelligence enabled healthcare: a hype, hope or harm, J. Family Med. Prim. Care 8 (11) (2019) 3461–3464.
[123] E.J. Emanuel, R.M. Wachter, Artificial intelligence in health care: will the value match the hype? JAMA 321 (23) (2019) 2281–2282.
[124] J. Sheffer, AI in healthcare: less hype, better data, Biomed. Instrum. Technol. 53 (2) (2019) 82–83.
[125] L. Hood, M. Flores, A personal view on systems medicine and the emergence of proactive P4 medicine: predictive, preventive, personalized and participatory, New Biotechnol. 29 (6) (2012) 613–624.

[126] G.T. Sharrer, Personalized medicine: ethics for clinical trials, Methods Mol. Biol. 823 (2012) 35–48.
[127] S. Russell, et al., Robotics: ethics of artificial intelligence, Nature 521 (7553) (2015) 415–418.
[128] V. Fineschi, Editorial: personalized medicine: a positional point of view about precision medicine and clarity for ethics of public health, Curr. Pharm. Biotechnol. 18 (3) (2017) 192–193.
[129] G.T. Sharrer, Personalized medicine: ethical aspects, Methods Mol. Biol. 1606 (2017) 37–50.
[130] M. Shoaib, et al., Personalized medicine in a new genomic era: ethical and legal aspects, Sci. Eng. Ethics 23 (4) (2017) 1207–1212.
[131] J. Nabi, How bioethics can shape artificial intelligence and machine learning, Hast. Cent. Rep. 48 (5) (2018) 10–13.
[132] D. Gruson, et al., Data science, artificial intelligence, and machine learning: opportunities for laboratory medicine and the value of positive regulation, Clin. Biochem. 69 (2019) 1–7.
[133] O. El-Hassoun, et al., Artificial intelligence in service of medicine, Bratisl. Lek. Listy 120 (3) (2019) 218–222.
[134] H. Liyanage, et al., Artificial intelligence in primary health care: perceptions, issues, and challenges, Yearb. Med. Inform. 28 (1) (2019) 41–46.
[135] J. Morley, L. Floridi, An ethically mindful approach to AI for health care, Lancet 395 (10220) (2020) 254–255.
[136] C. Burr, M. Taddeo, L. Floridi, The ethics of digital well-being: a thematic review, Sci. Eng. Ethics 26 (2020) 2313–2343, https://doi.org/10.1007/s11948-020-00175-8.
[137] N.M. Safdar, J.D. Banja, C.C. Meltzer, Ethical considerations in artificial intelligence, Eur. J. Radiol. 122 (2020) 108768.
[138] E.W. Kluge, Artificial intelligence in healthcare: ethical considerations, Health Manage. Forum 33 (1) (2020) 47–49.
[139] T.Y.A. Liu, N.M. Bressler, Controversies in artificial intelligence, Curr. Opin. Ophthalmol. 31 (5) (2020) 324–328.
[140] J.P. Gilbert, et al., A call for an ethical framework when using social media data for artificial intelligence applications in public health research, Can. Commun. Dis. Rep. 46 (6) (2020) 169–173.
[141] J. Atherton, Development of the electronic health record, Virtual Mentor 13 (3) (2011) 186–189.
[142] D.E. Adkins, Machine learning and electronic health records: a paradigm shift, Am. J. Psychiatry 174 (2) (2017) 93–94.
[143] F.M. Bublitz, et al., Disruptive technologies for environment and health research: an overview of artificial intelligence, blockchain, and internet of things, Int. J. Environ. Res. Public Health 16 (20) (2019) 3847.
[144] G. Leeming, J. Ainsworth, D.A. Clifton, Blockchain in health care: hype, trust, and digital health, Lancet 393 (10190) (2019) 2476–2477.
[145] S. Conard, Best practices in digital health literacy, Int. J. Cardiol. 292 (2019) 277–279.
[146] M.P. McBee, C. Wilcox, Blockchain technology: principles and applications in medical imaging, J. Digit. Imaging 33 (2020) 726–734, https://doi.org/10.1007/s10278-019-00310-3.
[147] A. Margheri, et al., Decentralised provenance for healthcare data, Int. J. Med. Inform. 141 (2020) 104197.
[148] M. Kim, et al., Design of secure protocol for cloud-assisted electronic health record system using blockchain, Sensors (Basel) 20 (10) (2020) 2913.

[149] D. Tith, et al., Application of blockchain to maintaining patient records in electronic health record for enhanced privacy, scalability, and availability, Healthc. Inform. Res. 26 (1) (2020) 3–12.

[150] A.H. Mayer, C.A. da Costa, R.D.R. Righi, Electronic health records in a blockchain: a systematic review, Health Informatics J. 26 (2) (2020) 1273–1288.

[151] H. Tang, J.H. Ng, Googling for a diagnosis—use of Google as a diagnostic aid: internet based study, BMJ 333 (7579) (2006) 1143–1145.

[152] L.A. Celi, et al., Bridging the health data divide, J. Med. Internet Res. 18 (12) (2016) e325.

[153] A.N. Chester, et al., Patient-targeted Googling and social media: a cross-sectional study of senior medical students, BMC Med. Ethics 18 (1) (2017) 70.

[154] C. Zimmer, et al., Use of daily Internet search query data improves real-time projections of influenza epidemics, J. R. Soc. Interface 15 (147) (2018) 20180220, https://doi.org/10.1098/rsif.2018.0220.

[155] M.G. Sanal, et al., Artificial intelligence and deep learning: the future of medicine and medical practice, J. Assoc. Physicians India 67 (4) (2019) 71–73.

[156] A.K. Triantafyllidis, A. Tsanas, Applications of machine learning in real-life digital health interventions: review of the literature, J. Med. Internet Res. 21 (4) (2019) e12286.

[157] L.J. Kricka, History of disruptions in laboratory medicine: what have we learned from predictions? Clin. Chem. Lab. Med. 57 (3) (2019) 308–311.

[158] Y. Luo, W. Li, S. Qiu, Anomaly detection based latency-aware energy consumption optimization for iot data-flow services, Sensors (Basel) 20 (1) (2019) 122, https://doi.org/10.3390/s20010122.

[159] B. Ma, et al., Muscle fatigue detection and treatment system driven by internet of things, BMC Med. Inform. Decis. Mak. 19 (Suppl. 7) (2019) 275.

[160] J.P. Rajan, et al., Fog computing employed computer aided cancer classification system using deep neural network in internet of things based healthcare system, J. Med. Syst. 44 (2) (2019) 34.

[161] P. Rajan Jeyaraj, E.R.S. Nadar, Atrial fibrillation classification using deep learning algorithm in Internet of Things-based smart healthcare system, Health Informatics J. (2019). 1460458219891384.

[162] G. Kyriakopoulos, et al., Internet of things (IoT)-enabled elderly fall verification, exploiting temporal inference models in smart homes, Int. J. Environ. Res. Public Health 17 (2020) 408.

[163] R.L. Fritz, G. Dermody, A nurse-driven method for developing artificial intelligence in "smart" homes for aging-in-place, Nurs. Outlook 67 (2) (2019) 140–153.

[164] S. Topolski, J. Sturmberg, Validation of a non-linear model of health, J. Eval. Clin. Pract. 20 (6) (2014) 1026–1035.

[165] H. Ben Hassen, W. Dghais, B. Hamdi, An E-health system for monitoring elderly health based on Internet of Things and Fog computing, Health. Inf. Sci. Syst. 7 (1) (2019) 24.

[166] L. Rampasek, A. Goldenberg, TensorFlow: biology's gateway to deep learning? Cell Syst. 2 (1) (2016) 12–14.

[167] M. Alzantot, et al., RSTensorFlow: GPU enabled tensorflow for deep learning on commodity android devices, MobiSys 2017 (2017) 7–12.

[168] D. Mulfari, A. Palla, L. Fanucci, Embedded systems and tensorflow frameworks as assistive technology solutions, Stud. Health Technol. Inform. 242 (2017) 396–400.

[169] S. Jiang, et al., Solving fourier ptychographic imaging problems via neural network modeling and TensorFlow, Biomed. Opt. Express 9 (7) (2018) 3306–3319.

[170] K. Wongsuphasawat, et al., Visualizing dataflow graphs of deep learning models in tensorflow, IEEE Trans. Vis. Comput. Graph. 24 (1) (2018) 1–12.

[171] F. Laporte, J. Dambre, P. Bienstman, Highly parallel simulation and optimization of photonic circuits in time and frequency domain based on the deep-learning framework PyTorch, Sci. Rep. 9 (1) (2019) 5918.

[172] K.M. Chen, et al., Selene: a PyTorch-based deep learning library for sequence data, Nat. Methods 16 (4) (2019) 315–318.

[173] K.T. Schutt, et al., SchNetPack: a deep learning toolbox for atomistic systems, J. Chem. Theory Comput. 15 (1) (2019) 448–455.

[174] M.S. Kohn, et al., IBM's health analytics and clinical decision support, Yearb. Med. Inform. 9 (2014) 154–162.

[175] E. Harper, Can big data transform electronic health records into learning health systems? Stud. Health Technol. Inform. 201 (2014) 470–475.

[176] F. Cheng, Z. Zhao, Machine learning-based prediction of drug-drug interactions by integrating drug phenotypic, therapeutic, chemical, and genomic properties, J. Am. Med. Inform. Assoc. 21 (e2) (2014) e278–e286.

[177] N. Rifai, et al., Disruptive innovation in laboratory medicine, Clin. Chem. 61 (9) (2015) 1129–1132.

[178] H.L. Semigran, et al., Evaluation of symptom checkers for self diagnosis and triage: audit study, BMJ 351 (2015) h3480.

[179] A.J. Elliot, et al., Internet-based remote health self-checker symptom data as an adjuvant to a national syndromic surveillance system, Epidemiol. Infect. 143 (16) (2015) 3416–3422.

[180] M. Ramos-Casals, et al., Google-driven search for big data in autoimmune geoepidemiology: analysis of 394,827 patients with systemic autoimmune diseases, Autoimmun. Rev. 14 (8) (2015) 670–679.

[181] N. Peek, R. Marin Morales, M. Peleg, Artificial intelligence in medicine AIME 2013, Artif. Intell. Med. 65 (1) (2015) 1–3.

[182] E. Topol, Digital medicine: empowering both patients and clinicians, Lancet 388 (10046) (2016) 740–741.

[183] L. Powley, et al., Are online symptoms checkers useful for patients with inflammatory arthritis? BMC Musculoskelet. Disord. 17 (1) (2016) 362.

[184] M. Limb, NHS announces online symptom checker, BMJ 354 (2016) i4905.

[185] M.J. Khoury, M.F. Iademarco, W.T. Riley, Precision public health for the era of precision medicine, Am. J. Prev. Med. 50 (3) (2016) 398–401.

[186] W. Diprose, N. Buist, Artificial intelligence in medicine: humans need not apply? N. Z. Med. J. 129 (1434) (2016) 73–76.

[187] R. Miotto, et al., Deep patient: an unsupervised representation to predict the future of patients from the electronic health records, Sci. Rep. 6 (2016) 26094.

[188] A. Torkamani, et al., High-definition medicine, Cell 170 (5) (2017) 828–843.

[189] M. Recht, R.N. Bryan, Artificial intelligence: threat or boon to radiologists? J. Am. Coll. Radiol. 14 (11) (2017) 1476–1480.

[190] K.A. Mikk, H.A. Sleeper, E.J. Topol, The pathway to patient data ownership and better health, JAMA 318 (15) (2017) 1433–1434.

[191] M.L. Ackerman, T. Virani, B. Billings, Digital mental health—innovations in consumer driven care, Nurs. Leadersh. (Tor. Ont) 30 (3) (2017) 63–72.

[192] Artificial intelligence use in healthcare growing fast, J. AHIMA 88 (6) (2017) 76. https://pubmed.ncbi.nlm.nih.gov/29424995/.

[193] B.F. King Jr., Guest editorial: discovery and artificial intelligence, AJR Am. J. Roentgenol. 209 (6) (2017) 1189–1190.

[194] Z. Huang, J.M. Juarez, X. Li, Data mining for biomedicine and healthcare, J. Healthc. Eng. 2017 (2017) 7107629.
[195] J. Sensmeier, Harnessing the power of artificial intelligence, Nurs. Manag. 48 (11) (2017) 14–19.
[196] D. Bonderman, Artificial intelligence in cardiology, Wien. Klin. Wochenschr. 129 (23–24) (2017) 866–868.
[197] C. Krittanawong, Healthcare in the 21st century, Eur. J. Intern. Med. 38 (2017) e17.
[198] M.D. Kohli, R.M. Summers, J.R. Geis, Medical image data and datasets in the era of machine learning-whitepaper from the 2016 C-MIMI meeting dataset session, J. Digit. Imaging 30 (4) (2017) 392–399.
[199] G. Sharma, A. Carter, Artificial intelligence and the pathologist: future frenemies? Arch. Pathol. Lab. Med. 141 (5) (2017) 622–623.
[200] M.J. Fritzler, et al., The utilization of autoantibodies in approaches to precision health, Front. Immunol. 9 (2018) 2682.
[201] D. Grapov, et al., Rise of deep learning for genomic, proteomic, and metabolomic data integration in precision medicine, OMICS 22 (10) (2018) 630–636.
[202] K.W. Johnson, et al., Artificial intelligence in cardiology, J. Am. Coll. Cardiol. 71 (23) (2018) 2668–2679.
[203] A.M. Williams, et al., Artificial intelligence, physiological genomics, and precision medicine, Physiol. Genomics 50 (4) (2018) 237–243.
[204] S.A. Dugger, A. Platt, D.B. Goldstein, Drug development in the era of precision medicine, Nat. Rev. Drug Discov. 17 (3) (2018) 183–196.
[205] B.D. Modena, et al., Advanced and accurate mobile health tracking devices record new cardiac vital signs, Hypertension 72 (2) (2018) 503–510.
[206] S.R. Steinhubl, et al., Virtual care for improved global health, Lancet 391 (10119) (2018) 419.
[207] K.A. Mikk, H.A. Sleeper, E. Topol, Patient data ownership-reply, JAMA 319 (9) (2018) 935–936.
[208] R.A. Kellogg, J. Dunn, M.P. Snyder, Personal omics for precision health, Circ. Res. 122 (9) (2018) 1169–1171.
[209] K.J. Karczewski, M.P. Snyder, Integrative omics for health and disease, Nat. Rev. Genet. 19 (5) (2018) 299–310.
[210] V.B. Kolachalama, P.S. Garg, Machine learning and medical education, NPJ Digit. Med. 1 (2018) 54.
[211] T. Panch, P. Szolovits, R. Atun, Artificial intelligence, machine learning and health systems, J. Glob. Health 8 (2) (2018) 020303.
[212] J. Goldhahn, V. Rampton, G.A. Spinas, Could artificial intelligence make doctors obsolete? BMJ 363 (2018) k4563.
[213] F. Pesapane, M. Codari, F. Sardanelli, Artificial intelligence in medical imaging: threat or opportunity? Radiologists again at the forefront of innovation in medicine, Eur. Radiol. Exp. 2 (1) (2018) 35.
[214] A.E. Chung, et al., Health and fitness apps for hands-free voice-activated assistants: content analysis, JMIR Mhealth Uhealth 6 (9) (2018) e174.
[215] B. Wahl, et al., Artificial intelligence (AI) and global health: how can AI contribute to health in resource-poor settings? BMJ Glob. Health 3 (4) (2018) e000798.
[216] Y. Lee, et al., Deep learning in the medical domain: predicting cardiac arrest using deep learning, Acute Crit. Care 33 (3) (2018) 117–120.
[217] A. Morsy, Can AI truly transform health care?: a recent IEEE pulse on stage forum offers some perspective, IEEE Pulse 9 (4) (2018) 18–20.

[218] B. Mesko, G. Hetenyi, Z. Gyorffy, Will artificial intelligence solve the human resource crisis in healthcare? BMC Health Serv. Res. 18 (1) (2018) 545.
[219] P. Baird, Going big data, with caution, Biomed. Instrum. Technol. 52 (s2) (2018) 39–40.
[220] L. Tarassenko, P. Watkinson, Artificial intelligence in health care: enabling informed care, Lancet 391 (10127) (2018) 1260.
[221] K. Dreyer, B. Allen, Artificial intelligence in health care: brave new world or golden opportunity? J. Am. Coll. Radiol. 15 (4) (2018) 655–657.
[222] Y. Berlyand, et al., How artificial intelligence could transform emergency department operations, Am. J. Emerg. Med. 36 (8) (2018) 1515–1517.
[223] The Lancet, Artificial intelligence in health care: within touching distance, Lancet 390 (10114) (2018) 2739.
[224] M. Kayaalp, Patient privacy in the era of big data, Balkan Med. J. 35 (1) (2018) 8–17.
[225] R. Miotto, et al., Deep learning for healthcare: review, opportunities and challenges, Brief. Bioinform. 19 (6) (2018) 1236–1246.
[226] N. Noorbakhsh-Sabet, et al., Artificial intelligence transforms the future of health care, Am. J. Med. 132 (7) (2019) 795–801, https://doi.org/10.1016/j.amjmed.2019.01.017.
[227] A.C. Berry, et al., Online symptom checker diagnostic and triage accuracy for HIV and hepatitis C, Epidemiol. Infect. 147 (2019) e104.
[228] G.Z. Papadakis, et al., Deep learning opens new horizons in personalized medicine, Biomed. Rep. 10 (4) (2019) 215–217.
[229] M.E. Matheny, D. Whicher, S. Thadaney Israni, Artificial intelligence in health care: a report from the national academy of medicine, JAMA 323 (6) (2020) 509–510, https://doi.org/10.1001/jama.2019.21579.
[230] A. Blasiak, J. Khong, T. Kee, CURATE.AI: optimizing personalized medicine with artificial intelligence, SLAS Technol. (2019). 2472630319890316.
[231] K.H. Pettersen, Artificial intelligence will change the health services, Tidsskr. Nor. Laegeforen. 139 (14) (2019), https://doi.org/10.4045/tidsskr.19.0479.
[232] M. Sujan, P. Scott, K. Cresswell, Digital health and patient safety: technology is not a magic wand, Health Informatics J. (2019). 1460458219876183.
[233] M. Pillai, et al., Using artificial intelligence to improve the quality and safety of radiation therapy, J. Am. Coll. Radiol. 16 (9 Pt B) (2019) 1267–1272.
[234] P. Mistry, Artificial intelligence in primary care, Br. J. Gen. Pract. 69 (686) (2019) 422–423.
[235] Amisha, et al., Overview of artificial intelligence in medicine, J. Family Med. Prim. Care 8 (7) (2019) 2328–2331.
[236] T. Truong, et al., A framework for applied AI in healthcare, Stud. Health Technol. Inform. 264 (2019) 1993–1994.
[237] T. Mattsson, Editorial: digitalisation and artificial intelligence in European healthcare, Eur. J. Health Law (2019) 1–4.
[238] D. Chen, et al., Deep learning and alternative learning strategies for retrospective real-world clinical data, NPJ Digit. Med. 2 (2019) 43.
[239] B. Chin-Yee, R. Upshur, Three problems with big data and artificial intelligence in medicine, Perspect. Biol. Med. 62 (2) (2019) 237–256.
[240] D.C. Hague, Benefits, pitfalls, and potential bias in health care AI, N. C. Med. J. 80 (4) (2019) 219–223.
[241] E. Racine, W. Boehlen, M. Sample, Healthcare uses of artificial intelligence: challenges and opportunities for growth, Healthc. Manage. Forum 32 (5) (2019) 272–275.
[242] S. Kothari, et al., Artificial Intelligence (AI) and rheumatology: a potential partnership, Rheumatology (Oxford) 58 (11) (2019) 1894–1895.

[243] C. Kuziemsky, et al., Role of artificial intelligence within the telehealth domain, Yearb. Med. Inform. 28 (1) (2019) 35–40.
[244] T.R. Clancy, Artificial intelligence and nursing: the future is now, J. Nurs. Adm. 50 (3) (2020) 125–127.
[245] O.A. Paiva, L.M. Prevedello, The potential impact of artificial intelligence in radiology, Radiol. Bras. 50 (5) (2017) V–VI.
[246] M. Kohli, et al., Implementing machine learning in radiology practice and research, AJR Am. J. Roentgenol. 208 (4) (2017) 754–760.
[247] M.L. Giger, Machine learning in medical imaging, J. Am. Coll. Radiol. 15 (3 Pt B) (2018) 512–520.
[248] SFR-IA Group; CERF; French Radiology Community, Artificial intelligence and medical imaging 2018: French Radiology Community white paper, Diagn. Interv. Imaging 99 (11) (2018) 727–742.
[249] M.P. McBee, et al., Deep learning in radiology, Acad. Radiol. 25 (11) (2018) 1472–1480.
[250] P. Lakhani, et al., Machine learning in radiology: applications beyond image interpretation, J. Am. Coll. Radiol. 15 (2) (2018) 350–359.
[251] L. Saba, et al., The present and future of deep learning in radiology, Eur. J. Radiol. 114 (2019) 14–24.
[252] Y. Kobayashi, M. Ishibashi, H. Kobayashi, How will "democratization of artificial intelligence" change the future of radiologists? Jpn. J. Radiol. 37 (1) (2019) 9–14.
[253] M.P. Recht, et al., Integrating artificial intelligence into the clinical practice of radiology: challenges and recommendations, Eur. Radiol. 30 (6) (2020) 3576–3584, https://doi.org/10.1007/s00330-020-06672-5.
[254] J. Sogani, et al., Artificial intelligence in radiology: the ecosystem essential to improving patient care, Clin. Imaging 59 (1) (2020) A3–A6.
[255] G. Xie, et al., Artificial intelligence in nephrology: how can artificial intelligence augment nephrologists' intelligence? Kidney Dis. (Basel) 6 (1) (2020) 1–6.
[256] M. Mitchell, L. Kan, Digital technology and the future of health systems, Health Syst. Reform 5 (2) (2019) 113–120.
[257] E.H. Shortliffe, M.J. Sepúlveda, Clinical decision support in the era of artificial intelligence, JAMA 320 (21) (2018) 2199–2200.
[258] J.D. Siegel, T.A. Parrino, Computerized diagnosis: implications for clinical education, Med. Educ. 22 (1) (1988) 47–54.
[259] G. Banks, Artificial intelligence in medical diagnosis: the INTERNIST/CADUCEUS approach, Crit. Rev. Med. Inform. 1 (1) (1986) 23–54.
[260] R.C. Parker, R.A. Miller, Creation of realistic appearing simulated patient cases using the INTERNIST-1/QMR knowledge base and interrelationship properties of manifestations, Methods Inf. Med. 28 (4) (1989) 346–351.
[261] M. Pradhan, G. Provan, M. Henrion, Experimental analysis of large belief networks for medical diagnosis, Proc. Annu. Symp. Comput. Appl. Med. Care (1994) 775–779.
[262] S. Weiss, et al., A model-based method for computer-aided medical decision-making, Artif. Intell. 11 (1–2) (1978) 145–172.
[263] K.R. Siegersma, et al., Artificial intelligence in cardiovascular imaging: state of the art and implications for the imaging cardiologist, Neth. Hear. J. 27 (9) (2019) 403–413.
[264] A. Hirschmann, et al., Artificial intelligence in musculoskeletal imaging: review of current literature, challenges, and trends, Semin. Musculoskelet. Radiol. 23 (3) (2019) 304–311.
[265] P.M. Johnson, M.P. Recht, F. Knoll, Improving the speed of MRI with artificial intelligence, Semin. Musculoskelet. Radiol. 24 (1) (2020) 12–20.

[266] F.J. Gilbert, S.W. Smye, C.B. Schonlieb, Artificial intelligence in clinical imaging: a health system approach, Clin. Radiol. 75 (1) (2020) 3–6.
[267] J.S. Smolen, et al., Reference sera for antinuclear antibodies. II. Further definition of antibody specificities in international antinuclear antibody reference sera by immunofluorescence and western blotting, Arthritis Rheum. 40 (3) (1997) 413–418.
[268] G. Wieringa, Teaching the pony new tricks: competences for specialists in laboratory medicine to meet the challenges of disruptive innovation, Clin. Chem. Lab. Med. 57 (3) (2019) 398–402.
[269] C. Krittanawong, The rise of artificial intelligence and the uncertain future for physicians, Eur. J. Intern. Med. 48 (2018) e13–e14.
[270] M. Prastawa, E. Bullitt, G. Gerig, Simulation of brain tumors in MR images for evaluation of segmentation efficacy, Med. Image Anal. 13 (2) (2009) 297–311.
[271] S. Keil, et al., RECIST and WHO criteria evaluation of cervical, thoracic and abdominal lymph nodes in patients with malignant lymphoma: manual versus semi-automated measurement on standard MDCT slices, Röfo 181 (9) (2009) 888–895.
[272] O. Kubassova, et al., A computer-aided detection system for rheumatoid arthritis MRI data interpretation and quantification of synovial activity, Eur. J. Radiol. 74 (3) (2010) e67–e72.
[273] A.R. Crowley, et al., Measuring bone erosion and edema in rheumatoid arthritis: a comparison of manual segmentation and RAMRIS methods, J. Magn. Reson. Imaging 33 (2) (2011) 364–371.
[274] F. Lopez Segui, et al., Teleconsultations between patients and healthcare professionals in primary care in catalonia: the evaluation of text classification algorithms using supervised machine learning, Int. J. Environ. Res. Public Health 17 (3) (2020) 1093, https://doi.org/10.3390/ijerph17031093.
[275] M. Jones, F. DeRuyter, J. Morris, The digital health revolution and people with disabilities: perspective from the United States, Int. J. Environ. Res. Public Health 17 (2) (2020) 381.
[276] A.A. Seyhan, C. Carini, Are innovation and new technologies in precision medicine paving a new era in patients centric care? J. Transl. Med. 17 (1) (2019) 114.
[277] A.H. Sapci, H.A. Sapci, Innovative assisted living tools, remote monitoring technologies, artificial intelligence-driven solutions, and robotic systems for aging societies: systematic review, JMIR Aging 2 (2) (2019) e15429.
[278] A. Galderisi, et al., Continuous glucose monitoring linked to an artificial intelligence risk index: early footprints of intraventricular hemorrhage in preterm neonates, Diabetes Technol. Ther. 21 (3) (2019) 146–153.
[279] S. Kato, et al., Effectiveness of lifestyle intervention using the internet of things system for individuals with early type 2 diabetes mellitus, Intern. Med. 59 (1) (2020) 45–53.
[280] R. Hemdani, et al., The COVID-19 outbreak: a game-changer in reinforcing the use of telemedicine in dermatology? Skinmed 18 (3) (2020) 187–188.
[281] E. Menardi, et al., Telemedicine during COVID-19 pandemic, J. Arrhythm. 36 (4) (2020) 804–805.
[282] J. Bokolo Anthony, Use of telemedicine and virtual care for remote treatment in response to COVID-19 pandemic, J. Med. Syst. 44 (7) (2020) 132.
[283] A.S. Chaudhari, et al., Utility of deep learning super-resolution in the context of osteoarthritis MRI biomarkers, J. Magn. Reson. Imaging 51 (3) (2020) 768–779, https://doi.org/10.1002/jmri.26872.
[284] S.H. Park, et al., What should medical students know about artificial intelligence in medicine? J. Educ. Eval. Health Prof. 16 (2019) 18.

[285] A.F. Leite, et al., Radiomics and machine learning in oral healthcare, Proteomics Clin. Appl. (2020) e1900040.
[286] S. Gyftopoulos, et al., Artificial intelligence in musculoskeletal imaging: current status and future directions, AJR Am. J. Roentgenol. 213 (3) (2019) 506–513.
[287] J. Martorell-Marugan, et al., Deep learning in omics data analysis and precision medicine, in: H. Husi (Ed.), Computational Biology [Internet], Codon Publications, Brisbane (AU), 2019. Chapter 3.
[288] J.E. Van Eyk, M.P. Snyder, Precision medicine: role of proteomics in changing clinical management and care, J. Proteome Res. 18 (1) (2019) 1–6.
[289] F. Wang, A. Preininger, AI in health: state of the art, challenges, and future directions, Yearb. Med. Inform. 28 (1) (2019) 16–26.
[290] P. Butow, E. Hoque, Using artificial intelligence to analyse and teach communication in healthcare, Breast 50 (2020) 49–55.
[291] J. Corral-Acero, et al., The 'Digital Twin' to enable the vision of precision cardiology, Eur. Heart J. (2020), https://doi.org/10.1093/eurheartj/ehaa159, ehaa159.
[292] A. Turing, Computing machinery and intelligence, Mind LIX (236) (1950) 433–460.
[293] M. Harris, et al., A systematic review of the diagnostic accuracy of artificial intelligence-based computer programs to analyze chest X-rays for pulmonary tuberculosis, PLoS One 14 (9) (2019) e0221339.
[294] J. Kedra, et al., Current status of use of big data and artificial intelligence in RMDs: a systematic literature review informing EULAR recommendations, RMD Open 5 (2) (2019) e001004.
[295] J.A. Berinstein, et al., The IBD SGI diagnostic test is frequently used by non-gastroenterologists to screen for inflammatory bowel disease, Inflamm. Bowel Dis. 24 (5) (2018) e18.
[296] K.C. Kalunian, et al., Measurement of cell-bound complement activation products enhances diagnostic performance in systemic lupus erythematosus, Arthritis Rheum. 64 (12) (2012) 4040–4047.
[297] T. Dervieux, et al., Validation of a multi-analyte panel with cell-bound complement activation products for systemic lupus erythematosus, J. Immunol. Methods 446 (2017) 54–59.
[298] D.J. Wallace, et al., Randomised prospective trial to assess the clinical utility of multianalyte assay panel with complement activation products for the diagnosis of SLE, Lupus Sci. Med. 6 (1) (2019) e000349.
[299] J. Mossell, et al., The avise lupus test and cell-bound complement activation products aid the diagnosis of systemic lupus erythematosus, Open Rheumatol. J. 10 (2016) 71–80.
[300] N.J. Olsen, D.R. Karp, Finding lupus in the ANA haystack, Lupus Sci. Med. 7 (1) (2020) e000384.
[301] J.K.H. Andersen, et al., Neural networks for automatic scoring of arthritis disease activity on ultrasound images, RMD Open 5 (1) (2019) e000891.
[302] A. Haj-Mirzaian, et al., Role of artificial intelligence in assessment of peripheral joint MRI in inflammatory arthritis: a systematic review and meta-analysis, Am. Coll. Rheumatol. Meet. Abst. 1 (2020), 1547. In press.

Chapter 2

Precision medicine as an approach to autoimmune diseases

Marvin J. Fritzler[a] and Michael Mahler[b]
[a]Cumming School of Medicine, University of Calgary, Calgary, AB, Canada, [b]Inova Diagnostics, Inc., San Diego, CA, United States

Abbreviations

ACPA	anti-citrullinated peptide antibodies
AI	artificial intelligence
AIM	autoimmune inflammatory myopathy
CTC	circulating tumor cells
DMARD	disease-modifying anti-rheumatic drug
DNN	deep neural networks
EBM	evidence-based medicine
EKG	electrocardiogram
EMR	electronic medical records
FDR	first degree relatives
IA	inflammatory arthritis
MAAAA	multi-analyte arrays with analytic algorithms
ML	machine learning
PH	precision health
PM	precision medicine
RA	rheumatoid arthritis
SjS	Sjögren syndrome
SLE	systemic lupus erythematosus
SSc	systemic sclerosis
VBM	value-based medicine

1 Introduction

Despite tremendous advances in the field of medicine, several limitations persist and new challenges arise due to the recognition of new diseases. The emergence of "new" autoimmune diseases such as autoinflammatory syndromes

Precision Medicine and Artificial Intelligence. https://doi.org/10.1016/B978-0-12-820239-5.00007-3

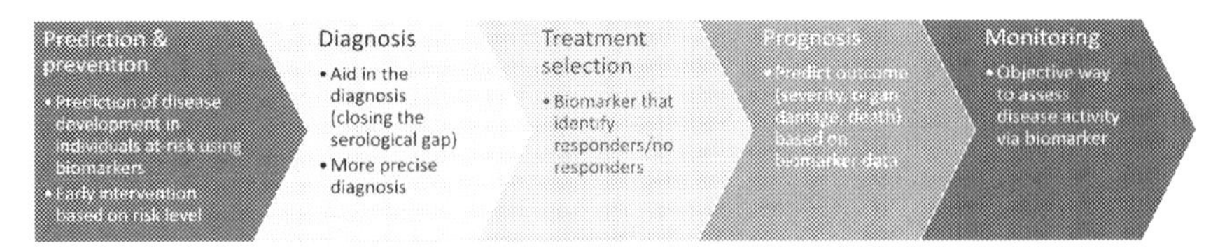

FIG. 1 Unmet needs in the journey of patients with chronic diseases. Management of a patient with a chronic condition is complex and involves several important decision points starting from disease prediction and prevention, the diagnosis, important treatment decisions, providing prognostic information to finally monitoring disease activity (remission and relapse) as well as treatment response and success.

"paranodopathies" [1, 2], re-emergence of drug-induced and drug-related immune response adverse events associated with the use of checkpoint inhibitors in cancer therapy [3, 4] are only a few examples. Even during the pandemic of the corona virus disease (COVID-19), parallels between the viral infection and autoimmune diseases might be useful to understand the pathogenesis of COVID-19 related complications [5–7]. In the last decade, many chronic conditions have been increasingly managed by using principles of Precision Medicine (PM). Obviously, much PM publicity has focused on cancer, heart disease and monogenic conditions, while other diseases increasingly embrace PM concepts. Since PM has yet to make major inroads in the field of autoimmunity, it is informative to learn from examples in the field of cancer [8] and apply those to autoimmunity. This chapter provides a high-level overview of some key elements and persisting challenges for PM approaches to autoimmune diseases (see Fig. 1), in many cases citing rheumatoid arthritis (RA) as example and as prototype disease to learn from.

2 From individualized to precision medicine

PM has been defined as "*a medical model that proposes the customization of healthcare, with medical decisions, treatments, practices, or products being tailored to the individual patient. In this model, diagnostic testing is often employed for selecting appropriate and optimal therapies based on the context of a patient's genetic content or other molecular or cellular analysis. Tools employed in precision medicine can include molecular diagnostics, imaging, and analytics*" (see https://en.wikipedia.org/wiki/Precision_medicine). A simplified version of PM based on patient stratification is shown in Fig. 2. While molecular diagnostics, imaging and other analytics may describe the key data inputs, it is instrumental to remember the importance of more traditional data points such as the complete medical history, physical exam and patient-provided inputs (see Table 1).

Although the origin of PM has been traced back more than a century [75], the term PM evolved from initial attempts to describe the concept of a more tailored approach to medicine. Other terms that have been used include, but are not limited to

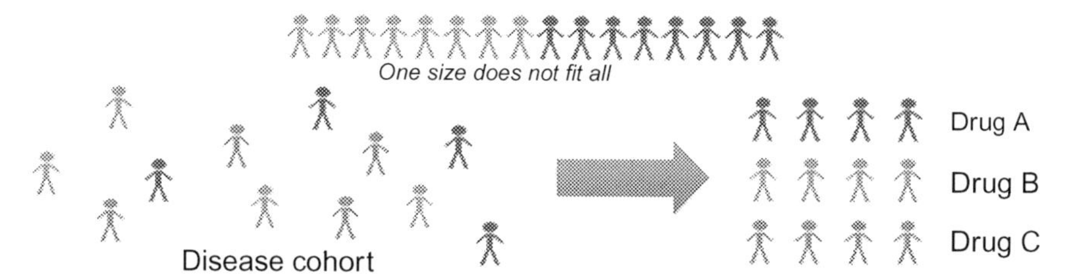

FIG. 2 Simplified concept of patient stratification as part of precision medicine. Using clinical parameters, biomarkers and other data sources, patient stratification has the potential to provide more homogeneous subsets that will benefit from specific treatments.

TABLE 1 Examples of precision medicine in autoimmune condition.

Disease area	Key focus	References
Rheumatoid arthritis	– Disease prediction and prevention – Patient stratification and prognosis – Management of disease – Imaging	[9–24]
Systemic lupus erythematosus	– Patient stratification and prognosis	[17, 25–40]
Systemic sclerosis	– Treatment selection – Cancer risk assessment – Molecular stratification (focus on autoantibodies and other biomarkers)	[41–49]
Autoimmune myositis	– Patient stratification and prognosis – Cancer risk assessment	[50–56]
Type 1 diabetes	– Patient remote monitoring – Life-style modifications – Prediction and prevention	[57–64]
Juvenile dermatomyositis	– Rapid response to therapy – Diminished side effects of therapies because of earlier treatment	[65]
Juvenile idiopathic arthritis	– Decreased side effects form methotrexate and other medications – Predicting and preventing uveitis	[66–68]
Primary biliary cholangitis	– Improve outcomes based on genotypic and molecular characteristics that correlate to specific phenotypic – Right drug, right patient at right time	[69–71]
Multiple sclerosis	– Disease prediction and prevention – Right drug, right patient at the right time	[72, 73]
Antiphospholipid syndrome/thrombosis	– Hydroxychloroquine to prevent thrombotic events	[74]

- Personalized medicine
- Personalized health
- Precision health

Although, the term PM is widely used and accepted, it may miss the mark because the definition of precision means that a given target can be hit repeatedly, even if it is not the correct target. In contrast, accuracy implies that hitting the right target (bulls eye) is achieved (see Fig. 3), but not necessarily for all attempts (limited precision). The ideal goal for PM is to achieve high precision and high accuracy, but if one had to choose, accuracy might be preferable. An additional limitation is the term "medicine" which might not include the concepts of disease prevention, prognosis or interests of public health [76, 77]. Consequently, some experts prefer the term "health" instead of "medicine." In general, there is strong consensus about the scope and the goals of PM. This consensus takes into consideration that the original concept to treat every patient differently was too ambitious and is now focused on homogeneous groups based on molecular taxonomy instead of individuals [78, 79].

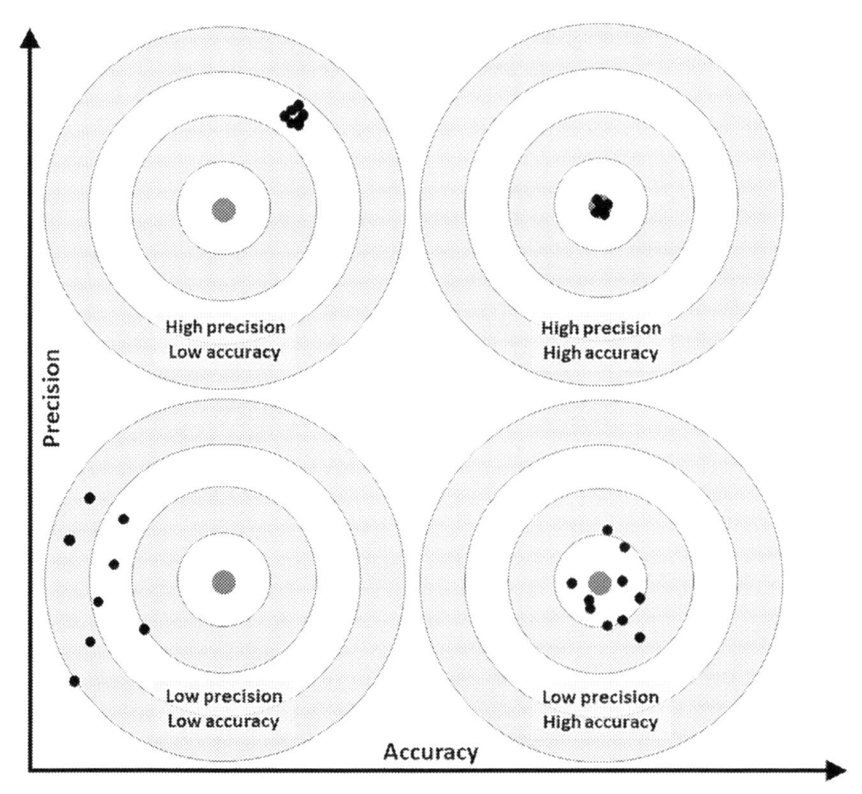

FIG. 3 Precision vs accuracy. Precision means to hit the same area over and over again. However, this does not mean hitting the correct area. In contrast, accuracy means to hit the right area, potentially not precisely.

3 Trends that promote precision medicine

With an explosion of basic and clinical information such as data from clinical trials and evidence-based therapeutic options, it has become increasingly difficult for health care providers to stay abreast of the constantly expanding field of biomarkers. In the United States it was reported that 12 million serious diagnostic errors occur in a single year, that most people will experience a diagnostic error in their lifetime and that more than one-third of medical-related deaths are preventable [80]. Another significant challenge is delay in arriving at a timely and accurate diagnosis. One outcome of these challenges is a sizable caseload of malpractice lawsuits [81]. It is anticipated that PM, coupled with artificial intelligence (AI) [82–85], machine learning (ML) and deep neural networks (DNN) [86], will help health care practitioners overcome these challenges by keeping them abreast of current medical evidence, predict individual risk of developing autoimmune conditions and provide an evidence-based approach to clinically complex situations [50, 87, 88].

Another trend that drives PM is a shift from the service-based to value-based medicine (VBM). Instead of reimbursement by payers (e.g., health insurance companies, government agencies) based on provided service (e.g., drugs prescribed), payments are based on outcome or efficacy of the intervention [89–91], an approach referred to "pay for performance." VBM starts with the best evidence-based data and converts it to patient value-based data, allowing clinicians to deliver higher quality patient care than by using evidence-based medicine (EBM) alone. The ultimate goals of VBM are improving quality of health care and efficient use of health care resources.

4 Challenges for therapeutics, medical interventions and clinical trials

4.1 Heterogeneity of autoimmune diseases

With remarkable advances in our understanding the pathogenesis and pathophysiology of autoimmune diseases has come a growing therapeutic pipeline that has provided numerous therapeutic options. However, when new clinical trials begin or even if the therapeutics achieve regulatory approval, they often fail to meet outcome targets and/or only a minority of patients respond. These failures have been, at least in part, attributed to patient compliance but most notably to the marked heterogeneity (e.g., clinical phenotypes) within diagnostic categories [92]. An illustrative example is belimumab which was approved by the Food and Drug Administration USA (FDA) for SLE in 2011 (and now for lupus nephritis). Remarkably, it was the first drug approved for that disease for over 50 years, and only after leveraging anti-dsDNA antibodies as a biomarker to select patients [25, 93–101]. In the constantly growing autoimmune diseases market, it is thought that small biotechnology companies can especially benefit from leveraging AI technologies for clinical trial design [102, 103]. A potential approach for pharmaceutical companies is the use of companion or complementary diagnostic assays, which have made huge impacts in the field of cancer treatment [104]. Details of companion or complementary diagnostic assays are provided in chapter 4 of this book.

4.2 Improved standard of care increases threshold for outcome measures

Improved standard of care raises the bar for clinical trials to achieve desirable endpoints. For example, in RA, a common endpoint of clinical trials includes the absence or presence of joint erosions. If RA patients are treated early in their disease course, joint erosions are less common, and because of this highly desirable outcome, large numbers of patients are included in the clinical trial (see Fig. 4).

5 Prevention of disease morbidity and mortality

The key to prevention of disease morbidity and mortality is establishing an early and accurate at-risk diagnosis. Since this implies that advanced disease has not yet evolved (i.e., joint erosions, kidney or lung disease) to the point where a patient may not meet classification or diagnostic criteria and, hence, by necessity there is a significant dependency on predictive and prognostic biomarkers. Disease prediction and prevention represents a complex field that requires consideration of several factors, such as:

- Financial aspects. There is extensive evidence that advanced disease (e.g., kidney or central nervous system disease in systemic lupus erythematosus (SLE) [105–107], interstitial lung disease or pulmonary arterial hypertension

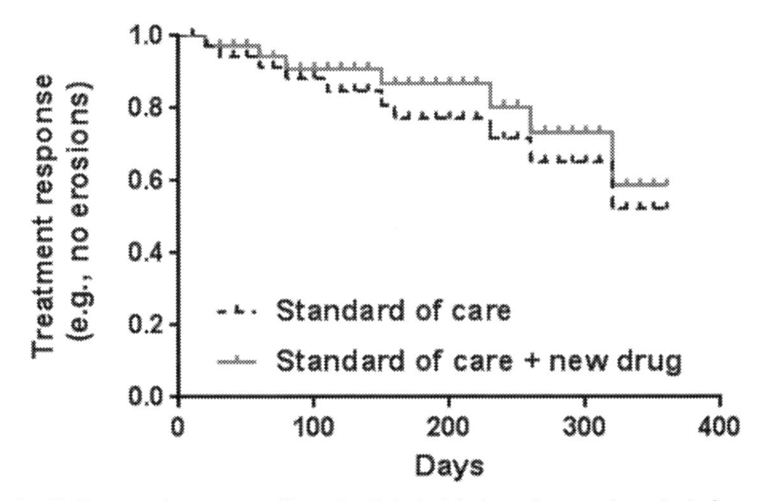

FIG. 4 Challenge to demonstrate efficacy in clinical trials due to improved standard of care illustrated using rheumatoid arthritis. During clinical trial design, the study endpoints have to be defined. In most clinical trials, a placebo group treated with standard of care is included to show superiority of the novel drug. With improved standard of care, this can often be challenging.

in systemic sclerosis (SSc) [108, 109], joint erosions and joint dysfunction in RA [110–112]) are attended by remarkably higher direct and indirect health care costs.
- Logistics. One of the challenges is to develop an evidence-based approach to identify individuals at risk of developing a disease. Whereas a number of studies are based on apparently healthy volunteers, it has yet to be established that this approach will be valid for larger populations of mixed ethnic (genetic variability) or geographic/environmental (epigenetic variability) backgrounds.
- Ethical perspectives. As noted above, some studies developing predictive, prognostic and preventive datasets are based on volunteers, prompting ethical concerns [113–116]. With the wide-spread use of electronic medical records (EMR) where large datasets on groups or individuals abound, the ethics of accessing or mining those datasets as approach to disease prevention has to conform with requirements of the Health Insurance Portability and Accountability Act (HIPAA) and/or have established biomedical legal or ethical support.

As a result of the above-mentioned factors, the target population for screening and prevention, as well as the scope of diseases under investigation has to be clearly defined.

Disease prevention for autoimmune diseases is clearly a major challenge and progress has been limited, but in other diseases approaches to prevention is rapidly being established. Some common examples include:

- Infectious diseases and the use of vaccines.
- Cancer prevention by mitigating toxin exposure and altering lifestyle (e.g., smoking cessation).
- Cardiovascular disease through the use of lipid lowering drugs and altering lifestyle (e.g., smoking cessation) [117].

As stated above, in autoimmunity, approaches to disease prevention are still evolving. Recently, several clinical trials are exploring the prevention of RA before the clinical onset, or before progression to irreversible joint damage [9, 10]. Similar approaches are also being applied to SLE [26, 27, 118], SSc [41–43] and juvenile dermatomyositis [65]. A key element for the success of disease prediction and prevention is education and participation of individuals at risk [119]. Although this seems easy, it might not be trivial. It is of outmost importance that the actionable risk is well explained and conveyed to individuals prompting them to take personal responsibility for their health and wellness [119–121].

In this context, it is noteworthy that significant progress has been made toward prediction and prevention of diabetes [57, 58, 122]. Type 1 diabetes (T1D) is a chronic T-cell mediated autoimmune disease characterized by destruction of pancreatic beta cells. Although new data have better defined the complex etiology underling the interrelation of genetic and environmental factors in the natural history of T1D, relevant pieces of the puzzle still are missing. Genetic predisposition is mainly associated with histocompatibility leukocyte antigen (HLA) alleles; however, recent data suggest that new as well as still unknown genes might better define the complex multigenetic risk of this disease. In addition to the genetic components, the concordance in familial aggregation in T1D indicates a pivotal role of environmental factors that facilitate autoreactivity and autoantibody production. A new, early stage of T1D has recently been proposed in which the detection of two or more autoantibodies in the blood, might identify children at increased risk of eventually developing T1D. In contrast to the improvements achieved by prediction models, primary, secondary and tertiary prevention have yet to achieve a safe and efficacious intervention strategies. In a phase 2 randomized, placebo-controlled, double-blind trial of teplizumab (an Fc receptor-nonbinding anti-CD3 monoclonal antibody) involving relatives of T1D patients who were at high risk for development of T1D, the hazard ratio for the diagnosis of T1D was 0.41 (95% confidence interval, 0.22–0.78; $P = 0.006$). The annualized rates of T1D diagnosis were 14.9% per year in the teplizumab group and 35.9% per year in the placebo group indicating that the preventive treatment reduced progression to clinical T1D in high-risk participants [58].

Other disease prevention areas include, but are not limited to, autoimmune skin disease [123], multiple sclerosis [124], antiphospholipid syndrome [125] as well as primary biliary cholangitis [125–133].

5.1 Modifiable and non-modifiable risk factors

The development of an autoimmune disease is based on certain risk factors that can be divided into two main categories: the modifiable and the non-modifiable. Especially the modifiable factors play in important role in the pre-clinical phase of autoimmunity which might be initiated by modification of the risk profile and is characterized by a break of B- and T-cell tolerance as part of the autoimmune processes leading to increasing tissue damage and significant morbidity [134]. More details are provided in chapters 7 and 10.

5.2 Prediction tools

Several risk factors have been identified that can be used to predict future development of autoimmune diseases [11, 135–137]. However, there is no single factor that provides sufficient value to predict: (a) precisely who will develop autoimmunity, (b) which autoimmune disease will emerge, and equally important (c) when the transition to clinically defined disease will occur. Consequently, it is likely that several risk factors will have to be combined in prediction models and scores. Furthermore, identifying patterns of biomarkers may help identify specific immunologic or inflammation pathways that can be targeted for prevention that are aligned with a given individuals "personal" trajectory of autoimmunity development. For example (and hypothetically), identifying within Subject A that autoantibody X, cytokine Y and glycosylation status Z are abnormal may help inform when the individual might develop RA, as well as identify a specific therapeutic target (e.g., cytokine Y blockade) for prevention.

5.3 Education of populations at risk for autoimmune diseases

Models to predict autoimmune diseases are emerging and important; however, educating individuals regarding their specific risk for future disease represents an equally important aspect of efforts to early treatment and disease prevention. Different approaches have been considered and novel tools have become available that might help to facilitate the education of at-risk individuals. In particular, a trial was conducted among unaffected first-degree relatives (FDR) in which subjects were randomized to receive either a personalized RA risk tool that linked genetics, demographics, biomarkers (ACPA and RF) and behaviors or others who received standard information about RA [138]. In this approach the application of next-generation sequencing technology to PM in cancer (e.g., the Tumor Biomarker Committee of the Chinese Society of Clinical Oncology risk tool) was shown to be successful in increasing the motivation toward positive health behavior and lifestyle changes in at-risk individuals. Similarly, individuals who received personalized risk education demonstrated a significant improved knowledge based on FDR of RA patients, which might represent the first steps toward abrogating RA-risk-related lifestyle. A recent study showed a significantly improved motivation to change RA risk-related behaviors over 12 months when individuals received a personalized education [135]. Further, subjects receiving their personalized RA risk score were more likely to be reassured about anxiety related to their RA risk as opposed to those in the comparison arm of the study. Therefore, it is possible to tailor these personalized interventions to at-risk individuals without harmful psychological repercussions. Taken together, there is an increasing body of evidence that involving patients, relatives and health care providers in decision making regarding prevention and management of RA can improve overall outcomes. However, major breakthroughs might not be achievable without significant incentives and penalties driven by health care payers [139, 140]. More details are provided in chapter 10 of this book.

5.4 Prevention of autoimmunity: Simple, safe and potentially beneficial approaches

Reduction of RA development or progression can be improved by relatively simple approaches. For example, smoking has been demonstrated to represent a risk factor for the development of ACPA-positive RA, a more aggressive form of the disease [141]. Therefore, it should be incumbent on all rheumatologist to motivate RA patients to stop smoking and potentially family care physicians to motivate FDR to consider smoking cessation, which would also provide other benefits such as reduction of cancer risk and economic benefits for former smokers. In addition, dietary supplements such as omega 3 fatty acids and vitamin D supplements have been considered valuable for attenuation of several diseases [142, 143]. In addition, a recent study has indicated that sleep duration might be related to transition from pre-clinical phase to clinically apparent SLE [28]. This finding must be confirmed and validated. Similar approaches and caveats apply to diet and other life-style choices. To conclude this brief overview, several simple and safe measures can be applied that likely will reduce (not eliminate) the risk of developing autoimmune disease (for details, see also chapters 7 and 10).

5.5 Clinical trials for the prevention of autoimmune diseases

In RA, significant improvements have been made in achieving an early diagnosis and timely treatment [78]. The early identification of patients in the pre-clinical phase of RA is of high importance because it became evident during the last decade that early intervention can prevent joint damage in patients with RA [11, 137]. In addition to improving the diagnosis of RA in individuals with clinically-apparent inflammatory arthritis (IA), the elucidation of the natural history of RA during which circulating RA-related biomarker profiles can be elevated years prior to the first clinically-apparent IA has opened the door to considerations of RA prevention [78, 144–147]. Indeed, elevation of ACPA (with or without RF) in individuals without clinically-apparent IA are key inclusion factors in several clinical trials that have completed, or are underway [10, 12, 13, 148].

Several other studies are currently testing the ability of a variety of agents such as glucocorticoids, hydroxychloroquine, abatacept and statins to delay or halt the development of IA/RA. All of the completed or ongoing RA prevention trials (see Table 2) utilize the concept of the "window of opportunity" to prevent or delay the clinical impact of disease, where basically intervening at the right time in RA development can lead to substantial future benefit, including patient outcomes as well as health economic benefits [14, 149–154]. As in the PRAIRI trial [10], these studies relied primarily on blood-based autoantibody biomarkers to precisely identify individuals who are at high-risk for future RA.

The results from these studies will need to be examined carefully to determine how they will ultimately impact future prevention clinical trials and more cogently, clinical care of RA. The design of next generation of prevention trials need to address both the optimal timing of intervention in pre-RA and identifying risk-based, multi-level interventions.

A significant challenge to identify the right time for treatment is the patient-specific transition and trajectory from pre-clinical to clinical RA (Fig. 1). In some patients, the pre-clinical phase encompasses several years whereas in others the conversion to complete RA might be over a much shorter time period. Factors triggering, contributing, or accelerating this transition are not fully understood. Additional information will be available from other prevention

TABLE 2 Overview of big data approach for precision medicine.

Data type	Potential source	Comments
Demographic data	Electronic medical record (EMR), patient provided	Data can be obtained either from EMR, or more increasingly via patients (e.g., in phone applications)
Environmental data	GPS, patient provided, jurisdictional monitoring agencies	Pollution, air quality, exposure to toxic substances and xenobiotics
Biomarker	Wearable devices, laboratory tests, imaging	Can provide data on disease pathology, disease activity, prognostics
Clinical assessment	Clinician, EMR	Input errors need to be minimized

trials as well as other studies of the natural history of RA. In turn, this information will hopefully shed light on evidence-based approaches to identify the precise interventional agent to match the biologic profile of each patient so that prevention can be optimized.

Moreover, the willingness to participate and comply with recommended therapies or intervention by individuals at risk to develop autoimmune conditions depends on several factors including patient perceptions and preferences that may be related to the safety and tolerability profile of the respective drugs (both short-term and long-term) [155–159]. In this respect, drugs which have been used for many years will clearly have an advantage based on the wealth of safety data available (e.g., hydroxychloroquine). Other aspects include the costs and the accessibility to the drug. In addition, the long-term efficiency, in other words the ability to introduce long term drug-free remission, should be considered. Lastly, the administration format of the therapeutic (oral *vs.* injection) will also have an impact. Depending on the cost and safety profile of the treatment, different accuracy levels of the prediction model are required (see Fig. 2).

6 Mobile devices and web-based software applications and the future of medicine

6.1 Symptom checker

The diagnosis of any disease typically begins with the recognition of key signs and symptoms exhibited by the affected individual. With the broad availability of the internet, more and more individuals are seeking online help and advice, which includes searches on or "consultations" with Dr. Google, Wikipedia, IBM's (Big Blue) or "Dr. Watson," Alexa, Cortana, Siri, Bixby and other social media platforms [160–167]. It has been estimated that more than one third of American adults use the internet to diagnose medical conditions although this has raised significant concerns from clinicians. Consequently, mobile phone and web-based software applications such as symptom-checker-apps have been developed that can provide a probability of a disease based on combinations of symptoms and demographic aspects [168–173]. A study by Harvard Medical School demonstrated varying accuracies of the systems [169]. A recent small study evaluating the utility of such a web-based tool for the diagnosis of RA suggested that online advice for rheumatic diseases can be inappropriate and that the diagnoses suggested are frequently inaccurate [168]. For example, social media recommendations to seek emergency advice may prompt inappropriate health care utilization. Hence, further optimization of "apps" and related diagnostic interfaces is needed and will likely improve online decision support. However, utilizing mobile apps to collect real world data for research and clinical purpose has already demonstrated potential to improve clinical trials and management of RA patients. There are several accessories such as wearable devices (e.g., smart watches) that can record real-time, big data in "apps" to provide more complete insights into the disease progression and to enrich clinical decision making [174–176].

From the Apple Watch's EKG capabilities to new continuous glucose monitoring systems, wearable medical technologies have a wide range of potential applications in health care. A number of mobile applications are now available for rheumatic disease patients [177, 178]. Although the increasing availability of these "apps" may enable people with rheumatic diseases to more effectively and efficiently self-manage their health [179], an approach to ensuring appropriate development and evaluation of "apps" has been limited. To date, most "apps" are for RA with the three main focus areas of pain, fatigue, and physical activity being recorded. Another issue is that the useful lifetime of these "apps" is limited as some are no longer in use or readily available. Moreover, it was found that very few studies showed improvement of disease outcome measures though use off the "apps" [177]. Hence, because of the increasing use mobile "apps" for self-management, continuous improvement as well as optimized standards and quality assurance of existing and new "apps" are still required.

6.2 Wearable technologies

As for other aspects of PM, progress in autoimmunity is limited, but is slowly gaining momentum. Learning from other clinical areas will likely help to facilitate advances in autoimmunity. An example of wearable technologies that might enable PM includes Walking Data from Wearables Predicting Alzheimer's Disease [85]. Alzheimer's disease patients commonly have altered walking mechanics, or gait. Gait speed, symmetry, and stride length are typically impaired in patients with the disease, and their walking speed is much more variable. This can be detected via clinical assessment, with the physician observing and timing patients as they walk a certain distance. Alternatively, these patients could be monitored through portable equipment such as sensors within smartphones, watches, and other wearables to provide accurate data regarding the patient's gait and thereby providing a way to continuously monitor patients walking habits. This information in turn could be enhanced even further with contact sensors in a shoe or sock that provide pressure readings. Another example of a wearable device is one that precisely detects cancer cells in blood. Such a wearable device can continuously monitor for circulating tumor cells (CTCs) in the blood [180]. These cancer cells are typically obtained via blood samples to provide a biomarker for treatment, but a wrist-worn prototype could potentially screen patients' blood for a few hours to obtain only the CTCs of interest. As another example, excess sun exposure can cause flares in SLE, triggering symptoms such as severe skin rashes, joint pain, weakness, and fatigue. There are a number of mobile and wearable devices that can help SLE patients [181], accurately track sun exposure. The UV sensor paired with smartphone measures and managing radiation exposure advising the user when they reached tolerable limits. Another example of wearable technology paired with AI as an aid in the management of autoimmune conditions is an activity tracker for RA and axial spondyloarthritis [182, 183].

6.3 Future of rheumatology?

Burmester [15] recently presented a hypothetical case of a 44-year-old female patient who uncovers her diagnosis by talking to an AI-based voice recognition system that inquires about symptoms and assesses further information via the internet revealing that her physical activity significantly dropped from an average of 8500 to 4700 steps as recorded by a wearable device. This information triggers a referral to a supermarket (big box store) to acquire joint imaging and an 'omics' biomarker profile. Based on the AI algorithmic analysis of most significant information, including a positive RF, ACPA and anti-Ro60 antibody results, HLA shared epitope and reduced enzyme activity for metabolism of nonsteroidal drugs, a diagnosis of RA is established with 99% probability. Next, using a computer system she is educated about four treatment options with detailed explanation of predicted responses, side effects and other risks. She opts for the most aggressive treatment including a biologic, a disease-modifying antirheumatic drug (DMARD) and chimeric antigen receptor (CAR) T-cell therapy. This selection finally triggers a referral to a rheumatologist who now can spend essential time with the patient. Although there are several gaps and concerns related to the described hypothetical case, some aspects might help to overcome current limitations of patient care in rheumatology, which includes access of patients to rheumatologists and time that can be shared between patient and clinician (e.g., rheumatologists) [85, 184–186]. This will ultimately reduce potential delays in diagnosis and will enable earlier treatment. Mobile and wearable technologies offer a convenient means of monitoring many physiological features, presenting a multitude of medical solutions [174, 175, 187, 188]. Not only are these devices easy for the consumer to use, but they offer real-time data for physicians.

7 Importance of biomarkers

Biomarkers are defined as indicators of a definable biological or pathological process that may be used in diagnosis or to monitor therapy [189]. The National Institutes of Health Working Group has categorized biomarkers as those providing data and/or evidence pertaining to disease susceptibility and/or risk, diagnosis, monitoring, prediction and prognosis, pharmacodynamics and therapeutic responses and therapeutic safety [189, 190] (see also Figure 1).

In the realm of laboratory medicine, biomarkers are often generalized to the 'omics' [29]:

- Genomics/metagenomics
- Proteomics
- Lipidomic and glycomics
- Metabolomics
- Metalomics
- Microbiomics

With increasing availability and use of multi-analyte arrays with analytic algorithms (MAAAA) [30, 78, 191] and gene chips, the application of "omics" to PM has the potential to provide large data sets for individual patients that require interpretative skills and accuracy that can likely be only achieved through the use of AI, ML augmented ML, and DNN. Using these approaches, significant insight in complex autoimmune diseases such as RA [16–18, 145, 147, 192, 193], SLE [17, 29–37, 78, 194–201], SSc [44–46, 202, 203], AIM [204, 205] and SjS [206] is of increasing importance in order to facilitate actionable clinical decision making [207, 208].

8 Health economics

On the backdrop of aging populations, pandemics and the rising prevalence of chronic disease, the demand for more accessible, efficient and effective health care is rising in many countries. With limited health care resources (especially when put in the context of a pandemic), it is important that new approaches and innovative technologies are used in a way that provides the most value for patients. Since the management of autoimmune diseases is attended by remarkable health care expenditures, there is a significant opportunity to achieve meaningful savings. As discussed earlier in this chapter, a compelling reason to focus on early disease diagnosis and morbidity prevention is the high economic burden that comes with the care of advanced organ involvement in autoimmune diseases. To optimize the impact PM on autoimmune diseases, increased investments on health and disease prevention are required. One increasingly important aspect regarding health economics in autoimmune diseases is to intervene early or to even prevent disease progression to end-organ compromise and failure. This will require screening of individuals at risk. When it comes to population screening (or screening of at-risk populations), health economics becomes an important factor in the equation. Since the global health care expenditures as a proportion of the gross domestic product are constantly increasing and reaching non-sustainable thresholds in many jurisdictions, health economic studies of direct and indirect costs will be needed to demonstrate that investments in screening for pre-autoimmune conditions and early interventions or therapies are attended by incrementally improved clinical outcomes. Several studies have compared the cost per response of different treatment options in autoimmune conditions, especially in RA. As outlined above, such aspects will become more and more important following the trend toward value *vs.* service-based health care systems [209–211]. When it comes to prediction and prevention of autoimmune diseases, reduction of health care costs and negative socioeconomic impacts associated with such conditions seems achievable. Promoting early disease prevention or remission could decrease the health care burden of autoimmune diseases by tempering the need for long-term costly treatments, diminishing disability and improving quality of life.

9 Conclusion

Application of PM to autoimmune diseases is a compelling approach to improve patient care and outcomes while reducing associated health care costs. Early diagnosis and treatment decision will be instrumental in achieving the goals of PM which rely on precise, accurate and reliable biomarkers to support clinical assessment and patient provided information. Due to regulatory requirements and need for validation, slow but steady progress can be expected toward achieving the goals purported by PM approaches.

Competing interests

M. Mahler is employed at Inova Diagnostics, an autoimmune diagnostics company. M.J. Fritzler is a consultant to and has received honoraria from Inova Diagnostics.

References

[1] A. Uncini, J.M. Vallat, Autoimmune nodo-paranodopathies of peripheral nerve: the concept is gaining ground, J. Neurol. Neurosurg. Psychiatry 89 (6) (2018) 627–635.

[2] N. Garg, et al., Conduction block in immune-mediated neuropathy: paranodopathy versus axonopathy, Eur. J. Neurol. 26 (8) (2019) 1121–1129.

[3] L.H. Calabrese, C. Calabrese, L.C. Cappelli, Rheumatic immune-related adverse events from cancer immunotherapy, Nat. Rev. Rheumatol. 14 (10) (2018) 569–579.

[4] L. Calabrese, X. Mariette, The evolving role of the rheumatologist in the management of immune-related adverse events (irAEs) caused by cancer immunotherapy, Ann. Rheum. Dis. 77 (2) (2018) 162–164.

[5] P. Sarzi-Puttini, et al., COVID-19, cytokines and immunosuppression: what can we learn from severe acute respiratory syndrome? Clin. Exp. Rheumatol. 38 (2) (2020) 337–342.

[6] F. Ferro, et al., COVID-19: the new challenge for rheumatologists, Clin. Exp. Rheumatol. 38 (2) (2020) 175–180.

[7] E.G. Favalli, et al., COVID-19 infection and rheumatoid arthritis: faraway, so close! Autoimmun. Rev. 20 (2020) 102523.
[8] X. Zhang, et al., Application of next-generation sequencing technology to precision medicine in cancer: joint consensus of the Tumor Biomarker Committee of the Chinese Society of Clinical Oncology, Cancer Biol. Med. 16 (1) (2019) 189–204.
[9] D.M. Gerlag, J.M. Norris, P.P. Tak, Towards prevention of autoantibody-positive rheumatoid arthritis: from lifestyle modification to preventive treatment, Rheumatology (Oxford) 55 (4) (2016) 607–614.
[10] D.M. Gerlag, et al., Effects of B-cell directed therapy on the preclinical stage of rheumatoid arthritis: the PRAIRI study, Ann. Rheum. Dis. 78 (2) (2019) 179–185.
[11] K.D. Deane, Preclinical rheumatoid arthritis and rheumatoid arthritis prevention, Curr. Rheumatol. Rep. 20 (8) (2018) 50.
[12] K. Mankia, A. Di Matteo, P. Emery, Prevention and cure: the major unmet needs in the management of rheumatoid arthritis, J. Autoimmun. 110 (2019) 102399.
[13] D. Alpizar-Rodriguez, A. Finckh, Is the prevention of rheumatoid arthritis possible? Clin. Rheumatol. 39 (5) (2020). 1393-1389.
[14] K. Mankia, P. Emery, A new window of opportunity in rheumatoid arthritis: targeting at-risk individuals, Curr. Opin. Rheumatol. 28 (3) (2016) 260–266.
[15] G.R. Burmester, Rheumatology 4.0: big data, wearables and diagnosis by computer, Ann. Rheum. Dis. 77 (7) (2018) 963–965.
[16] L.A. Trouw, R.E. Toes, Rheumatoid arthritis: autoantibody testing to predict response to therapy in RA, Nat. Rev. Rheumatol. 12 (10) (2016) 566–568.
[17] T.L. Wampler Muskardin, et al., Lessons from precision medicine in rheumatology, Mult. Scler. 26 (5) (2020). https://doi.org/10.1177/1352458519884249.
[18] K.D. Deane, T.T. Cheung, Rheumatoid arthritis prevention: challenges and opportunities to change the paradigm of disease management, Clin. Ther. 41 (7) (2019) 1235–1239.
[19] H.H. Chang, et al., A molecular signature of preclinical rheumatoid arthritis triggered by dysregulated PTPN22, JCI Insight 1 (17) (2016) e90045.
[20] A. Finckh, D. Alpizar-Rodriguez, P. Roux-Lombard, Value of biomarkers in the prevention of rheumatoid arthritis, Clin. Pharmacol. Ther. 102 (4) (2017) 585–587.
[21] M. Mahler, Population-based screening for ACPAs: a step in the pathway to the prevention of rheumatoid arthritis? Ann. Rheum. Dis. 76 (11) (2017) e42.
[22] L.E. Burgers, et al., Brief report: clinical trials aiming to prevent rheumatoid arthritis cannot detect prevention without adequate risk stratification: a trial of methotrexate versus placebo in undifferentiated arthritis as an example, Arthritis Rheumatol. 69 (5) (2017) 926–931.
[23] H. Radner, et al., 2017 EULAR recommendations for a core data set to support observational research and clinical care in rheumatoid arthritis, Ann. Rheum. Dis. 77 (4) (2018) 476–479.
[24] D. Luo, et al., Mobile apps for individuals with rheumatoid arthritis: a systematic review, J. Clin. Rheumatol. 25 (3) (2019) 133–141.
[25] R. Felten, et al., Advances in the treatment of systemic lupus erythematosus: from back to the future, to the future and beyond, Joint Bone Spine 86 (4) (2019) 429–436.
[26] N.J. Olsen, et al., Study of anti-malarials in incomplete lupus erythematosus (SMILE): study protocol for a randomized controlled trial, Trials 19 (1) (2018) 694.
[27] M.Y. Choi, et al., Preventing the development of SLE: identifying risk factors and proposing pathways for clinical care, Lupus 25 (8) (2016) 838–849.
[28] K.A. Young, et al., Less than 7 hours of sleep per night is associated with transitioning to systemic lupus erythematosus, Lupus 27 (9) (2018) 1524–1531.
[29] W. Song, et al., Advances in applying of multi-omics approaches in the research of systemic lupus erythematosus, Int. Rev. Immunol. (2020) 1–11.

[30] M.J. Fritzler, M. Mahler, Redefining systemic lupus erythematosus—SMAARTT proteomics, Nat. Rev. Rheumatol. 14 (8) (2018) 451–452.
[31] G. Barturen, M.E. Alarcon-Riquelme, SLE redefined on the basis of molecular pathways, Best Pract. Res. Clin. Rheumatol. 31 (3) (2017) 291–305.
[32] M.J. Lewis, et al., Autoantibodies targeting TLR and SMAD pathways define new subgroups in systemic lupus erythematosus, J. Autoimmun. 91 (2018) 1–12.
[33] Q. Luo, et al., Circular RNAs hsa_circ_0000479 in peripheral blood mononuclear cells as novel biomarkers for systemic lupus erythematosus, Autoimmunity 194 (2020) 1–10.
[34] Y. Kotliarov, et al., Broad immune activation underlies shared set point signatures for vaccine responsiveness in healthy individuals and disease activity in patients with lupus, Nat. Med. 26 (4) (2020) 618–629.
[35] N. Bona, et al., Oxidative stress, inflammation and disease activity biomarkers in lupus nephropathy, Lupus 29 (3) (2020) 311–323.
[36] M.A. Smith, et al., SLE plasma profiling identifies unique signatures of lupus nephritis and discoid lupus, Sci. Rep. 9 (1) (2019) 14433.
[37] D.S. Pisetsky, Evolving story of autoantibodies in systemic lupus erythematosus, J. Autoimmun. 110 (6) (2019) 102356.
[38] M. Barbhaiya, et al., Influence of alcohol consumption on the risk of systemic lupus erythematosus among women in the nurses' health study cohorts, Arthritis Care Res. (Hoboken) 69 (3) (2017) 384–392.
[39] K.A. Young, et al., Screening characteristics for enrichment of individuals at higher risk for transitioning to classified SLE, Lupus 28 (5) (2019) 597–606.
[40] M.Y. Choi, M.J. Fritzler, Autoantibodies in SLE: prediction and the p value matrix, Lupus 28 (11) (2019) 1285–1293.
[41] C.A. Mecoli, A.A. Shah, More than skin deep: bringing precision medicine to systemic sclerosis, Arthritis Rheumatol. 72 (3) (2020) 383–385.
[42] P.J. Wermuth, et al., Existing and novel biomarkers for precision medicine in systemic sclerosis, Nat. Rev. Rheumatol. 14 (7) (2018) 421–432.
[43] P.J. Wermuth, S. Piera-Velazquez, S.A. Jimenez, Identification of novel systemic sclerosis biomarkers employing aptamer proteomic analysis, Rheumatology (Oxford) 57 (10) (2018) 1698–1706.
[44] J. Blagojevic, et al., Classification, categorization and essential items for digital ulcer evaluation in systemic sclerosis: a DeSScipher/European Scleroderma Trials and Research group (EUSTAR) survey, Arthritis Res. Ther. 21 (1) (2019) 35.
[45] S.I. Nihtyanova, et al., Using autoantibodies and cutaneous subset to develop outcome-based disease classification in systemic sclerosis, Arthritis Rheumatol. 72 (3) (2020) 465–476.
[46] A.L. Herrick, et al., Patterns and predictors of skin score change in early diffuse systemic sclerosis from the European Scleroderma Observational Study, Ann. Rheum. Dis. 77 (4) (2018) 563–570.
[47] A.A. Shah, et al., Evaluation of cancer-associated myositis and scleroderma autoantibodies in breast cancer patients without rheumatic disease, Clin. Exp. Rheumatol. 35 (Suppl. 106(4)) (2017) 71–74.
[48] T. Igusa, et al., Autoantibodies and scleroderma phenotype define subgroups at high-risk and low-risk for cancer, Ann. Rheum. Dis. 77 (8) (2018) 1179–1186.
[49] R.L. Smeets, et al., Diagnostic profiles for precision medicine in systemic sclerosis; stepping forward from single biomarkers towards pathophysiological panels, Autoimmun. Rev. 19 (5) (2020) 102515.
[50] M. Mahler, B. Rossin, O. Kubassova, Augmented versus artificial intelligence for stratification of patients with myositis, Ann. Rheum. Dis. 79 (12) (2020) e162.

[51] A. Aussy, O. Boyer, N. Cordel, Dermatomyositis and immune-mediated necrotizing myopathies: a window on autoimmunity and cancer, Front. Immunol. 8 (2017) 992.
[52] M. Mahler, M.J. Fritzler, Detection of myositis-specific antibodies: additional notes, Ann. Rheum. Dis. 78 (5) (2019) e45.
[53] B. Stuhlmuller, et al., Disease specific autoantibodies in idiopathic inflammatory myopathies, Front. Neurol. 10 (2019) 438.
[54] M. Best, et al., Use of anti-transcriptional intermediary factor-1 gamma autoantibody in identifying adult dermatomyositis patients with cancer: a systematic review and meta-analysis, Acta Derm. Venereol. 99 (3) (2019) 256–262.
[55] C. Anquetil, et al., Myositis-specific autoantibodies, a cornerstone in immune-mediated necrotizing myopathy, Autoimmun. Rev. 18 (3) (2019) 223–230.
[56] M. Mahler, et al., Standardisation of myositis-specific antibodies: where are we today? Ann. Rheum. Dis. (2019). Aug. 3, 21603.
[57] F. Chiarelli, C. Giannini, M. Primavera, Prediction and prevention of type 1 diabetes in children, Clin. Pediatr. Endocrinol. 28 (3) (2019) 43–57.
[58] K.C. Herold, et al., An anti-CD3 antibody, teplizumab, in relatives at risk for type 1 diabetes, N. Engl. J. Med. 381 (7) (2019) 603–613.
[59] S. Abhari, et al., Artificial intelligence applications in type 2 diabetes mellitus care: focus on machine learning methods, Healthc. Inform. Res. 25 (4) (2019) 248–261.
[60] R. Singla, et al., Artificial intelligence/machine learning in diabetes care, Indian J. Endocrinol. Metab. 23 (4) (2019) 495–497.
[61] S. Ashrafzadeh, O. Hamdy, Patient-driven diabetes care of the future in the technology era, Cell Metab. 29 (3) (2019) 564–575.
[62] A. Galderisi, et al., Continuous glucose monitoring linked to an artificial intelligence risk index: early footprints of intraventricular hemorrhage in preterm neonates, Diabetes Technol. Ther. 21 (3) (2019) 146–153.
[63] D.T. Broome, C.B. Hilton, N. Mehta, Policy implications of artificial intelligence and machine learning in diabetes management, Curr. Diab. Rep. 20 (2) (2020) 5.
[64] S. Kato, et al., Effectiveness of lifestyle intervention using the internet of things system for individuals with early type 2 diabetes mellitus, Intern. Med. 59 (1) (2020) 45–53.
[65] J. Martorell-Marugan, et al., Deep learning in omics data analysis and precision medicine, in: H. Husi (Ed.), Computational Biology, Codon Publications, Brisbane (AU), 2019. Chapter 3.
[66] L.B. Ramsey, et al., Association of SLCO1B1 *14 allele with poor response to methotrexate in juvenile idiopathic arthritis patients, ACR Open Rheumatol. 1 (1) (2019) 58–62.
[67] J. Roszkiewicz, E. Smolewska, In the pursuit of methotrexate treatment response biomarker in juvenile idiopathic arthritis-are we getting closer to personalised medicine? Curr. Rheumatol. Rep. 19 (4) (2017) 19.
[68] J. Roszkiewicz, K. Orczyk, E. Smolewska, Tocilizumab in the treatment of systemic-onset juvenile idiopathic arthritis—single-centre experience, Reumatologia 56 (5) (2018) 279–284.
[69] L. Cristoferi, et al., Prognostic models in primary biliary cholangitis, J. Autoimmun. 95 (2018) 171–178.
[70] L. Cristoferi, et al., Individualizing care: management beyond medical therapy, Clin. Liver Dis. 22 (3) (2018) 545–561.
[71] V. Ronca, et al., Precision medicine in primary biliary cholangitis, J. Dig. Dis. 20 (7) (2019) 338–345.
[72] R. Seccia, et al., Considering patient clinical history impacts performance of machine learning models in predicting course of multiple sclerosis, PLoS One 15 (3) (2020) e0230219.

[73] G. Bose, M.S. Freedman, Precision medicine in the multiple sclerosis clinic: selecting the right patient for the right treatment, Mult. Scler. 26 (5) (2020) 540–547.
[74] E. Kravvariti, et al., The effect of hydroxychloroquine on thrombosis prevention and antiphospholipid antibody levels in primary antiphospholipid syndrome: a pilot open label randomized prospective study, Autoimmun. Rev. 19 (4) (2020) 102491.
[75] M.K. Konstantinidou, et al., Are the origins of precision medicine found in the corpus hippocraticum? Mol. Diagn. Ther. 21 (6) (2017) 601–606.
[76] S.R. Snyder, et al., Generic cost-effectiveness models: a proof of concept of a tool for informed decision-making for public health precision medicine, Public Health Genomics 21 (5–6) (2018) 217–227.
[77] M.J. Khoury, M.F. Iademarco, W.T. Riley, Precision public health for the era of precision medicine, Am. J. Prev. Med. 50 (3) (2016) 398–401.
[78] M.J. Fritzler, et al., The utilization of autoantibodies in approaches to precision health, Front. Immunol. 9 (2018) 2682.
[79] A.J. Elliot, et al., Internet-based remote health self-checker symptom data as an adjuvant to a national syndromic surveillance system, Epidemiol. Infect. 143 (16) (2015) 3416–3422.
[80] P.W. Yoon, et al., Potentially preventable deaths from the five leading causes of death—United States, 2008–2010, MMWR Morb. Mortal. Wkly Rep. 63 (17) (2014) 369–374.
[81] G.D. Schiff, et al., Diagnostic error in medicine: analysis of 583 physician-reported errors, Arch. Intern. Med. 169 (20) (2009) 1881–1887.
[82] N. Rifai, et al., Disruptive innovation in laboratory medicine, Clin. Chem. 61 (9) (2015) 1129–1132.
[83] E.J. Topol, The big medical data miss: challenges in establishing an open medical resource, Nat. Rev. Genet. 16 (5) (2015) 253–254.
[84] S. Jha, E.J. Topol, Information and artificial intelligence, J. Am. Coll. Radiol. 15 (3 Pt B) (2018) 509–511.
[85] E.J. Topol, High-performance medicine: the convergence of human and artificial intelligence, Nat. Med. 25 (1) (2019) 44–56.
[86] G.Z. Papadakis, et al., Deep learning opens new horizons in personalized medicine, Biomed. Rep. 10 (4) (2019) 215–217.
[87] A. Rajkomar, J. Dean, I. Kohane, Machine learning in medicine, N. Engl. J. Med. 380 (14) (2019) 1347–1358.
[88] A. Rajkomar, et al., Automatically charting symptoms from patient-physician conversations using machine learning, JAMA Intern. Med. 179 (6) (2019) 836–838.
[89] M.B. Rosenthal, Beyond pay for performance—emerging models of provider-payment reform, N. Engl. J. Med. 359 (12) (2008) 1197–1200.
[90] R.K. Smoldt, D.A. Cortese, Pay-for-performance or pay for value? Mayo Clin. Proc. 82 (2) (2007) 210–213.
[91] K.K. Kondo, et al., Implementation processes and pay for performance in healthcare: a systematic review, J. Gen. Intern. Med. 31 (Suppl. 1) (2016) 61–69.
[92] R. Goldberg, S. Kauffman, E.J. Topol, Study design and the drug development process, JAMA 311 (19) (2014) 2023.
[93] W. Stohl, D.M. Hilbert, The discovery and development of belimumab: the anti-BLyS-lupus connection, Nat. Biotechnol. 30 (1) (2012) 69–77.
[94] C. Ding, Belimumab, an anti-BLyS human monoclonal antibody for potential treatment of inflammatory autoimmune diseases, Expert. Opin. Biol. Ther. 8 (11) (2008) 1805–1814.

[95] D.J. Wallace, et al., A phase II, randomized, double-blind, placebo-controlled, dose-ranging study of belimumab in patients with active systemic lupus erythematosus, Arthritis Rheum. 61 (9) (2009) 1168–1178.
[96] R. Furie, et al., A phase III, randomized, placebo-controlled study of belimumab, a monoclonal antibody that inhibits B lymphocyte stimulator, in patients with systemic lupus erythematosus, Arthritis Rheum. 63 (12) (2011) 3918–3930.
[97] W. Stohl, et al., Belimumab reduces autoantibodies, normalizes low complement levels, and reduces select B cell populations in patients with systemic lupus erythematosus, Arthritis Rheum. 64 (7) (2012) 2328–2337.
[98] R.F. van Vollenhoven, et al., Belimumab in the treatment of systemic lupus erythematosus: high disease activity predictors of response, Ann. Rheum. Dis. 71 (8) (2012) 1343–1349.
[99] N.B. Kandala, et al., Belimumab: a technological advance for systemic lupus erythematosus patients? Report of a systematic review and meta-analysis, BMJ Open 3 (7) (2013) e002852.
[100] I. Cavazzana, et al., Autoantibodies' titre modulation by anti-BlyS treatment in systemic lupus erythematosus, Lupus 28 (9) (2019) 1074–1081.
[101] C. Anjo, et al., Effectiveness and safety of belimumab in patients with systemic lupus erythematosus in a real-world setting, Scand. J. Rheumatol. 48 (6) (2019) 469–473.
[102] R.A. Moscicki, P.K. Tandon, Drug-development challenges for small biopharmaceutical companies, N. Engl. J. Med. 376 (5) (2017) 469–474.
[103] M. Woo, An AI boost for clinical trials, Nature 573 (7775) (2019) S100–S102.
[104] D.S. Pisetsky, B.H. Rovin, P.E. Lipsky, New perspectives in rheumatology: biomarkers as entry criteria for clinical trials of new therapies for systemic lupus erythematosus: the example of antinuclear antibodies and anti-DNA, Arthritis Rheumatol. 69 (3) (2017) 487–493.
[105] M.R.W. Barber, A.E. Clarke, Socioeconomic consequences of systemic lupus erythematosus, Curr. Opin. Rheumatol. 29 (5) (2017) 480–485.
[106] M.R.W. Barber, et al., Economic evaluation of lupus nephritis in the systemic lupus international collaborating clinics inception cohort using a multistate model approach, Arthritis Care Res. 70 (9) (2018) 1294–1302.
[107] M.R.W. Barber, et al., Economic evaluation of damage accrual in an international SLE inception cohort using a multi-state model approach, Arthritis Care Res. (Hoboken) 72 (12) (2020) 1800–1808.
[108] D.E. Furst, et al., Annual medical costs and healthcare resource use in patients with systemic sclerosis in an insured population, J. Rheumatol. 39 (12) (2012) 2303–2309.
[109] J. Lopez-Bastida, et al., Social/economic costs and health-related quality of life in patients with scleroderma in Europe, Eur. J. Health Econ. 17 (Suppl. 1) (2016) 109–117.
[110] A. Hresko, T.C. Lin, D.H. Solomon, Medical care costs associated with rheumatoid arthritis in the US: a systematic literature review and meta-analysis, Arthritis Care Res. (Hoboken) 70 (10) (2018) 1431–1438.
[111] G.M. Oderda, et al., The potential impact of monitoring disease activity biomarkers on rheumatoid arthritis outcomes and costs, Per. Med. 15 (4) (2018) 291–301.
[112] C.K. Porter, et al., Cohort profile of a US military population for evaluating pre-disease and disease serological biomarkers in rheumatoid and reactive arthritis: rationale, organization, design, and baseline characteristics, Contemp. Clin. Trials Commun. 17 (2020) 100522.
[113] S. Vos, et al., Ethical considerations for modern molecular pathology, J. Pathol. 246 (4) (2018) 405–414.
[114] V. Fineschi, Editorial: personalized medicine: a positional point of view about precision medicine and clarity for ethics of public health, Curr. Pharm. Biotechnol. 18 (3) (2017) 192–193.

[115] G.T. Sharrer, Personalized medicine: ethical aspects, Methods Mol. Biol. 1606 (2017) 37–50.
[116] M. Shoaib, et al., Personalized medicine in a new genomic era: ethical and legal aspects, Sci. Eng. Ethics 23 (4) (2017) 1207–1212.
[117] G.G. Repetti, et al., Novel therapies for prevention and early treatment of cardiomyopathies, Circ. Res. 124 (11) (2019) 1536–1550.
[118] J.A. James, et al., Hydroxychloroquine sulfate treatment is associated with later onset of systemic lupus erythematosus, Lupus 16 (6) (2007) 401–409.
[119] L. Hood, M. Flores, A personal view on systems medicine and the emergence of proactive P4 medicine: predictive, preventive, personalized and participatory, New Biotechnol. 29 (6) (2012) 613–624.
[120] G. Gibson, et al., PART of the WHOLE: a case study in wellness-oriented personalized medicine, Yale J. Biol. Med. 88 (4) (2015) 397–406.
[121] E.T. Juengst, M.L. McGowan, Why does the shift from "personalized medicine" to "precision health" and "wellness genomics" matter? AMA J. Ethics 20 (9) (2018) E881–E890.
[122] C.J. Rosen, J.R. Ingelfinger, Traveling down the long road to type 1 diabetes mellitus prevention, N. Engl. J. Med. 381 (7) (2019) 666–667.
[123] S. Tavakolpour, et al., Pathogenic and protective roles of cytokines in pemphigus: a systematic review, Cytokine 129 (2020) 155026.
[124] M.A. Islam, S. Kundu, R. Hassan, Gene therapy approaches in an autoimmune demyelinating disease: multiple sclerosis, Curr. Gene Ther. 19 (6) (2020) 376–385.
[125] M. Blank, et al., Prevention of experimental antiphospholipid syndrome and endothelial cell activation by synthetic peptides, Proc. Natl. Acad. Sci. USA 96 (9) (1999) 5164–5168.
[126] Y. Shoenfeld, Primary biliary cirrhosis and autoimmune rheumatic diseases: prediction and prevention, Isr. J. Med. Sci. 28 (2) (1992) 113–116.
[127] Y. Bar-Dayan, et al., Aspirin for prevention of myocardial infarction. A double-edged sword, Ann. Med. Interne. (Paris) 148 (6) (1997) 430–433.
[128] Y. Sherer, Y. Shoenfeld, Immunomodulation for treatment and prevention of atherosclerosis, Autoimmun. Rev. 1 (1–2) (2002) 21–27.
[129] D. Shepshelovich, Y. Shoenfeld, Prediction and prevention of autoimmune diseases: additional aspects of the mosaic of autoimmunity, Lupus 15 (3) (2006) 183–190.
[130] H. Torres-Aguilar, et al., Tolerogenic dendritic cells in autoimmune diseases: crucial players in induction and prevention of autoimmunity, Autoimmun. Rev. 10 (1) (2010) 8–17.
[131] J. Damoiseaux, et al., Autoantibodies 2015: from diagnostic biomarkers toward prediction, prognosis and prevention, Autoimmun. Rev. 14 (6) (2015) 555–563.
[132] N.R. Rose, Prediction and prevention of autoimmune disease: a personal perspective, Ann. N. Y. Acad. Sci. 1109 (2007) 117–128.
[133] N.R. Rose, Prediction and prevention of autoimmune disease in the 21st century: a review and preview, Am. J. Epidemiol. 183 (5) (2016) 403–406.
[134] V. Malmstrom, A.I. Catrina, L. Klareskog, The immunopathogenesis of seropositive rheumatoid arthritis: from triggering to targeting, Nat. Rev. Immunol. 17 (1) (2017) 60–75.
[135] J.A. Sparks, et al., Personalized risk estimator for rheumatoid arthritis (PRE-RA) family study: rationale and design for a randomized controlled trial evaluating rheumatoid arthritis risk education to first-degree relatives, Contemp. Clin. Trials 39 (1) (2014) 145–157.
[136] J.A. Ford, et al., Impact of cyclic citrullinated peptide antibody level on progression to rheumatoid arthritis in clinically tested CCP-positive patients without RA, Arthritis Care Res. (Hoboken) 71 (12) (2019) 1583–1592.
[137] K.D. Deane, V.M. Holers, The natural history of rheumatoid arthritis, Clin. Ther. 41 (7) (2019) 1256–1269.

[138] J.A. Sparks, et al., Disclosure of personalized rheumatoid arthritis risk using genetics, biomarkers, and lifestyle factors to motivate health behavior improvements: a randomized controlled trial, Arthritis Care Res. (Hoboken) 70 (6) (2018) 823–833.
[139] J.R. Trosman, et al., Decision making on medical innovations in a changing health care environment: insights from accountable care organizations and payers on personalized medicine and other technologies, Value Health 20 (1) (2017) 40–46.
[140] I. Akhmetov, R.V. Bubnov, Innovative payer engagement strategies: will the convergence lead to better value creation in personalized medicine? EPMA J. 8 (1) (2017) 5–15.
[141] X. Liu, et al., Impact and timing of smoking cessation on reducing risk for rheumatoid arthritis among women in the Nurses' Health Studies, Arthritis Care Res. (Hoboken) 71 (7) (2019) 914–924.
[142] D. Gallo, et al., Immunomodulatory effect of vitamin D and its potential role in the prevention and treatment of thyroid autoimmunity: a narrative review, J. Endocrinol. Investig. 43 (4) (2020) 413–429.
[143] R. Illescas-Montes, et al., Vitamin D and autoimmune diseases, Life Sci. 233 (2019) 116744.
[144] J. Shi, et al., Anti-carbamylated protein antibodies are present in arthralgia patients and predict the development of rheumatoid arthritis, Arthritis Rheum. 65 (4) (2013) 911–915.
[145] M.K. Verheul, et al., Triple positivity for anti-citrullinated protein autoantibodies, rheumatoid factor, and anti-carbamylated protein antibodies conferring high specificity for rheumatoid arthritis: implications for very early identification of at-risk individuals, Arthritis Rheumatol. 70 (11) (2018) 1721–1731.
[146] K.D. Deane, et al., Genetic and environmental risk factors for rheumatoid arthritis, Best Pract. Res. Clin. Rheumatol. 31 (1) (2017) 3–18.
[147] L.B. Kelmenson, et al., Timing of elevations of autoantibody isotypes prior to diagnosis of rheumatoid arthritis, Arthritis Rheumatol. 72 (2) (2020) 251–261.
[148] A.P. Cope, Considerations for optimal trial design for rheumatoid arthritis prevention studies, Clin. Ther. 41 (7) (2019) 1299–1311.
[149] K. Raza, et al., Timing the therapeutic window of opportunity in early rheumatoid arthritis: proposal for definitions of disease duration in clinical trials, Ann. Rheum. Dis. 71 (12) (2012) 1921–1923.
[150] L. Hunt, M. Buch, The 'therapeutic window' and treating to target in rheumatoid arthritis, Clin. Med. (Lond.) 13 (4) (2013) 387–390.
[151] J.A. van Nies, et al., What is the evidence for the presence of a therapeutic window of opportunity in rheumatoid arthritis? A systematic literature review, Ann. Rheum. Dis. 73 (5) (2014) 861–870.
[152] K. Raza, A. Filer, The therapeutic window of opportunity in rheumatoid arthritis: does it ever close? Ann. Rheum. Dis. 74 (5) (2015) 793–794.
[153] C.M. Coffey, et al., Evidence of diagnostic and treatment delay in seronegative rheumatoid arthritis: missing the window of opportunity, Mayo Clin. Proc. 94 (11) (2019) 2241–2248.
[154] L.E. Burgers, K. Raza, A.H. van der Helm-van Mil, Window of opportunity in rheumatoid arthritis—definitions and supporting evidence: from old to new perspectives, RMD Open 5 (1) (2019) e000870.
[155] G.S. Hazlewood, et al., Treatment preferences of patients with early rheumatoid arthritis: a discrete-choice experiment, Rheumatology (Oxford) 55 (11) (2016) 1959–1968.
[156] G.S. Hazlewood, Measuring patient preferences: an overview of methods with a focus on discrete choice experiments, Rheum. Dis. Clin. N. Am. 44 (2) (2018) 337–347.
[157] A. Loyola-Sanchez, et al., Qualitative study of treatment preferences for rheumatoid arthritis and pharmacotherapy acceptance: indigenous patient perspectives, Arthritis Care Res. (Hoboken) 72 (4) (2020) 544–552.

[158] G.S. Hazlewood, et al., Patient preferences for maintenance therapy in Crohn's disease: a discrete-choice experiment, PLoS One 15 (1) (2020) e0227635.

[159] C. Durand, et al., Patient preferences for disease-modifying antirheumatic drug treatment in rheumatoid arthritis: a systematic review, J. Rheumatol. 47 (2) (2020) 176–187.

[160] H. Tang, J.H. Ng, Googling for a diagnosis—use of Google as a diagnostic aid: internet based study, BMJ 333 (7579) (2006) 1143–1145.

[161] J.P. D'Auria, Googling for health information, J. Pediatr. Health Care 26 (4) (2012) e21–e23.

[162] M. Ramos-Casals, et al., Google-driven search for big data in autoimmune geoepidemiology: analysis of 394,827 patients with systemic autoimmune diseases, Autoimmun. Rev. 14 (8) (2015) 670–679.

[163] A. Watad, et al., Readability of wikipedia pages on autoimmune disorders: systematic quantitative assessment, J. Med. Internet Res. 19 (7) (2017) e260.

[164] S. Doyle-Lindrud, Watson will see you now: a supercomputer to help clinicians make informed treatment decisions, Clin. J. Oncol. Nurs. 19 (1) (2015) 31–32.

[165] M.J. Fritzler, Choosing wisely: review and commentary on anti-nuclear antibody (ANA) testing, Autoimmun. Rev. 15 (3) (2016) 272–280.

[166] W.G. Dixon, K. Michaud, Using technology to support clinical care and research in rheumatoid arthritis, Curr. Opin. Rheumatol. 30 (3) (2018) 276–281.

[167] M. Kaminski, I. Loniewski, W. Marlicz, "Dr. Google, I am in Pain"-global internet searches associated with pain: a retrospective analysis of Google trends data, Int. J. Environ. Res. Public Health 17 (3) (2020) E954.

[168] L. Powley, et al., Are online symptoms checkers useful for patients with inflammatory arthritis? BMC Musculoskelet. Disord. 17 (1) (2016) 362.

[169] H.L. Semigran, et al., Evaluation of symptom checkers for self diagnosis and triage: audit study, BMJ 351 (2015) h3480.

[170] L.J. Bisson, et al., How accurate are patients at diagnosing the cause of their knee pain with the help of a web-based symptom checker? Orthop. J. Sports Med. 4 (2) (2016). 2325967116630286.

[171] T. Morita, et al., The potential possibility of symptom checker, Int. J. Health Policy Manag. 6 (10) (2017) 615–616.

[172] C. Shen, et al., Accuracy of a popular online symptom checker for ophthalmic diagnoses, JAMA Ophthalmol. 137 (6) (2019) 693.

[173] B.M. Davies, C.F. Munro, M.R. Kotter, A novel insight into the challenges of diagnosing degenerative cervical myelopathy using web-based symptom checkers, J. Med. Internet Res. 21 (1) (2019) e10868.

[174] J. Dunn, R. Runge, M. Snyder, Wearables and the medical revolution, Per. Med. 15 (5) (2018) 429–448.

[175] D. Witt, et al., Windows into human health through wearables data analytics, Curr. Opin. Biomed. Eng. 9 (2019) 28–46.

[176] Y. Song, J. Min, W. Gao, Wearable and implantable electronics: moving toward precision therapy, ACS Nano 13 (11) (2019) 12280–12286.

[177] A. Najm, et al., Mobile health apps for self-management of rheumatic and musculoskeletal diseases: systematic literature review, JMIR Mhealth Uhealth 7 (11) (2019) e14730.

[178] A. Najm, et al., EULAR points to consider for the development, evaluation and implementation of mobile health applications aiding self-management in people living with rheumatic and musculoskeletal diseases, RMD Open 5 (2) (2019) e001014.

[179] P. van Riel, et al., Patient self-management and tracking: a European experience, Rheum. Dis. Clin. N. Am. 45 (2) (2019) 187–195.

[180] T.H. Kim, et al., A temporary indwelling intravascular aphaeretic system for in vivo enrichment of circulating tumor cells, Nat. Commun. 10 (1) (2019) 1478.
[181] J. Turner, et al., A review on the ability of smartphones to detect ultraviolet (UV) radiation and their potential to be used in UV research and for public education purposes, Sci. Total Environ. 706 (2020) 135873.
[182] E. Fortune, et al., Activity level classification algorithm using SHIMMER wearable sensors for individuals with rheumatoid arthritis, Conf. Proc. IEEE Eng. Med. Biol. Soc. 2011 (2011) 3059–3062.
[183] L. Gossec, et al., Detection of flares by decrease in physical activity, collected using wearable activity trackers in rheumatoid arthritis or axial spondyloarthritis: an application of machine learning analyses in rheumatology, Arthritis Care Res. (Hoboken) 71 (10) (2019) 1336–1343.
[184] K.A. Mikk, H.A. Sleeper, E.J. Topol, The pathway to patient data ownership and better health, JAMA 318 (15) (2017) 1433–1434.
[185] K.A. Mikk, H.A. Sleeper, E. Topol, Patient data ownership-reply, JAMA 319 (9) (2018) 935–936.
[186] N. Foulquier, P. Redou, A. Saraux, How health information technologies and artificial intelligence may help rheumatologists in routine practice, Rheumatol. Ther. 6 (2) (2019) 135–138.
[187] B.D. Modena, et al., Advanced and accurate mobile health tracking devices record new cardiac vital signs, Hypertension 72 (2) (2018) 503–510.
[188] E. Topol, Digital medicine: empowering both patients and clinicians, Lancet 388 (10046) (2016) 740–741.
[189] Biomarkers Definitions Working Group, Biomarkers and surrogate endpoints: preferred definitions and conceptual framework, Clin. Pharmacol. Ther. 69 (3) (2001) 89–95.[190]In BEST (Biomarkers, EndpointS, and other Tools) Resource, Silver Spring, MD, 2016.In BEST (Biomarkers, EndpointS, and other Tools) Resource, Silver Spring, MD, 2016.
[191] N.J. Olsen, M.Y. Choi, M.J. Fritzler, Emerging technologies in autoantibody testing for rheumatic diseases, Arthritis Res. Ther. 19 (1) (2017) 172.
[192] L.A. Trouw, M. Mahler, Closing the serological gap: promising novel biomarkers for the early diagnosis of rheumatoid arthritis, Autoimmun. Rev. 12 (2) (2012) 318–322.
[193] J. Shi, et al., The specificity of anti-carbamylated protein antibodies for rheumatoid arthritis in a setting of early arthritis, Arthritis Res. Ther. 17 (2015) 339.
[194] Z. Huang, et al., MALDI-TOF MS combined with magnetic beads for detecting serum protein biomarkers and establishment of boosting decision tree model for diagnosis of systemic lupus erythematosus, Rheumatology (Oxford) 48 (6) (2009) 626–631.
[195] H. Zhang, et al., B cell-related circulating microRNAs with the potential value of biomarkers in the differential diagnosis, and distinguishment between the disease activity and lupus nephritis for systemic lupus erythematosus, Front. Immunol. 9 (2018) 1473.
[196] W.F. Lee, et al., Biomarkers associating endothelial dysregulation in pediatric-onset systemic lupus erythematous, Pediatr. Rheumatol. Online J. 17 (1) (2019) 69.
[197] Q. Luo, et al., Identification of circular RNAs hsa_circ_0044235 and hsa_circ_0068367 as novel biomarkers for systemic lupus erythematosus, Int. J. Mol. Med. 44 (4) (2019) 1462–1472.
[198] Y. Wang, et al., Novel biomarkers containing citrullinated peptides for diagnosis of systemic lupus erythematosus using protein microarrays, Clin. Exp. Rheumatol. 37 (6) (2019) 929–936.
[199] H.I. Brunner, et al., Urine biomarkers of chronic kidney damage and renal functional decline in childhood-onset systemic lupus erythematosus, Pediatr. Nephrol. 34 (1) (2019) 117–128.

[200] L.C.V. Alves, et al., Evaluation of potential biomarkers for the diagnosis and monitoring of systemic lupus erythematosus using the cytometric beads array (CBA), Clin. Chim. Acta 499 (2019) 16–23.

[201] J. Kong, et al., Potential protein biomarkers for systemic lupus erythematosus determined by bioinformatics analysis, Comput. Biol. Chem. 83 (2019) 107135.

[202] S.I. Nihtyanova, C.P. Denton, Autoantibodies as predictive tools in systemic sclerosis, Nat. Rev. Rheumatol. 6 (2) (2010) 112–116.

[203] S. Peytrignet, et al., Changes in disability and their relationship with skin thickening, in diffuse and limited cutaneous systemic sclerosis: a retrospective cohort study, Scand. J. Rheumatol. 48 (3) (2019) 230–234.

[204] K. Mariampillai, et al., Development of a new classification system for idiopathic inflammatory myopathies based on clinical manifestations and myositis-specific autoantibodies, JAMA Neurol. 75 (12) (2018) 1528–1537.

[205] N. Wesner, et al., Anti-RNP antibodies delineate a subgroup of myositis: a systematic retrospective study on 46 patients, Autoimmun. Rev. 19 (3) (2020) 102465.

[206] J.A. James, et al., Unique Sjogren's syndrome patient subsets defined by molecular features, Rheumatology (Oxford) 59 (2020) 860–868.

[207] G. Barturen, et al., Moving towards a molecular taxonomy of autoimmune rheumatic diseases, Nat. Rev. Rheumatol. 14 (2) (2018) 75–93.

[208] G. Barturen, et al., Moving towards a molecular taxonomy of autoimmune rheumatic diseases, Nat. Rev. Rheumatol. 14 (3) (2018) 180.

[209] C. Douglas, et al., The HackensackUMC value-based care model: building essentials for value-based purchasing, Nurs. Adm. Q. 40 (1) (2016) 51–59.

[210] I. Badash, et al., Redefining health: the evolution of health ideas from antiquity to the era of value-based care, Cureus 9 (2) (2017) e1018.

[211] S. Gentry, P. Badrinath, Defining health in the era of value-based care: lessons from England of relevance to other health systems, Cureus 9 (3) (2017) e1079.

Chapter 3

Biomarker and data science as integral part of precision medicine

Carlos Melus, Brenden Rossin, Mary Ann Aure, and Michael Mahler
Inova Diagnostics, Inc., San Diego, CA, United States

Abbreviations

ACR	American College of Rheumatology
AI	artificial intelligence
AID	autoimmune disease
AUC	area under the curve
CCP	cyclic citrullinated protein/peptide
CRP	C-reactive protein
DA	disease activity
ML	machine learning
MS	mass-spectrometry
PBC	primary biliary cholangitis
PCA	principal component analysis
PM	precision medicine
PMAT	particle-based multi-analyte panel
PM/Scl	polymyositis/scleroderma
PPV	positive predictive value
PSA	prostate-specific antigen
RA	rheumatoid arthritis
RNP	ribonucleoprotein
ROC	receiver operating characteristic
sCP	serum calprotectin
SLE	systemic lupus erythematosus
TNR	True Negative Rate
TPR	true positive rate

Precision Medicine and Artificial Intelligence. https://doi.org/10.1016/B978-0-12-820239-5.00006-1

1 Introduction

Biomarkers are important tools for patient care throughout the patient's journey and have been used for more than a century [1]. This chapter aims to provide an overview of biomarkers in autoimmunity starting with definitions for the individual groups or classes of biomarkers. It will also elaborate on the journey from discovery to validation and clinical implementation of biomarkers (Fig. 1). Lastly, the chapter provides statistical methods and visualization tools for data interpretation, benchmarking and biomarker selection.

2 Definitions of biomarkers

Despite the tremendous value, there is significant confusion about the fundamental definitions and concepts of biomarkers in research and clinical practice [2–4]. Several years ago, during a joint workshop of the US Food and Drug Administration (FDA) and the National Institutes of Health (NIH), it became clear that consensus definitions of biomarkers are needed. A joint task force was therefore formed to forge common definitions and to make them publicly available through a continuously updated online document—the "Biomarkers, EndpointS, and other Tools" (BEST) resource. Several subtypes of biomarkers have been defined according to their putative applications. Importantly, a single biomarker may meet multiple criteria for different uses, but it is important to develop evidence for each definition (intended use). Thus, while definitions may overlap, they also have clear distinguishing features that specify particular uses. An overview of different types of biomarkers is visualized in Fig. 2 and further described in Tables 1 and 2.

2.1 Diagnostic biomarkers

A diagnostic biomarker detects the presence or absence of a given disease (or helps to do so) at the timepoint of patient's assessment [1, 2, 4]. In autoimmunity, diagnostic biomarkers are mostly used as an aid in the diagnosis and

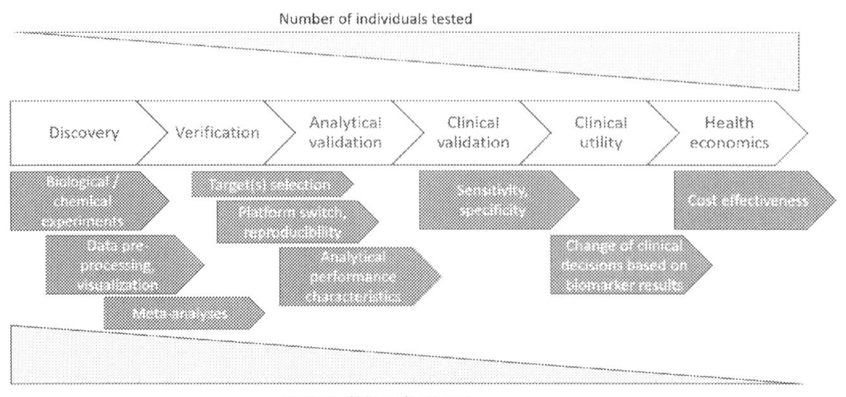

FIG. 1 The biomarker development process from research to clinical use. Biomarker discovery is a long process that undergoes several phases from discovery to clinical utilization.

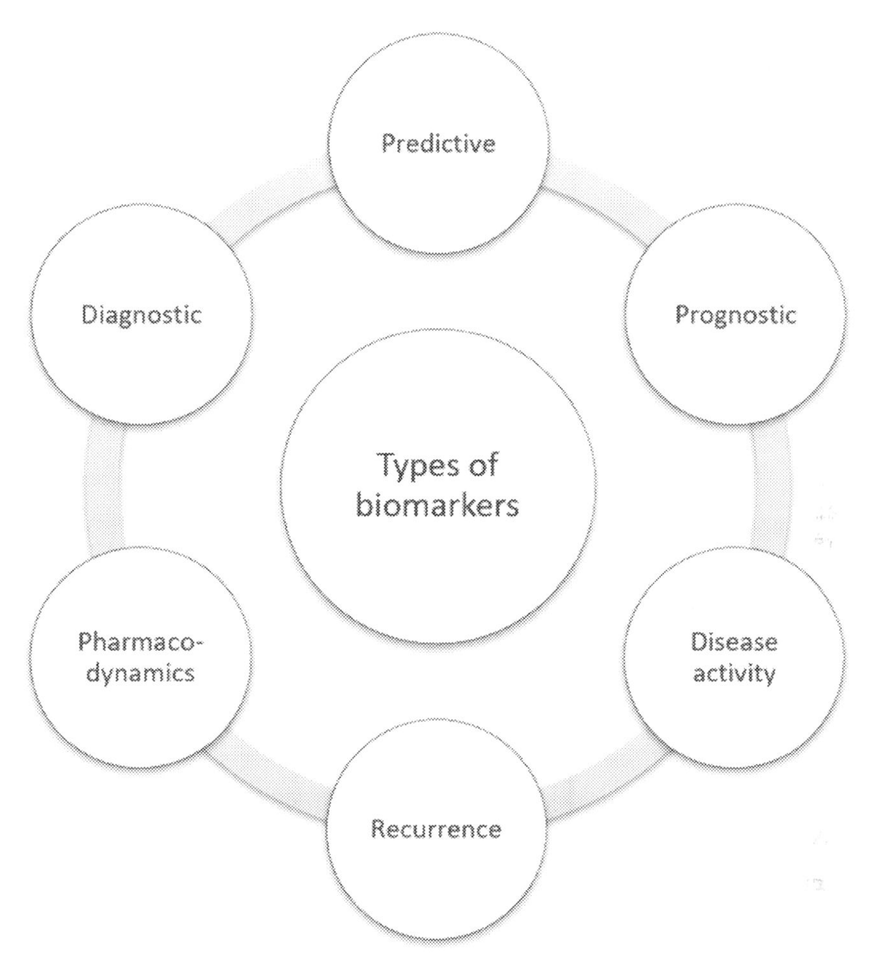

FIG. 2 Types of biomarkers. Biomarkers were historically used for diagnosis but more recently have been increasingly used for clinical decision support in other phases of the patient's journey.

are not diagnostic by itself (only in conjunction with clinical assessment). The weight the individual biomarker provides for the diagnosis varies significantly. For example, anti-Sm antibodies due to their high disease specificity carry a much higher weight compared to anti-Ro60 antibodies [5]. This is also reflected by the inclusion of anti-Sm, but not anti-Ro60 in the classification criteria for systemic lupus erythematosus (SLE) [6–9]. As we move into the era of precision medicine (PM) [1], this type of biomarker will evolve considerably. Such biomarkers may be used not only to identify people with a disease, but to redefine the classification of the disease.

2.2 Disease monitoring biomarkers

Biomarkers that can be measured sequentially to assess the status of a medical condition (e.g., flare or remission) it can be considered as a monitoring

TABLE 1 Types of biomarkers with examples from autoimmunity.

Method	Comment	Examples in autoimmunity
Diagnostic	Can help in the diagnosis; aid in the diagnosis, mostly not diagnostic by itself	Anti-Sm antibodies for SLE, anti-CCP antibodies for RA
Prognostic	Identifies likelihood of clinical event (e.g., death) independent of therapy received	Anti-MDA5 antibodies
Predictive	May identify individuals with a more or less favorable response to treatment	Anti-PAD4 antibody positive individuals show favorable outcome after treatment escalation
Monitoring	Changes with DA and provides objective measure for DA	sCP anti-dsDNA antibodies, anti-C1q antibodies
Risk factors	Can be detected before onset of disease	Several, e.g., most ANAs
Digital	Digital biomarkers are the most recent form of biomarkers Might represent a powerful group of markers that can help to monitor health	Mobility assessment using wearable devices in rheumatoid arthritis

TABLE 2 Overview of different types of biomarkers.

Biomarker type	Matrix	Benefit	Challenge/pitfall
Antibodies, autoantibodies	Serum/plasma/ whole blood	Very stable, serum is a common matrix	Often epiphenomenon (not mechanistic marker)
Serum protein	Serum/plasma/ whole blood	Potentially mechanistic marker (if involved in pathogenesis)	Stability strongly depends on the protein characteristics
Genomic markers	Serum/plasma/ whole blood; saliva	Can identify risk genes	Autoimmune disease often multi-factorial
Proteomic markers	Serum/plasma/ whole blood		
Metabolomic markers	Serum/plasma/ whole blood		
Gene expression (mRNA)	Serum/plasma/ whole blood	Can provide insights in activated disease related pathways	Stability of RNA can be challenging

biomarker. Monitoring is a broad concept and consequently there is overlap with other categories of biomarkers as described below. In autoimmunity, several biomarkers are used for diagnosis as well as to assess disease activity. One specific example are anti-dsDNA antibodies that aid in the diagnosis of SLE and also in the assessment of flares and remission [10–13]. Depending on the assay and geography, this might represent an off-label use since this application might not be covered in the intended use statement of the assay (directional insert). However, there is ample evidence available that depending on the assay used, significant correlation between anti-dsDNA antibody levels and disease activity can be observed [12]. Another more recent example is the use of serum calprotectin (sCP) as a marker for disease activity (DA) in rheumatoid arthritis (RA) [14–16]. More and more data becomes available that sCP might be superior to C-reactive protein (CRP) in autoimmunity due to the higher specificity [16]. Another "biomarker" which represents a composition of several individual markers assembled in a score, can be used to assess DA in RA [17–68]. This marker has been evaluated in several publications and has recently been recognized by the American College of Rheumatology (ACR) for the assessment of DA in RA. An interesting aspect is how monitoring biomarkers are being used. In most cases, the changes in biomarker measurements and not the actual measure is utilized.

When considering DA of chronic conditions such as autoimmunity, patient assessment can lead to significant subjectivity that can alter or impact the precision of the disease activity index. Biomarkers as surrogate for the measurement of DA are less subjective and therefore often more reproducible. In such situations, digital assessment might provide opportunities to reduce variability in data assessment [69].

2.3 Pharmacodynamic/response biomarkers

When the level of a biomarker changes in response to treatment, it is often referred to as pharmacodynamic/response biomarker. This type of biomarker is extraordinarily useful both in clinical practice and early drug development. Although frequently used in other diseases (e.g., diabetes), not many examples for response biomarkers are applied routinely in patients with autoimmune diseases.

One recent example is sCP which might provide useful information both in terms of treatment selection and monitoring of treatment response in patients with RA [14–16]. This type of biomarker is related to companion and complementary biomarkers which are discussed in a different chapter of this book.

2.4 Risk markers/factors

Historically, predictive biomarkers and risk factors have been combined. For many diseases, risk factors are determined as part of disease prevention. A well-known example is prostate-specific antigen (PSA) which has been used for screening of prostate cancer for decades [70]. Although autoantibodies are known to precede disease onset for many years [71–75] and provide significant positive predictive value, they are rarely used in clinical practice to predict and prevent autoimmune conditions. However, the field is slowly moving in this direction [76–79]. Important aspects to consider for risk markers/factors is when/who to test and most importantly what to do with the information provided by the test [1, 78–80]. Without clearly defined pathways from biomarker results, minimal clinical utility can be demonstrated.

2.5 Predictive biomarkers

A predictive biomarker is defined by the finding that the presence or change in a biomarker predicts an individual or group of individuals more likely to experience a favorable or unfavorable effect from the exposure to a medical product or environmental agent [2]. Validating such an intended use requires a rigorous approach to clinical studies. Ideally, patients with or without the biomarker are randomized to one of two or more treatments (or a placebo comparator) and differences in outcome as function of treatment are significantly related to the difference in presence, absence, or level of the biomarker. Proof of a reliable predictive biomarker thus represents a "high hurdle" to clear. However, predictive biomarkers are important for enrichment strategies in the design and conduct of clinical trials. Especially in the pre-registration phase of development, focusing enrolment on participants with elevated levels of a predictive biomarker enables a clearer signal of the efficacy in the enrolled individuals. Using predictive biomarkers for enrichment is a more targeted approach than using prognostic biomarkers, which can be used to increase event rates but not to select specific patients who are more likely to respond or not respond to therapy [80]. A typical example is RA in which standard of care significantly improved over the past decades. Therefore achieving clinical endpoint in the placebo group can be challenging. In this case, biomarkers that can identify individuals that will more likely reach clinical endpoint (e.g., joint erosions) might be useful for clinical trials design (reduction of trial size). Examples of such biomarkers include anti-CarP [81–84] antibodies.

There are many examples for predictive biomarkers in autoimmune disease (AID) such as autoantibodies in myositis. For example, anti-MDA5 antibodies are associated with rapidly progressing lung diseases. Consequently, knowing this autoantibody is key for proper management of patient with this form of myositis [85]. Once identified, anti-MDA5 positive individuals can be subjected to more aggressive and targeted treatment which often leads to better outcome.

2.6 Prognostic biomarkers

A prognostic biomarker is used to define the probability of a clinical complication, such as disease recurrence or disease progression. Although this distinction is not uniformly accepted, prognostic biomarkers should be differentiated from susceptibility/risk biomarkers, which deal with association with the transition from healthy state to disease [86]. Furthermore, they are distinguished from predictive biomarkers, which identify factors associated with the effect of intervention or exposure. There are several biomarkers in autoimmune diseases that provide prognostic information. For example, anti-CarP antibodies predict future development of erosive disease in RA patients [81, 82]. Additionally, several autoantibodies in myositis are associated with cancer development [85].

2.7 Digital biomarkers

A very recent group of biomarkers are summarized as digital biomarkers. Due to their recent genesis, the definition is still evolving. Digital biomarkers are defined as objective, quantifiable physiological and behavioral data that are collected and measured by means of digital devices such as portables, wearables, implantables, or digestibles. The data collected is typically used to explain, influence, and/or predict health-related outcomes.

3 Pre-analytical aspects of biomarkers

The pre-analytical aspects of biomarkers are often overlooked in the scientific field but are key for successful translation from research to clinical practice. This includes the sample matrix (e.g., serum/plasma) as well as the stability of the marker in the matrix. Consequently, those aspects also influence the ease of use of a biomarker in clinical routine. Biomarkers that can be measured in a matrix that is well-established (serum/plasma, whole blood, urine, stool, sputum, plaque extract, cell culture media, tissue lysate, microvesicles, cell lysate, cerebrospinal fluid, saliva, etc.) in a certain disease will have advantages over biomarkers that require a novel sample matrix that is not routinely used.

4 Benefits and pitfalls of biomarkers

Non-invasive biomarkers can provide a wide range of benefits and can represent a viable surrogate for invasive procedures such as biopsies which can be associated with safety concerns for the patient. In addition, biomarkers are often more cost effective versus invasive procedures which has positive impacts on health economic outcomes. However, biomarkers can also come with drawbacks and limitations including analytical and clinical inaccuracy. Therefore, biomarkers might be used to help triage patients, but often require confirmation.

5 Biomarker discovery

Biomarker discovery represents a complex and long process encompassing several sequential phases (Fig. 1). Typically, the process starts with the careful selection of clinical samples for the initial discovery work which is key for future success. In most approaches, the number of targets studied during the different phases decreases while the number of samples increases. In addition, the test system or the assay platform is often switch to accomodate the increasing number of samples. The reason behind platform changes is that the discovery platform which allows to screen a selected number of specimens for multiple of markers is not suitable for validation or clinical use. It is not uncommon that the platform is changed more than once. Driving factors for the platform selection include volume requirement for the samples, number of analytes that can be tested as well as turnaround time and costs. Platform changes are not simple and might change the behavior and performance of biomarkers which often lead to failure in the validation phase.

The introduction of technologies such as mass spectrometry (MS) and protein and DNA arrays, combined with our understanding of the human genome, has led to renewed interest in the discovery of novel biomarkers. Furthermore, modern technologies provide the means by which new, single biomarkers could be discovered through use of reasonable hypotheses and novel analytical strategies. Despite the current optimism, a number of important limitations to the discovery of novel single biomarkers have been identified, including study design bias, and artifacts related to the collection and storage of samples. Despite the fact that new technologies and strategies often fail to identify well established biomarkers [87–89] and show a bias toward the identification of high-abundance markers, these technological advances have the capacity to revolutionize biomarker discovery. An overview of discovery methods is provided in Table 3.

5.1 Gene-expression profiling

Genomic microarrays experienced a tremendous growth in the application of gene-expression profiling which can provide insights into the pathogenesis, and the discovery of a large number of diagnostic autoimmune biomarkers. Despite the potential, these technologies for the discovery of diagnostic biomarkers have not yet delivered many significant novel biomarkers in the space of autoimmunity.

5.2 Mass-spectrometry-based proteomic profiling

Proteomic-pattern profiling is a more recent approach to biomarker discovery. Mass-spectrometry-based methods of proteomic analysis have improved and include more-advanced technology that brings higher mass accuracy, higher detection capability, and shorter cycling times, thereby enabling increased throughput and more-reliable data. MS can be utilized separately or couple with other methods. Based on the tremendous advances in MS, neat serum samples can be profiled for serum protein/peptides.

In spite of the optimism regarding this approach, a number of important limitations have been identified. These shortcomings include bias from artifacts related to the clinical sample collection and storage, the inherent qualitative nature of MS, failure to identify well-established biomarkers, bias when identifying high-abundance molecules within the serum, and disagreement between peaks

TABLE 3 Biomarker discovery technologies and platforms (focused on autoimmunity).

Method	Benefit	Challenge/limitation
Mass spectrometry	High density method; information obtained from MS can be used to switch platform and to develop, e.g., capture immunoassay	• Prone to identify artifacts; MS not widely available or expensive as routine method
B-cell epitope mapping	Can identify peptide sequence that will allow for improved consistency in supply chain	• Target antigen has to be known
Random peptide arrays	Can identify novel targets	• Prone to artifacts
Planar peptide arrays	Relatively easy to perform; can be cost effective	• Limited number of variants; quality of peptides (no control)
Protein arrays	High density method; often full-length antigens	• Coating concentration and folding of proteins often unclear • Platform switch require which can change characteristics of biomarker
Phage or bacterial display	High density method	• Often platform shift required

generated by different research laboratories. Another limitation concerns possible bioinformatic artifacts. Despite tremendous potential and advances of this technology, no product has yet reached the clinic and no independent validation studies have been published. Guideline-developing organizations and expert panels do not currently recommend serum proteomic profiling for clinical use.

Many autoantibodies have been matched to their corresponding autoantigenic target by a combination of western-blot and MS [90, 91]. Following this approach, cell extracts are size-p fractionated via gel electrophoresis and further analyzed by immunoprecipitation or western-blot using positive samples. Reactive protein bands are then extracted and submitted for MS testing.

5.3 Peptidomics

The low-molecular-weight plasma or serum proteome has been the focus of recent attempts to find novel biomarkers. Peptidomic profiling might represent nothing more than peptides cleaved during coagulation or functions inherent to plasma or serum, including immune modulation, inflammatory response and protease inhibition. Many of the aforementioned caveats associated with MS-based protein profiling technologies also apply to peptidomics.

5.4 Biomarker family approach

The family concept is frequently used in cancer research and diagnostic. Although less frequently, biomarkers in autoimmunity are also characterized using the family approach. For example, antibodies targeting the protein arginine deiminase (PAD) enzymes in patients with RA have been explored based on the family of PAD enzymes. Antibodies to PAD4, followed by anti-PAD3 and anti-PAD2 antibodies have been described. Just recently, PAD1 and PAD6 have also been identified as autoantigenic targets [92]. Since many of known autoantigens are part of macromolecular complexes, studying the immune response to members of those complex has led to the identification of novel autoantigens [93]. A few examples cover the ribonucleoprotein (RNP) [94] or the polymyositis/scleroderma (PM/Scl) complex [93,94] as well as the centromere antigens [94,95].

5.5 Secreted protein approach

The identification of secreted proteins in tissues or other biological fluids does not necessarily imply that the proteins will be detectable in the sera of patients. Serum-based diagnostic tests that measure circulating proteins depend on the stability of the protein, its clearance, its association with other serum proteins and the extent of post-translational modifications. In autoimmunity, serum proteins are routinely measured which includes CRP and more recently also the Vectra DA as well as sCP [14–16, 48].

5.6 Protein arrays

Protein arrays can be used for several applications including protein-protein, protein/peptide, protein/DNA interaction studies as well as autoantibody discovery. Protein arrays using part of the human genome expression libraries are mostly provided as planar arrays, but can also be present in form of bead-based arrays. One example of an autoantibody that has been discovered using a bead-based library are antibodies to BICD2 as a marker for systemic sclerosis [96]. In contrast, two novel autoantibodies that have shown value as an aid in the diagnosis of primary biliary cholangitis (PBC) were successfully discovered using a planar protein array [97–100].

5.7 Epitope mapping as discovery strategy

For antibodies as biomarkers, epitopes (the binding site of antibodies on antigens) can be used to develop biomarkers. As a simple rule, epitopes can be distinguished into linear (continuous) as well as conformational epitopes [94]. Mapping of linear epitopes is generally more straightforward. For this purpose, a wide variety of methods can be applied including phage or bacterial display as well as solid phase peptide arrays [94]. All of those approaches can start from different target sources including:

- random peptide libraries
- gene/protein approach

Identified linear epitopes can provide multiple benefits including higher specificity, higher sensitivity (linked to specificity; lower cut-off), improved precision as well as consistency (supply chain benefits) [93]. On the other side, mapping for conformational epitopes requires more complex strategies, often combining experimental components as well as computational aspects. A key example for a biomarker discovery for autoimmune diseases is the discovery of anti-CCP antibodies [82, 101]. This biomarker has changed the diagnosis and management of RA. For the discovery, a random peptide array synthesized on beads was successfully deployed which lead to the first generation of cyclic citrullinated protein/peptide (CCP). Later on, the antigen was further optimized which led to improved performance of the second-generation assay. Subsequently, the third generation was established which was developed using peptide engineering and not using peptide libraries [82]. An example of a more directed approach is the discovery of a linear epitope on Rpp38 as a marker for systemic sclerosis [102].

5.8 Different phases of biomarker discovery

As discussed above, biomarker discovery is carried out in several phases with data review as gate approach. The ultimate goal is to balance time and overall costs. In general, the initial stage includes testing of highly characterized patient samples for as many targets as possible followed by data analysis and data mining. In case candidate biomarkers can be found, the number of samples (target condition and controls) is expanded. While in some cases, the discovery platform is used for this second phase, very often the platform is changed to a platform that allows for high throughput testing. After successful completion of this verification, a more formal validation is performed that mostly is carried out on the target platform of the clinical application.

5.9 Platform switch

The platform used for biomarker discovery is rarely the same as the technology used for clinical testing. A specific example is MS which is an excellent discovery platform, but despite significant evolution has not reached broad adoption in routine diagnostics. Therefore, the knowledge gained from the discovery work is frequently used to develop or source antibody pairs to detect circulating proteins or peptides using a capture type immunoassay [103].

The same applies to the discovery of kelch-like peptide (KL-p) which has been discovered using planar protein arrays [97]. Later on, this novel biomarker has been validated using ELISA. Due to the challenges associated with the recombinant protein, linear epitope mapping was performed and a short linear peptide was successfully derived [98]. Finally, this peptide was used to develop an ELISA as well as a particle-based multi-analyte panel (PMAT) on the Aptiva system (Inova Diagnostics) [100].

6 Biomarker verification and validation

One goal is to define a method for validation that assures that the biomarker can be measured reliably, precisely, and repeatably at a low cost. All too often, assays are not validated, engendering misleading assumptions about the biomarker's value. The use of statistical tools such as receiver-operating characteristic (ROC) curves has enabled diagnostic biomarker evaluation. However, a common problem is the absence of a historical standard for defining the presence or absence of the disease or condition (also referred to as golden standard). Furthermore, decision thresholds and clinical utility are becoming important measures for assessing the value of biomarkers for clinical application. In the future, proof that a biomarker adds information about diagnosis may be necessary but not sufficient. Rather, the key question will be whether the additional information is substantial enough to lead to a change in clinical decision-making [1]. Statistics for evaluating this issue, such as the net reclassification index, are evolving. Researchers involved in early preclinical biomarker research would be well served to understand how the biomarker will eventually be evaluated.

7 Statistical approaches and big data

Statistics and more and more machine learning (ML) play a central role during the discovery of biomarkers. Below, some of the key elements are presented.

7.1 Visualization and relevance assessment

Data visualization is the representation of data or information in a graph, chart, or other visual format (Table 4, Figs. 3–7). It communicates relationships of the data with images. Data visualization is especially important when it comes to identifying candidate biomarkers (Fig. 3). It helps identifying patterns and trends that otherwise might go unnoticed. Identifying correlations between two or more variables can be difficult without a visualization.

TABLE 4 Overview of visualization tools.

Tool	Main use	Strength	Weakness
t-SNE	Used to reduce dimensionality of datasets with large number of variables Probabilistic technique		Sometimes different runs with the same hyperparameters may produce different results hence multiple plots must be observed before making any assessment with t-SNE t-SNE does not scale well with size of dataset
PCA	Used to reduce dimensionality of datasets with large number of variables Mathematical technique	Scales well with size of dataset	PCA will not be able to interpret the complex polynomial relationship between features while t-SNE is made to capture exactly that
ROC curve	Assessment of the discrimination ability of continues variables using binary reference. Can be used as single ROC curve or as comparative analysis of ~ 2–10 different curves	Independent from threshold of variable (biomarker) Should be used over Precision recall curve when there are roughly equal numbers of observations for each class	
AUC	Single value used to interpret ROC analysis	Easy to understand	Does not consider clinically relevant area
Precision recall curve	Assessment of the discrimination ability of continues variables using binary reference. Less frequently used vs ROC curves	Should be used over ROC curves when there is a moderate to large class imbalance	Not widely used in diagnostic field

Continued

TABLE 4 Overview of visualization tools—cont'd

Tool	Main use	Strength	Weakness
Venn diagram	Overlap between biomarkers using binary outcomes (most useful for 2–4 variables). Size of circles can help to visualize the number per category	Simple and easy to read	Does not allow display on large number of variables May exclude important datapoints that do not have results for all variables/groups being compared
Upset plot	Overlap between biomarkers using binary outcomes (more appropriate vs Venn diagrams for 4 or more variables)	Easy to interpret when evaluating 4 or more variables	Not widely used in diagnostic field
Volcano plot	Identification of biomarkers during discovery	Ability to visualize large number of datapoints for biomarker discovery	
Correlogram	Quantitative correlation between variables. Most useful for about 2–10 different variables (biomarkers)		Most useful for distributed datasets
Clustergram	Useful tool to understand relationships between variables	Can be used with large number of variables Very visual	
Correlation matrix	Can be used to summarize correlograms	Provides high level summary for quantitative or qualitative agreements	Does not provide details

AUC, area under the curve; *PCA*, principle component analysis; *ROC*, receiver operating characteristic; *t-SNE*, t-distributed stochastic neighboring entities.

A volcano plot is a scatterplot combining a measure of statistical significance from a statistical test (e.g., a *P*-value from an ANOVA model) with the magnitude of the change, enabling quick visual identification of the data points (proteins, genes, etc.) that display large magnitude changes that are also statistically

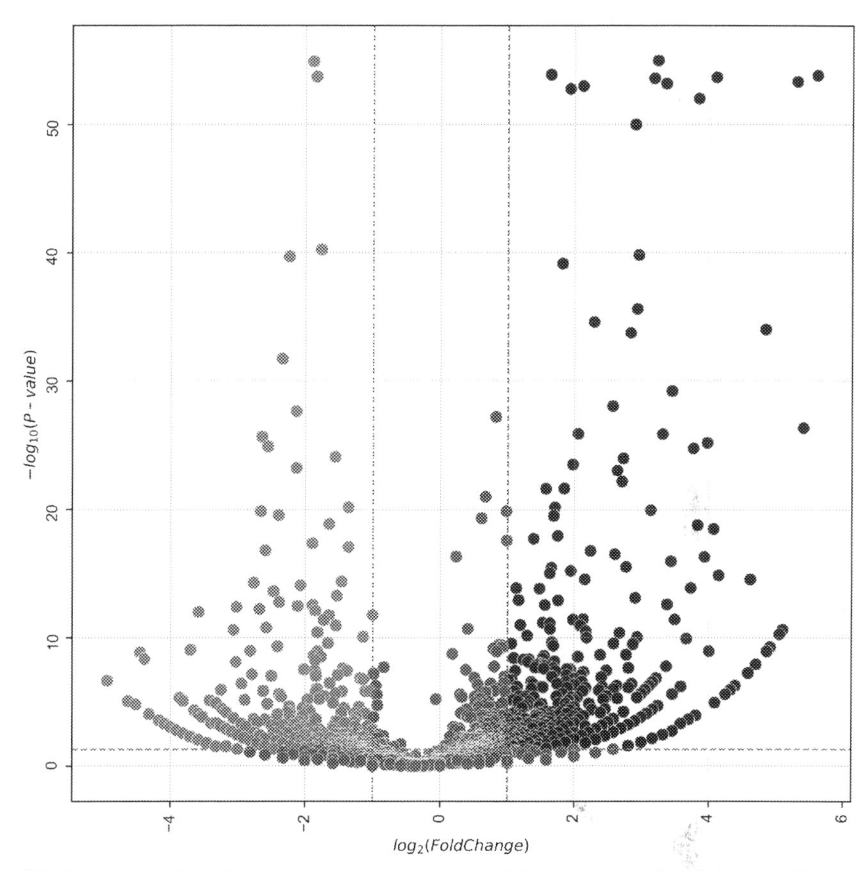

FIG. 3 Volcano plot. Volcano plot is a scatterplot combining a measure of statistical significance from a statistical test (e.g., a *P*-value from an ANOVA model) with the magnitude of the change, enabling quick visual identification of the data points (proteins, genes, etc.) that display large magnitude changes that are also statistically significant. These plots are increasingly common in biology and chemistry studies such as genomics, proteomics, and metabolomics, where there are usually many thousands of replicate data points. Plot created in Python 3.7 using Matplotlib 3.1.1 (https://matplotlib.org/), one of the best-known and widely-used libraries for creating visualizations in Python.

significant. These plots are increasingly common in biology and chemistry studies such as genomics, proteomics, and metabolomics, where there are usually many thousands of replicate data points (Fig. 3).

One of the classic options to graph correlation and distribution between two magnitudes is the scatter plot. It can be used to check if there is any relationship between two variables. Correlation coefficients (Pearson, Spearman, Kendall, etc.) can be calculated to measure the strength of the relationship (Fig. 4).

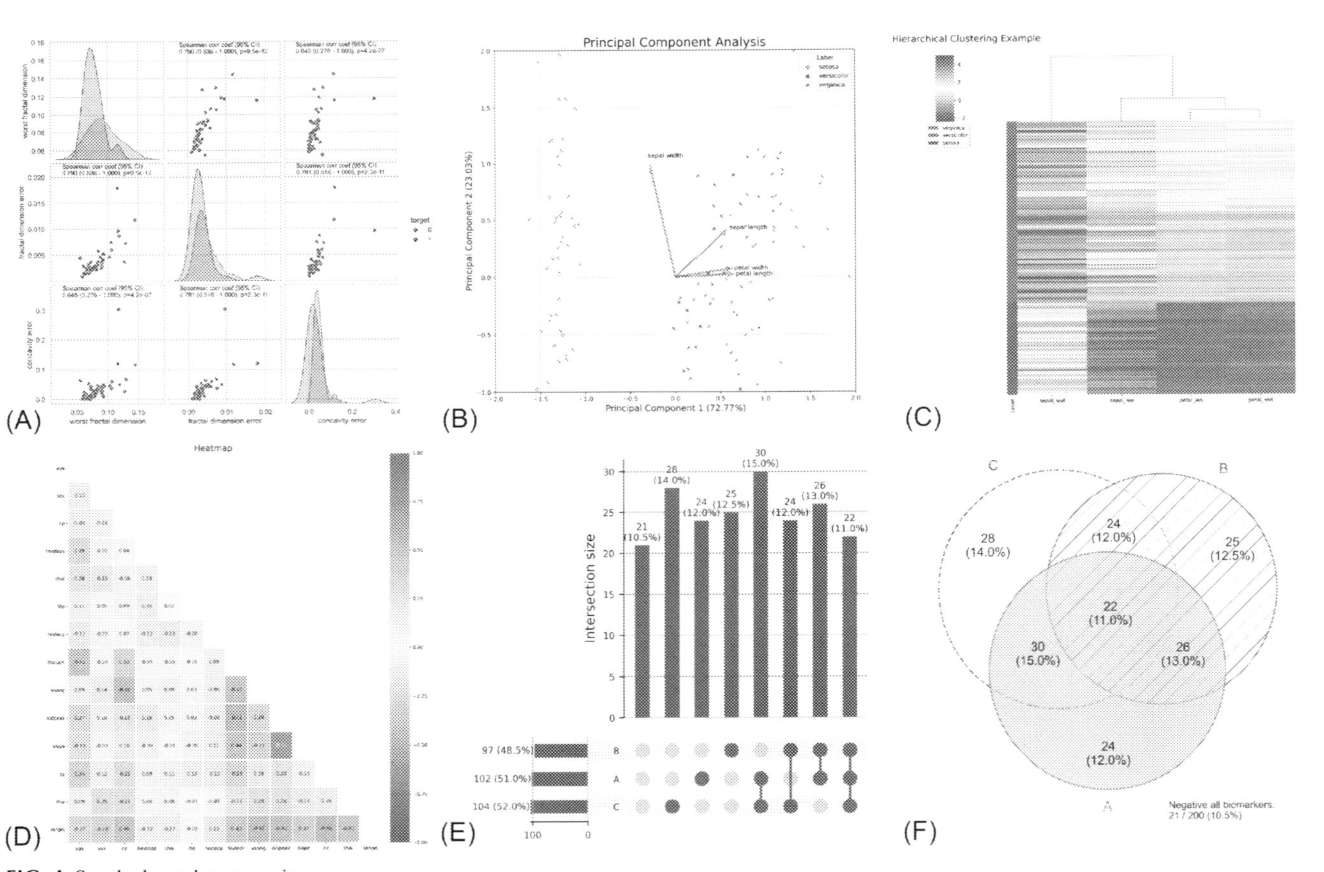

FIG. 4 See the legend on opposite page.

7.1.1 Multi-variate analysis/large datasets

When it comes to large datasets data visualization becomes even more critical in order to comprehend and interpret results. Here, dimensionality reduction can also be helpful when working with many features in a dataset. It is difficult to visualize more than four features in a plot, yet it is relatively common to work with datasets including tens of biomarker (features) per sample. Dimensionality reduction processes such as Principal Component Analysis (PCA) map data to a lower dimensional (2D/3D) space in a way that variance in the dataset is maximized in the representation (Fig. 4).

A correlogram is a matrix of scatterplots that help analyze the relationship between pairs of magnitudes (biomarkers) in a dataset. The diagonal of the matrix usually provides a representation of the distribution of the values of each variable by means of a histogram or density plot.

Venn diagrams use binary datasets to visualize qualitative overlap between features/biomarkers (Fig. 4). Modern approaches use the size of the circles to illustrate the number of cases in each category (example in [104]). The main limitation of this plot type is it becomes very difficult to create and interpret when there are more than three variables. In such situations, UpSet plots represent a useful alternative to Venn diagrams, especially for larger number of variables/features (biomarkers). UpSet plots enable representation of simultaneous elements in a set and their membership. The number of intersections can also become a problem when the number of sets increases.

7.2 The statistical trap and correction methods

The more markers are being tested, the higher the likelihood of finding potential candidates. This can be corrected by either increasing the threshold for statistical correlation and/or by statistical correction methods.

FIG. 4 Visualization tools for multiple biomarker results. Panel (A) displays a correlogram which is a matrix of scatterplots to help analyze the relationship between pairs of biomarkers in a dataset with representation of the distribution of the values of each variable by means of a histogram. In order to analyze the relationship between many biomarkers, dimensionality reduction is key to understand data. Here principal component (PCA) plots (B) and (C) clustergrams help to visualize data complexity. The correlation matrix (D) represents a popular tool to compare features or parameters for correlation. In (E) UpSet plots also visualize set overlaps, but are more readable than Venn diagrams, especially for larger number of sets. Venn diagrams (F) can be especially useful to classify and compare patients according to the result of each test. These plots were created in Python 3.6 using Matplotlib 3.1.1, and Seaborn 0.10.1 (http://seaborn.pydata.org/), a data visualization library based on Matplotlib that is closely integrated with data and relationships between variables. The UpSet plot (E) was created using the UpSetPlot 0.4.0 library (https://github.com/jnothman/UpSetPlot), an implementation also based on Matplotlib, with contributions from some of the authors.

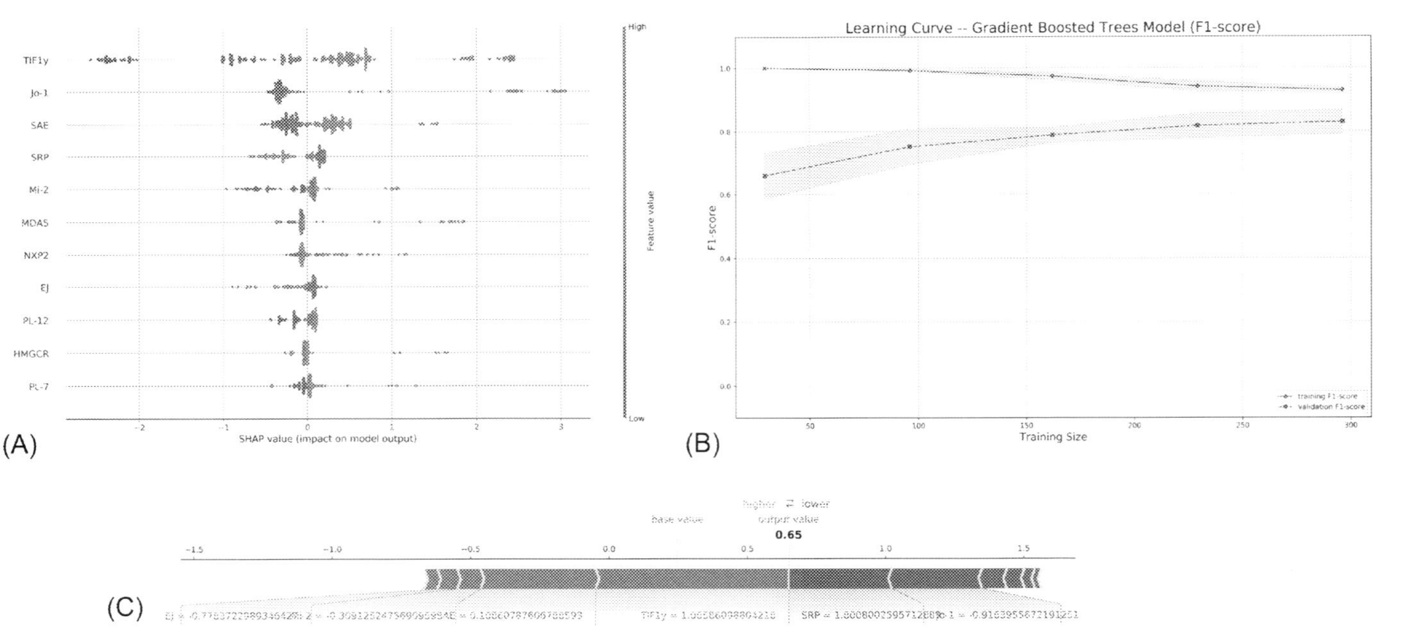

FIG. 5 Feature contribution of a classification model. In (A) the contribution of different features to a classifier is presented. This visualization tool is of particular value to understand machine learning models. In panel (B), the accuracy of the model in training and validation set is presented. With increasing data, performance in both sets should be equal. Significant difference often indicates lack of data points. In (C) shows how the individual features contributed to the decision for a patient. Plot created in Python 3.6 using Matplotlib 3.1.1 and Shapley 0.32.0 (https://github.com/slundberg/shap), a library to explain the contribution of inputs to the output of any machine learning model, based on game theory.

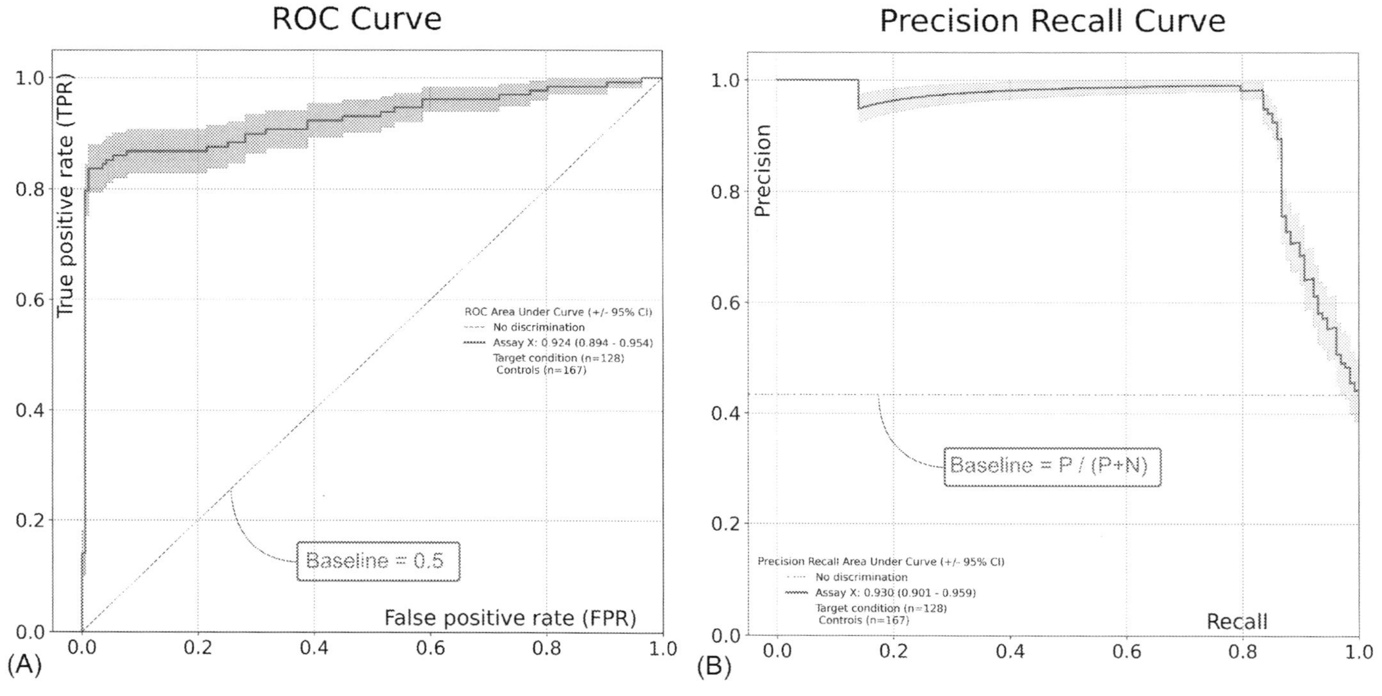

FIG. 6 Visualization of performance characteristics for diagnostic assays. In (A) a receiver operating characteristic (ROC) curve is illustrated. Panel (B) displays a precision recall curve. For both panels, the 95% confidence interval (CI) is indicated by the shaded area. The area under the curve (AUC) as a single parameter measure to describe the discrimination of the biomarker is illustrated. Plot created in Python 3.6 using Matplotlib 3.1.1.

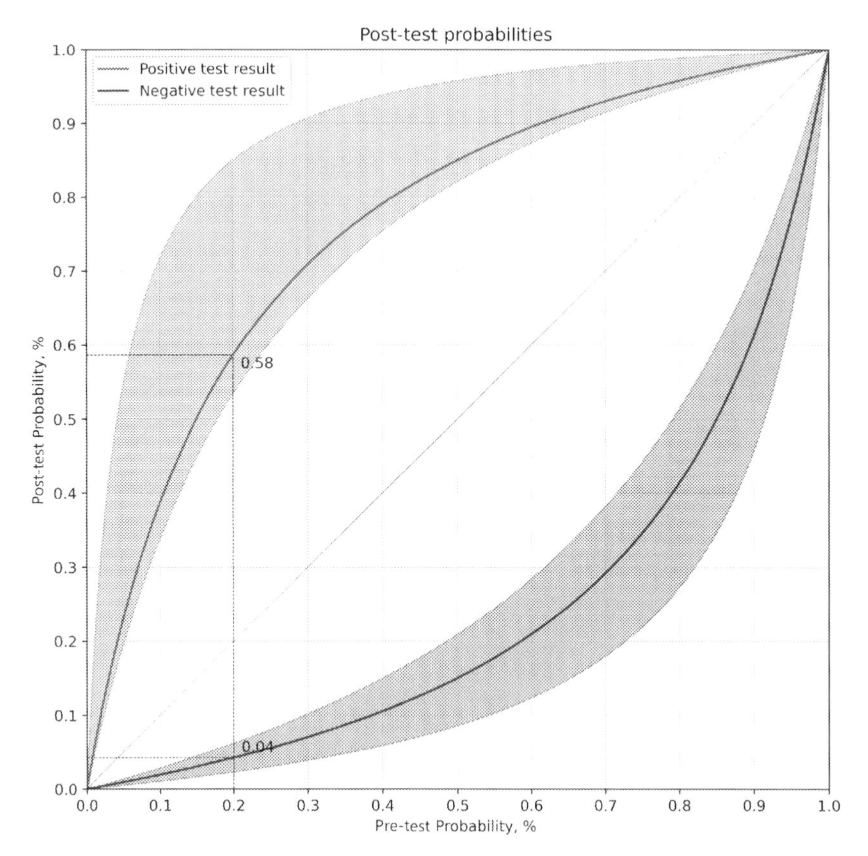

FIG. 7 Pre-test probability plot. Pre-test probability plots are used to describe the impact of a test result on probability of a certain condition. The 95% confidence interval (CI) is indicated using shaded areas. In this example [test with 85% sensitivity (95% CI 78–91%) and 85% specificity (95% CI 83–96%)] a positive test result changes the probability of the target condition from 20% to 58.0%. In contrast, a negative test result reduces the probability from 20% to 4%. Plot created in Python 3.6 using Matplotlib 3.1.1.

7.3 Methods to measure and describe clinical performance characteristics

Several approaches are being used to describe the clinical performance characteristics of an assay (Table 5). One key concept in clinical performance is the confusion matrix, which is a visualization of the model predictions versus the ground-truth results.

The confusion matrix enables calculating multiple performance metrics of a particular model (Table 5):

The most basic performance measurement, **accuracy**, is defined as the number of correct predictions over the total number of samples: $\frac{\text{true positive (TP)} + \text{true negative (TN)}}{TP + TN + \text{false negative (FN)} + \text{false positive (FP)}}$.

TABLE 5 Confusion matrix (also known as 2x2 tables).

		Predicted result	
		Positive	**Negative**
Actual result	Positive	True positive (TP)	False negative (FN)
	Negative	False positive (FP)	True negative (TN)

TABLE 6 Statistical methods to describe biomarker performance.

Statistical measure	General explanation	Strength	Weakness
Sensitivity	Statistical measure of how accurately a test correctly identifies diseased individuals	Does not depend on prevalence	Applies only to diseased persons
Specificity	Statistical measure of how well a test correctly identifies absence of the disease in question	Does not depend on prevalence	Applies only to non-diseased persons
Diagnostic efficiency	Combination of sensitivity and specificity	Does not depend on prevalence	Applies only to diseased and non-diseased persons
False negative (clinically)	Negative test result of a diseased individual	Does not depend on prevalence	Relies entirely on the cutoff used and the distribution accruing to test outcomes for TPs and FNs Interference; depends on type of test; applies only to diseased persons with negative tests; compromises the effectiveness of the screening

Continued

TABLE 6 Statistical methods to describe biomarker performance—cont'd

Statistical measure	General explanation	Strength	Weakness
False positive (clinically)	Positive test result of an individual without the disease in question	Does not depend on prevalence; depends on plausibility of substantiated hypothesis	Interference; depends on type of test; applies only to non-diseased persons with positive tests; cannot be excluded without confirmation
False negative (analytically)	Negative test result in the presence of the respective analyte	Does not depend on prevalence	Interference; depends on type of test; applies only to diseased persons with negative tests
False positive (analytically)	Positive test result in the absence of the respective analyte	Does not depend on prevalence	Interference; depends on type of test; applies only to non-diseased persons with positive tests
Positive predictive value	Ratio of true positive to combined true and false positives	Clinical relevance	Depends on prevalence
Negative predictive value	Ratio of true negatives to combined true and false negatives	Clinical relevance	Depends on prevalence
Positive likelihood ratio	The probability of a positive test results in patients with the disease divided by the probability of a positive test result in individuals without the disease.	Does not depend on prevalence	Applies to only positive tests
Negative likelihood ratio	The probability of a negative test result in patients with the disease divided by the probability of a negative test result in individuals without the disease.	Does not depend on prevalence	Applies to only negative tests

TABLE 6 Statistical methods to describe biomarker performance—cont'd

Statistical measure	General explanation	Strength	Weakness
Odds ratio (clinically)	A statistic that quantifies the strength of the association between two events. It represents the odds that an outcome will occur given a particular exposure, compared to the odds of the outcome occurring in the absence of that exposure	Does not depend on prevalence; combines sensitivity and specificity	Values false positive and false negative errors equally; not intuitive
Odds ratio (analytically)	A statistic that quantifies the strength of the association between two groups. It represents the odds that an outcome will occur given a particular analyte, compared to the odds of the outcome occurring in the absence of that analyte	Does not depend on prevalence; combines sensitivity and specificity	Values false positive and false negative errors equally; not intuitive
Area under the curve	The area under the receiver operating characteristic curves evaluate the ability of an assays at various threshold to discriminate two groups	Does not depend on prevalence; combines sensitivity and specificity	Lack of clinical interpretation

Accuracy can be misleading, especially in situations with class imbalance, where the prediction might be correct for the majority of the results and achieve a high classification, but still completely miss all the elements of the smaller population. A good example is fraud detection: fraudulent activities are extremely rare when compared to regular transactions. Simply predicting all transactions as "legitimate" would yield an accuracy close to 100%, but this model would not be able to detect any fraudulent transaction, actually making it useless as a fraud detection tool.

Sensitivity, also known as Recall or True Positive Rate (TPR) is defined as: $\frac{TP}{TP+FN}$. Regarding clinical tests, sensitivity measures the ability to correctly detect positive patients.

Specificity, or True Negative Rate (TNR) is defined as: $\frac{TN}{TN+FP}$. For a clinical test, specificity measures the ability to correctly identify healthy or disease controls as negative.

Sensitivity and specificity are the most commonly used parameters (Table 3). Despite their popularity, sensitivity and specificity have significant limitations. Both of them are calculated based on just one part of the population (positive or negative patients, respectively), and do not include any information about the rest of the population.

Receiver Operating Characteristic (**ROC**) curves (Fig. 6) are very commonly used in the performance analysis of biomarkers since they provide a very simple way to visualize the True Positive Rate (sensitivity) and False Positive Rate (1-specificity) of a given marker (at various thresholds), typically used to assess the performance of a biomarker. This type of analysis is independent from the cut-off of the assay and can also help to understand if poor performance of a biomarker or a diagnostic assay is attributed to incorrect cut-off value assignment.

Precision-Recall curves (Fig. 6) provide similar information compared to the ROC curve. Recall (True Positive Rate, or sensitivity) represented in the *y* axis of the ROC curve is plotted in the *x* axis of the precision-recall curve. Precision, defined as the proportion of positive predictions that are true positives: $\frac{TP}{TP+FP}$, is plotted on the *y* axis of the Precision-Recall curve. Precision-Recall curves highlights the number of false positives relative to the class size and are recommended when the dataset imbalance is moderate to large, which is a very common scenario in clinical studies. ROC curves represent better the total amount of false positives independent of what group they come up.

One of the main differences between ROC and Precision-Recall curves is the baseline (the non-discrimination area). In the ROC curve, the baseline is fixed, at 0.5 (50% probability, similar to a completely random event, like a coin toss), while in the Precision-Recall curve, the baseline is the ratio of positives (P) to total samples (positives and negatives): $\frac{P}{P+N}$. For balanced datasets, the value is 0.5, but for imbalanced distributions it can be much lower or higher, giving some additional perspective of the true performance of the assay [105].

The **F1 score** is another measure of a test's performance. It provides a balanced average of precision and recall, providing a more realistic measure of a test performance, especially if the prevalence is extremely low. A limitation of this score is it does not take true negatives into account and can be misleading for unbalanced datasets.

From the clinical perspective, performance characteristics have different implications on the likelihood of patients having a certain condition, especially if the diagnostic test is an aid in the diagnosis. Therefore, the translation of test result into probability of individual patients is important. A simple but effective concept is the visualization of pre-test and post-test probability [105–110].

7.4 Machine learning models

Combining features in ML models can provide improved performance vs individual biomarkers. However, understanding the contribution of features can be important for the user as well as for regulatory agencies (Fig. 5).

A standard ML workflow starts with collecting the data. It can be stored in text files (Comma Separated Values, CSV files), excel spreadsheets, or databases. Once data is available, it undergoes data processing. This step is used to "clean" the data and make it usable for the next steps. Some of the major activities during the data processing step are:

- Exploratory Data Analysis (EDA) is a common practice before starting to work with the data, evaluating the main characteristics of the data, frequently involving visualization of the dataset or some parts, to get a better understanding what steps are required to develop a model.
- One of the outputs of the initial EDA is the type of data available. To generate models with the ability to generalize well in different populations, data should be heterogenous for all parameters, and capture as much variance as possible in the input parameters of the model. Multiple studies contain one or many features that can be considered homogeneous due to the lack of diversity, resulting in potentially imbalanced or biased datasets.
- Normalization, scaling, and standardizing: when comparing measurements in different scales, values should be brought to a common scale without affecting the variance in the original data. Usually, data normalization results in improved model performance.
- Overfitting/underfitting: when a model is very accurate on the training data but does not work well when presented with new data, it is overfitted, and cannot generalize well. The opposite situation, underfitting, results in the model missing trends and patterns in the data, and not being able to generalize.
- Training/test split: one the best practices to avoid overfitting is splitting the data into, at least, two subsets, one used only for training and the other just

for testing the model. This avoids testing the predictive power of a model on data previously seen during the training process. Typically, 80% of the original dataset is randomly assigned to training and the remaining 20% to testing. If the dataset is sparse or the test cohort is too small, some of the classes might not be represented in the test set.
- Another good practice is visualizing the training progress, or the learning curve for a given model (Fig. 5). Such an exercise can provide some insight on the model performance as well as the ability to improve using additional data. Sometimes it is best to stop the modeling process until additional data is collected to obtain the expected performance.

8 From bench to bedside

Many biomarkers have been discovered and published in a broad range of scientific journals including high impact journals (examples in [111,112]). However, only a fraction has entered clinical application [3]. The underlying reasons for the "Death Valley" of biomarkers are manifold and include sensitivity or specificity challenges, reproducibility, stability, interference. In addition, economical aspects can also prevent biomarkers from entering clinical use.

9 Conclusions

Biomarkers play a critical role in various phases of patient care in autoimmunity. However, most biomarkers in AID are mostly used for diagnostic applications. Understanding the relationship between measurable biological processes and clinical outcomes is vital to expanding our portfolio of treatments for AID. With increasing knowledge and available data, AID biomarker use will be expanded to user phases of the patient's journey.

References

[1] M.J. Fritzler, et al., The utilization of autoantibodies in approaches to precision health, Front. Immunol. 9 (2018) 2682.

[2] Biomarkers Definitions Working Group, Biomarkers and surrogate endpoints: preferred definitions and conceptual framework, Clin. Pharmacol. Ther. 69 (3) (2001) 89–95.

[3] G. Poste, Bring on the biomarkers, Nature 469 (7329) (2011) 156–157.

[4] K. Strimbu, J.A. Tavel, What are biomarkers? Curr. Opin. HIV AIDS 5 (6) (2010) 463–466.

[5] M. Mahler, et al., Current concepts and future directions for the assessment of autoantibodies to cellular antigens referred to as anti-nuclear antibodies, J. Immunol. Res. 2014 (2014) 315179.

[6] M. Aringer, et al., 2019 European League Against Rheumatism/American College of Rheumatology Classification Criteria for systemic lupus erythematosus, Arthritis Rheumatol. 71 (9) (2019) 1400–1412.

[7] M. Aringer, et al., 2019 European League Against Rheumatism/American College of Rheumatology classification criteria for systemic lupus erythematosus, Ann. Rheum. Dis. 78 (9) (2019) 1151–1159.

[8] S.K. Tedeschi, et al., Multicriteria decision analysis process to develop new classification criteria for systemic lupus erythematosus, Ann. Rheum. Dis. 78 (5) (2019) 634–640.
[9] M. Aringer, N. Leuchten, S.R. Johnson, New criteria for lupus, Curr. Rheumatol. Rep. 22 (6) (2020) 18.
[10] C. Bentow, et al., International multi-center evaluation of a novel chemiluminescence assay for the detection of anti-dsDNA antibodies, Lupus 25 (8) (2016) 864–872.
[11] M. Mahler, et al., Performance characteristics of different anti-double-stranded DNA antibody assays in the monitoring of systemic lupus erythematosus, J. Immunol. Res. 2017 (2017) 1720902.
[12] E. Mummert, et al., The clinical utility of anti-double-stranded DNA antibodies and the challenges of their determination, J. Immunol. Methods 459 (2018) 11–19.
[13] C. Sjowall, et al., Two-parametric immunological score development for assessing renal involvement and disease activity in systemic lupus erythematosus, J. Immunol. Res. 2018 (2018) 1294680.
[14] E.C. de Moel, et al., Circulating calprotectin (S100A8/A9) is higher in rheumatoid arthritis patients that relapse within 12 months of tapering anti-rheumatic drugs, Arthritis Res. Ther. 21 (1) (2019) 268.
[15] M. Jarlborg, et al., Serum calprotectin: a promising biomarker in rheumatoid arthritis and axial spondyloarthritis, Arthritis Res. Ther. 22 (1) (2020) 105.
[16] M. Bach, et al., A neutrophil activation biomarker panel in prognosis and monitoring of patients with rheumatoid arthritis, Arthritis Rheumatol. 72 (1) (2020) 47–56.
[17] P.S. Eastman, et al., Characterization of a multiplex, 12-biomarker test for rheumatoid arthritis, J. Pharm. Biomed. Anal. 70 (2012) 415–424.
[18] J.R. Curtis, et al., Validation of a novel multibiomarker test to assess rheumatoid arthritis disease activity, Arthritis Care Res. 64 (12) (2012) 1794–1803.
[19] M.F. Bakker, et al., Performance of a multi-biomarker score measuring rheumatoid arthritis disease activity in the CAMERA tight control study, Ann. Rheum. Dis. 71 (10) (2012) 1692–1697.
[20] X. Zhao, et al., Pre-analytical effects of blood sampling and handling in quantitative immunoassays for rheumatoid arthritis, J. Immunol. Methods 378 (1–2) (2012) 72–80.
[21] J.W. Peabody, et al., Impact of rheumatoid arthritis disease activity test on clinical practice, PLoS One 8 (5) (2013) e63215.
[22] M. Centola, et al., Development of a multi-biomarker disease activity test for rheumatoid arthritis, PLoS One 8 (4) (2013) e60635.
[23] S. Hirata, et al., A multi-biomarker score measures rheumatoid arthritis disease activity in the BeSt study, Rheumatology (Oxford) 52 (7) (2013) 1202–1207.
[24] A.H. van der Helm-van Mil, et al., An evaluation of molecular and clinical remission in rheumatoid arthritis by assessing radiographic progression, Rheumatology (Oxford) 52 (5) (2013) 839–846.
[25] W. Li, et al., Impact of a multi-biomarker disease activity test on rheumatoid arthritis treatment decisions and therapy use, Curr. Med. Res. Opin. 29 (1) (2013) 85–92.
[26] R. van Vollenhoven, et al., Response to: 'MBDA: what is it good for?' by Yazici et al, Ann. Rheum. Dis. 73 (11) (2014) e73.
[27] I.M. Markusse, et al., A multibiomarker disease activity score for rheumatoid arthritis predicts radiographic joint damage in the BeSt study, J. Rheumatol. 41 (11) (2014) 2114–2119.
[28] Y. Yazici, C.J. Swearingen, MBDA: what is it good for? Ann. Rheum. Dis. 73 (11) (2014) e72.
[29] R.F. van Vollenhoven, et al., Brief report: enhancement of patient recruitment in rheumatoid arthritis clinical trials using a multi-biomarker disease activity score as an inclusion criterion, Arthritis Rheumatol. 67 (11) (2015) 2855–2860.

[30] K. Michaud, et al., Outcomes and costs of incorporating a multibiomarker disease activity test in the management of patients with rheumatoid arthritis, Rheumatology (Oxford) 54 (9) (2015) 1640–1649.
[31] S. Hirata, et al., A multi-biomarker disease activity score tracks clinical response consistently in patients with rheumatoid arthritis treated with different anti-tumor necrosis factor therapies: a retrospective observational study, Mod. Rheumatol. 25 (3) (2015) 344–349.
[32] K. Hambardzumyan, et al., Pretreatment multi-biomarker disease activity score and radiographic progression in early RA: results from the SWEFOT trial, Ann. Rheum. Dis. 74 (6) (2015) 1102–1109.
[33] R. Fleischmann, et al., Brief report: estimating disease activity using multi-biomarker disease activity scores in rheumatoid arthritis patients treated with abatacept or adalimumab, Arthritis Rheumatol. 68 (9) (2016) 2083–2089.
[34] B.I. Gavrila, C. Ciofu, V. Stoica, Biomarkers in rheumatoid arthritis, what is new? J. Med. Life 9 (2) (2016) 144–148.
[35] S. Hirata, Y. Tanaka, New assessment method in rheumatoid arthritis, Nihon Rinsho 74 (6) (2016) 931–937.
[36] S. Hirata, Y. Tanaka, Assessment of disease activity in rheumatoid arthritis by multi-biomarker disease activity (MBDA) score, Nihon Rinsho Meneki Gakkai Kaishi 39 (1) (2016) 37–41.
[37] K. Hambardzumyan, et al., Association of a multibiomarker disease activity score at multiple time-points with radiographic progression in rheumatoid arthritis: results from the SWEFOT trial, RMD Open 2 (1) (2016) e000197.
[38] S. Hirata, et al., Association of the multi-biomarker disease activity score with joint destruction in patients with rheumatoid arthritis receiving tumor necrosis factor-alpha inhibitor treatment in clinical practice, Mod. Rheumatol. 26 (6) (2016) 850–856.
[39] Y.C. Lee, et al., Multibiomarker disease activity score and C-reactive protein in a cross-sectional observational study of patients with rheumatoid arthritis with and without concomitant fibromyalgia, Rheumatology (Oxford) 55 (4) (2016) 640–648.
[40] J. Rech, et al., Prediction of disease relapses by multibiomarker disease activity and autoantibody status in patients with rheumatoid arthritis on tapering DMARD treatment, Ann. Rheum. Dis. 75 (9) (2016) 1637–1644.
[41] W. Li, et al., Relationship of multi-biomarker disease activity score and other risk factors with radiographic progression in an observational study of patients with rheumatoid arthritis, Rheumatology (Oxford) 55 (2) (2016) 357–366.
[42] W.G. Reiss, et al., Interpreting the multi-biomarker disease activity score in the context of tocilizumab treatment for patients with rheumatoid arthritis, Rheumatol. Int. 36 (2) (2016) 295–300.
[43] C.A.M. Bouman, et al., A multi-biomarker score measuring disease activity in rheumatoid arthritis patients tapering adalimumab or etanercept: predictive value for clinical and radiographic outcomes, Rheumatology (Oxford) 56 (6) (2017) 973–980.
[44] K. Hambardzumyan, et al., A multi-biomarker disease activity score and the choice of second-line therapy in early rheumatoid arthritis after methotrexate failure, Arthritis Rheumatol. 69 (5) (2017) 953–963.
[45] S. Krabbe, et al., Investigation of a multi-biomarker disease activity score in rheumatoid arthritis by comparison with magnetic resonance imaging, computed tomography, ultrasonography, and radiography parameters of inflammation and damage, Scand. J. Rheumatol. 46 (5) (2017) 353–358.
[46] G.M. Oderda, et al., The potential impact of monitoring disease activity biomarkers on rheumatoid arthritis outcomes and costs, Pers. Med. 15 (4) (2018) 291–301.

[47] N.M.T. Roodenrijs, et al., The multi-biomarker disease activity score tracks response to rituximab treatment in rheumatoid arthritis patients: a post hoc analysis of three cohort studies, Arthritis Res. Ther. 20 (1) (2018) 256.
[48] K. Bechman, et al., Flares in rheumatoid arthritis patients with low disease activity: predictability and association with worse clinical outcomes, J. Rheumatol. 45 (11) (2018) 1515–1521.
[49] M. Ghiti Moghadam, et al., Multi-biomarker disease activity score as a predictor of disease relapse in patients with rheumatoid arthritis stopping TNF inhibitor treatment, PLoS One 13 (5) (2018) e0192425.
[50] J.R. Curtis, et al., Biomarker-related risk for myocardial infarction and serious infections in patients with rheumatoid arthritis: a population-based study, Ann. Rheum. Dis. 77 (3) (2018) 386–392.
[51] J.R. Curtis, et al., Influence of obesity, age, and comorbidities on the multi-biomarker disease activity test in rheumatoid arthritis, Semin. Arthritis Rheum. 47 (4) (2018) 472–477.
[52] J.R. Curtis, et al., Adjustment of the multi-biomarker disease activity score to account for age, sex and adiposity in patients with rheumatoid arthritis, Rheumatology (Oxford) 58 (5) (2019) 874–883.
[53] D. Chernoff, et al., Determination of the minimally important difference (MID) in multi-biomarker disease activity (MBDA) test scores: impact of diurnal and daily biomarker variation patterns on MBDA scores, Clin. Rheumatol. 38 (2) (2019) 437–445.
[54] L.H. Calabrese, MBDA: a valuable tool for medical decision making, J. Rheumatol. 46 (12) (2019) 1642.
[55] M. Ghiti Moghadam, et al., Predictors of biologic-free disease control in patients with rheumatoid arthritis after stopping tumor necrosis factor inhibitor treatment, BMC Rheumatol. 3 (2019) 3.
[56] D.M. Boeters, et al., ACPA-negative RA consists of subgroups: patients with high likelihood of achieving sustained DMARD-free remission can be identified by serological markers at disease presentation, Arthritis Res. Ther. 21 (1) (2019) 121.
[57] J.R. Curtis, et al., Predicting risk for radiographic damage in rheumatoid arthritis: comparative analysis of the multi-biomarker disease activity score and conventional measures of disease activity in multiple studies, Curr. Med. Res. Opin. 35 (9) (2019) 1483–1493.
[58] M. Hagen, et al., Cost-effective tapering algorithm in patients with rheumatoid arthritis: combination of multibiomarker disease activity score and autoantibody status, J. Rheumatol. 46 (5) (2019) 460–466.
[59] J.R. Curtis, et al., Uptake and clinical utility of multibiomarker disease activity testing in the United States, J. Rheumatol. 46 (3) (2019) 237–244.
[60] T.M. Johnson, et al., Correlation of the multi-biomarker disease activity score with rheumatoid arthritis disease activity measures: a systematic review and meta-analysis, Arthritis Care Res. 71 (11) (2019) 1459–1472.
[61] F. Xie, et al., Tocilizumab and the risk of cardiovascular disease: direct comparison among biologic disease-modifying antirheumatic drugs for rheumatoid arthritis patients, Arthritis Care Res. 71 (8) (2019) 1004–1018.
[62] C.H. Brahe, et al., Predictive value of a multi-biomarker disease activity score for clinical remission and radiographic progression in patients with early rheumatoid arthritis: a post-hoc study of the OPERA trial, Scand. J. Rheumatol. 48 (1) (2019) 9–16.
[63] A.T. Masi, What further data are needed to value the MBDA score for measuring RA disease activity, comment on the article by Johnson TM et al, Arthritis Care Res. 72 (9) (2020) 1339–1340.
[64] M.H.Y. Ma, et al., A multi-biomarker disease activity score can predict sustained remission in rheumatoid arthritis, Arthritis Res. Ther. 22 (1) (2020) 158.

[65] T.M. Johnson, et al., Reply to: what further data are needed to value the MBDA score for measuring RA disease activity, comment on the article by Johnson TM et al, Arthritis Care Res. 72 (9) (2020). 1340-1340.

[66] T.A. Boyd, et al., Correlation of serum protein biomarkers with disease activity in psoriatic arthritis, Expert. Rev. Clin. Immunol. 16 (3) (2020) 335–341.

[67] A.T. Masi, R. Fleischmann, Does ACPA-negative RA consist of subgroups related to sustained DMARD-free remission and serological markers at disease presentation? Comment on article by Boeters DM et al, Arthritis Res. Ther. 22 (1) (2020) 17.

[68] O.G. Segurado, E.H. Sasso, Vectra DA for the objective measurement of disease activity in patients with rheumatoid arthritis, Clin. Exp. Rheumatol. 32 (5 Suppl. 85) (2014) S-29-34.

[69] V. Hamy, et al., Developing smartphone-based objective assessments of physical function in rheumatoid arthritis patients: the PARADE study, Digit. Biomark 4 (1) (2020) 26–43.

[70] M.K. David, S.W. Leslie, Prostate Specific Antigen (PSA), StatPearls, Treasure Island, FL, 2020.

[71] J.R. Kolfenbach, et al., Autoimmunity to peptidyl arginine deiminase type 4 precedes clinical onset of rheumatoid arthritis, Arthritis Rheum. 62 (9) (2010) 2633–2639.

[72] M.R. Arbuckle, et al., Development of anti-dsDNA autoantibodies prior to clinical diagnosis of systemic lupus erythematosus, Scand. J. Immunol. 54 (1–2) (2001) 211–219.

[73] M.R. Arbuckle, et al., Development of autoantibodies before the clinical onset of systemic lupus erythematosus, N. Engl. J. Med. 349 (16) (2003) 1526–1533.

[74] P.D. Burbelo, et al., Autoantibodies are present before the clinical diagnosis of systemic sclerosis, PLoS One 14 (3) (2019) e0214202.

[75] L.B. Kelmenson, et al., Timing of elevations of autoantibody isotypes prior to diagnosis of rheumatoid arthritis, Arthritis Rheumatol. 72 (2) (2020) 251–261.

[76] K.D. Deane, C.C. Striebich, V.M. Holers, Editorial: prevention of rheumatoid arthritis: now is the time, but how to proceed? Arthritis Rheumatol. 69 (5) (2017) 873–877.

[77] K.D. Deane, Preclinical rheumatoid arthritis and rheumatoid arthritis prevention, Curr. Rheumatol. Rep. 20 (8) (2018) 50.

[78] K.D. Deane, V.M. Holers, The natural history of rheumatoid arthritis, Clin. Ther. 41 (7) (2019) 1256–1269.

[79] K.D. Deane, T.T. Cheung, Rheumatoid arthritis prevention: challenges and opportunities to change the paradigm of disease management, Clin. Ther. 41 (7) (2019) 1235–1239.

[80] M. Mahler, et al., Precision medicine in the care of rheumatoid arthritis: focus on prediction and prevention of future clinically-apparent disease, Autoimmun. Rev. 19 (5) (2020) 102506.

[81] J. Shi, et al., Autoantibodies recognizing carbamylated proteins are present in sera of patients with rheumatoid arthritis and predict joint damage, Proc. Natl. Acad. Sci. USA 108 (42) (2011) 17372–17377.

[82] L.A. Trouw, M. Mahler, Closing the serological gap: promising novel biomarkers for the early diagnosis of rheumatoid arthritis, Autoimmun. Rev. 12 (2) (2012) 318–322.

[83] X. Jiang, et al., Anti-CarP antibodies in two large cohorts of patients with rheumatoid arthritis and their relationship to genetic risk factors, cigarette smoking and other autoantibodies, Ann. Rheum. Dis. 73 (10) (2014) 1761–1768.

[84] J. Shi, et al., Carbamylation and antibodies against carbamylated proteins in autoimmunity and other pathologies, Autoimmun. Rev. 13 (3) (2014) 225–230.

[85] M. Mahler, et al., Standardisation of myositis-specific antibodies: where are we today? Ann. Rheum. Dis. (2019), https://doi.org/10.1136/annrheumdis-2019-216003.

[86] M. Mahler, Population-based screening for ACPAs: a step in the pathway to the prevention of rheumatoid arthritis? Ann. Rheum. Dis. 76 (11) (2017) e42.

[87] M.J. Fritzler, M. Mahler, Redefining systemic lupus erythematosus—SMAARTT proteomics, Nat. Rev. Rheumatol. 14 (8) (2018) 451–452.
[88] G. Barturen, M.E. Alarcon-Riquelme, SLE redefined on the basis of molecular pathways, Best Pract. Res. Clin. Rheumatol. 31 (3) (2017) 291–305.
[89] M.J. Lewis, et al., Autoantibodies targeting TLR and SMAD pathways define new subgroups in systemic lupus erythematosus, J. Autoimmun. 91 (2018) 1–12.
[90] H. Gunawardena, et al., Autoantibodies to a 140-kd protein in juvenile dermatomyositis are associated with calcinosis, Arthritis Rheum. 60 (6) (2009) 1807–1814.
[91] Z. Betteridge, et al., Identification of a novel autoantigen eukaryotic initiation factor 3 associated with polymyositis, Rheumatology (Oxford) 59 (5) (2020) 1026–1030.
[92] L. Martinez-Prat, et al., OP0118 deciphering the anti-protein-arginine deiminase (PAD) response identifies PAD1 and PAD6 as novel autoantigens in rheumatoid arthritis, Ann. Rheum. Dis. 79 (Suppl. 1) (2020) 78–79.
[93] M. Mahler, M.J. Fritzler, The changing landscape of the clinical value of the PM/Scl autoantibody system, Arthritis Res. Ther. 11 (2) (2009) 106.
[94] M. Mahler, M. Bluthner, K.M. Pollard, Advances in B-cell epitope analysis of autoantigens in connective tissue diseases, Clin. Immunol. 107 (2) (2003) 65–79.
[95] M.J. Fritzler, et al., Historical perspectives on the discovery and elucidation of autoantibodies to centromere proteins (CENP) and the emerging importance of antibodies to CENP-F, Autoimmun. Rev. 10 (4) (2011) 194–200.
[96] M.J. Fritzler, et al., Bicaudal D2 is a novel autoantibody target in systemic sclerosis that shares a key epitope with CENP-A but has a distinct clinical phenotype, Autoimmun. Rev. 17 (3) (2018) 267–275.
[97] G.L. Norman, et al., Anti-kelch-like 12 and anti-hexokinase 1: novel autoantibodies in primary biliary cirrhosis, Liver Int. 35 (2) (2015) 642–651.
[98] G.L. Norman, et al., The prevalence of anti-hexokinase-1 and anti-kelch-like 12 peptide antibodies in patients with primary biliary cholangitis is similar in Europe and North America: a large International, Multi-Center Study, Front. Immunol. 10 (2019) 662.
[99] A. Reig, et al., Novel anti-hexokinase 1 antibodies are associated with poor prognosis in patients with primary biliary cholangitis, Am. J. Gastroenterol. 115 (10) (2020) 1634–1641.
[100] D. Villalta, et al., Evaluation of a novel extended automated particle-based multi-analyte assay for the detection of autoantibodies in the diagnosis of primary biliary cholangitis, Clin. Chem. Lab. Med. 58 (9) (2020) 1499–1507.
[101] G.A. Schellekens, et al., Citrulline is an essential constituent of antigenic determinants recognized by rheumatoid arthritis-specific autoantibodies, J. Clin. Invest. 101 (1) (1998) 273–281.
[102] M. Koenig, et al., Autoantibodies to a novel Rpp38 (Th/To) derived B-cell epitope are specific for systemic sclerosis and associate with a distinct clinical phenotype, Rheumatology (Oxford) 58 (10) (2019) 1784–1793.
[103] S. Hong, J.E. Lee, Y. Park, J.W. Choi, Identifying potential biomarkers related to pre-term delivery by proteomic analysis of amniotic fluid., Sci. Rep. 10 (1) (2020) 19648, https://doi.org/10.1038/s41598-020-76748-1.
[104] M.K. Verheul, et al., Triple positivity for anti-citrullinated protein autoantibodies, rheumatoid factor, and anti-carbamylated protein antibodies conferring high specificity for rheumatoid arthritis: implications for very early identification of at-risk individuals, Arthritis Rheumatol. 70 (11) (2018) 1721–1731.
[105] X. Bossuyt, Clinical performance characteristics of a laboratory test. A practical approach in the autoimmune laboratory, Autoimmun. Rev. 8 (7) (2009) 543–548.

[106] P. Vermeersch, D. Blockmans, X. Bossuyt, Use of likelihood ratios can improve the clinical usefulness of enzyme immunoassays for the diagnosis of small-vessel vasculitis, Clin. Chem. 55 (10) (2009) 1886–1888.
[107] X. Bossuyt, et al., Likelihood ratios as a function of antibody concentration for anti-cyclic citrullinated peptide antibodies and rheumatoid factor, Ann. Rheum. Dis. 68 (2) (2009) 287–289.
[108] M.M. Leeflang, P.M. Bossuyt, L. Irwig, Diagnostic test accuracy may vary with prevalence: implications for evidence-based diagnosis, J. Clin. Epidemiol. 62 (1) (2009) 5–12.
[109] P.M. Bossuyt, Clinical validity: defining biomarker performance, Scand. J. Clin. Lab. Investig. Suppl. 242 (2010) 46–52.
[110] N. Vermeulen, et al., Likelihood ratio for Crohn's disease as a function of anti-Saccharomyces cerevisiae antibody concentration, Inflamm. Bowel Dis. 16 (1) (2010) 5–6.
[111] H. Okamoto, Stimulatory autoantibodies to the PDGF receptor in scleroderma, N. Engl. J. Med. 355 (12) (2006) 1278 (author reply 1279).
[112] J.F. Classen, et al., Lack of evidence of stimulatory autoantibodies to platelet-derived growth factor receptor in patients with systemic sclerosis, Arthritis Rheum. 60 (4) (2009) 1137–1144.

Chapter 4

Companion and complementary diagnostics: A key to precision medicine

Jan Trøst Jørgensen
Dx-Rx Institute, Fredensborg, Denmark

1. Introduction

The goal of individualizing pharmacotherapy has been on the agenda of the health care providers for decades, and one of the key elements in this effort has been the principles of rational use of drugs or rational pharmacotherapy [1]. The essence of these principles is that the individual patient should receive medications appropriate to their clinical needs in order to optimize the benefit and minimize the harm. Already in the 1960s and 1970s, these principles were translated into "the right drug for the right patient in the right dose at the right time" [2, 3]. When we discuss precision or personalized medicine today, we still often use the same different "rights" to describe the concept. However, there is one major difference when we compare then with now, and that is the increase in our molecular understanding of the pathophysiology and the mechanisms of action of drugs, which is essential for the implementation of precision medicine. The importance of such an understanding has been recognized for decades, but we have only experienced substantial progress over the last couple of decades. The advances in molecular medicine and especially molecular diagnostics have slowly enabled us to start to practice a more individualized pharmacotherapy and develop drugs defined for subsets of patients. We have learned that most diseases are heterogeneous and in order to achieve a more effective pharmacotherapy drugs must be developed accordingly and here, the predictive biomarkers play a key role [1, 4].

The first time we systematically see predictive biomarker testing integrated in cancer drug development was in relation to the development of trastuzumab (Herceptin, Roche/Genentech) for treatment of metastatic HER2 positive breast cancer. In the 1980s, the US scientist Dennis J. Slamon discovered the link between amplification of the *HER2* gene and a poor prognosis of patients with breast cancer. In an article published in *Science* in 1987, he suggested that a specific antagonist should be the development, which could block the signaling from the HER2 receptor protein [5]. This antagonist later became the monoclonal antibody trastuzumab and when Genentech developed the drug, they also developed

Precision Medicine and Artificial Intelligence. https://doi.org/10.1016/B978-0-12-820239-5.00003-6

an immunohistochemistry (IHC) assay for the detection of the HER2 protein in the tumor tissue. When they initiated the clinical development of trastuzumab in 1990s, the IHC assay was used to preselect patients for treatment with the drug, and the different clinical trials conducted demonstrated a clear link between HER2 overexpression and the efficacy of trastuzumab [6]. By using this HER2 IHC assay for preselecting the patients, Genentech formed the basis for the prospective enrichment trial design as we know it today. In September 1998, the US Food and Drug Administration (FDA), through a new coordinated procedure, simultaneously granted an approval of trastuzumab and the HER2 IHC assay, named HercepTest (Dako). This simultaneous approval makes sense as the assay is an important treatment decision tool for the prescribing physician and must be available at the same time as the drug [7]. The HercepTest became the first companion diagnostics (CDx) assay linked to the use of a specific drug. Not only was the development of trastuzumab a great scientific and medical achievement but the parallel development of drug and diagnostic has served as an important inspiration for a number of other targeted cancer drugs developed subsequently [8].

Looking at the targeted cancer drugs that have obtained regulatory approval within the last couple of decades, a large part has been developed using the drug-diagnostic codevelopment model. Today, the number of drugs that are guided by a CDx assay is constantly increasing and currently, close to 40 drugs have such an assay linked to their use [9]. In this chapter, a brief introduction will be given to companion and complementary diagnostics with specific attention payed to the clinical development and the regulatory requirements. Since most of the advances of CDx have been made in oncology, examples are provided that likely will provide guidance for the field of autoimmune diseases which are the focus of this book.

2. Companion and complementary diagnostics

The term used to describe a predictive biomarker assay linked to a specific drug has varied over time. In the years following the US FDA approval of the HercepTest this type of assay was referred to as pharmacodiagnostics, theranostics, or pharmacogenetics [10]. Almost a decade later, the term companion diagnostic appeared for the first time in the literature, and subsequently became the preferred term [11].

The US FDA issued the first regulatory guidance document on CDx in 2014 and here, this type of assay was officially defined for the first time [12]. According to this definition, a CDx assay is an in vitro diagnostic (IVD) device that provides information that is essential for the safe and effective use of a corresponding therapeutic product. Furthermore, four areas are specified in which a CDx assay could be essential: (I) to identify patients who are most likely to benefit from the therapeutic product; (II) to identify patients likely to be at increased risk of serious adverse reactions as a result of treatment with the therapeutic product; (III) to monitor response to treatment with the therapeutic product for the purpose of adjusting treatment (e.g., schedule, dose, discontinuation) to achieve improved safety or effectiveness; and finally, (IV) to identify patients in the population for whom the therapeutic product has been adequately studied, and found safe and effective, i.e., there is insufficient information about the safety and effectiveness of the therapeutic product in any other population. Overall, the US FDA definition can be summarized as outcome prediction (efficacy and safety) as well as therapy monitoring. The fourth item mentioned in the definition can be regarded as a kind of disclaimer, which is related to the type of study design most often used for the clinical validation of the CDx assays; the enrichment study design. As the US FDA do in their definition, it is important to emphasize that clinical outcome data cannot be extrapolated to any other patient population than the one defined by the CDx assay [13]. According to the definition, a CDx assay is deemed essential for the safe and effective use of the corresponding therapeutic product, thus testing with this type of assay is mandatory and consequently included in the regulatory labeling for the drug under the section "Indications and Usage."

Until recently, no official definition of a CDx assay has been available in the European Union (EU). However, this has changed with the adoption of the new EU regulations on IVD medical devices by the European Commission and the European Parliament in 2017 [14]. According to the new regulations the definition of a CDx assay is similar to the one found in the 2014 guidance document issued by the US FDA. Overall, the new EU regulations state that a CDx assay is a device that is essential for the safe and effective use of a corresponding medicinal product. For both the US, the EU and other countries where similar regulations have been implemented, this must be seen as a major step forward in relation to improve patient safety when it comes to CDx assays. These types of assays are important treatment decision tools on which pharmacotherapeutic interventions are based, and an incorrect result from a CDx assay could lead to inappropriate patient management with possible serious consequences for the individual [13].

Beside the CDx assays, the US FDA have introduced a new class of assays meant to guide pharmacotherapy, which is complementary diagnostics (CoDx). This term is relatively new and was first used in relation to the regulatory approval of the immune checkpoint inhibitor nivolumab (Opdivo, BMS) for second-line treatment of non-squamous non-small cell lung cancer (NSCLC) in the autumn of 2015 [15–17]. So far, no guidelines have been issued on the subject and no official definition is currently available. However, from presentations, publications, etc. by the US FDA officials it can be deduced that a CoDx is an IVD device that has the ability to identify a biomarker-defined subset of patients with a different benefit-risk profile than the broader population for which a therapeutic product is indicated, but that is not a prerequisite for receiving the therapeutic product [17–19]. In contrast to CDx assays, the US FDA do not regard these assays essential for a safe and effective use of the corresponding therapeutic product, which is also reflected in the drug labeling, as no information about testing is required here [16–19]. Table 1 compare the most important regulatory differences between CDx and CoDx assays. Furthermore, Fig. 1 illustrates how these assays are supposed to be used in the treatment decision process for the individual patient [19].

TABLE 1 Regulatory status of companion and complementary diagnostics.

	Companion diagnostics	Complementary diagnostics
Definition	An IVD device that provides information that is essential for the safe and effective use of a corresponding therapeutic product	An IVD device that has the ability to identify a biomarker-defined subset of patients with a different benefit-risk profile than the broader population for which a therapeutic product is indicated, but that is not a prerequisite for receiving the therapeutic product[a]
Information in drug labeling (indications and usage)	Yes	No
FDA regulated product	Yes	Yes
EMA regulated product[b]	Yes	?[c]

[a]*Preliminary FDA definition [18].*
[b]*According to new European regulations on IVD medical devices CDx assays are Class C. The new regulation will enter into force in 2022 [20].*
[c]*CoDx is not mentioned in the new European regulations on IVD medical devices but this type of assay will likely be a regulated product [20].*

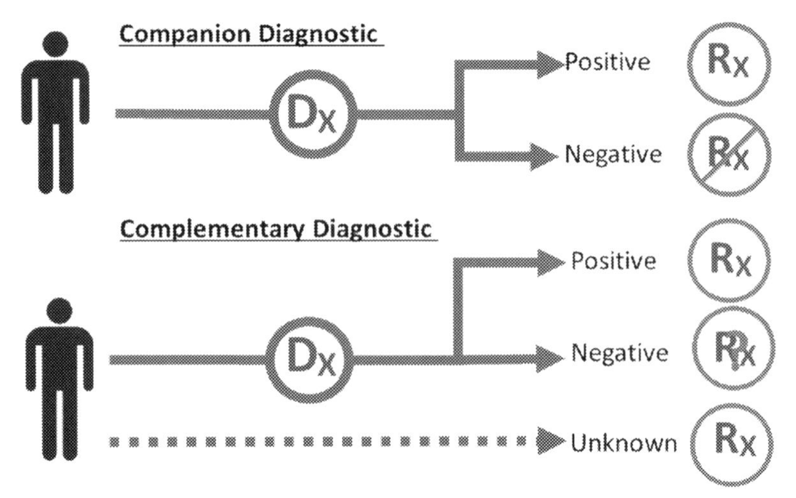

FIG. 1 The use of companion and complementary diagnostics [17]. *(Reprinted with permission from J.T. Jørgensen, Companion and complementary diagnostics—clinical and regulatory perspectives, Trends Cancer 2 (2016) 706–712, Elsevier.)*

3. Drug-diagnostic codevelopment

As described earlier in this chapter, an important incidence that formed the foundation for the current drug-diagnostic codevelopment model was the development of trastuzumab for treatment of women with metastatic breast cancer and the corresponding the HER2 IHC assay. Here, based on a solid molecular understanding of the pathogenesis of HER2 positive breast cancer, Dr. Slamon and colleagues suggested the development of an antagonist that could improve the therapeutic outcome for this group of patients [5]. For any drug development project, a molecular understanding of the pathophysiology and the mechanism of the drug is of paramount importance and in fact, this is the essence of the drug-diagnostic codevelopment model. Fig. 2 illustrates the traditional parallel development of drug and diagnostic with the three critical elements with respect to a CDx or CoDx assay development: (1) generation of a biomarker hypothesis, (2) analytical validation, and (3) clinical validation [21].

As described above, a molecular understanding is the key to the success of a drug-diagnostic codevelopment project and based on this knowledge the biomarker hypothesis is generated. The strength of such a hypothesis very much depends on how well these molecular mechanisms are understood, and sometimes more than one hypothesis is generated and one or all are subsequently tested using different prototype assays. The prototype assays are then used in early phase clinical trials in order to test the hypothesis on a link between biomarker status and clinical outcome following treatment with the investigational drug. If such a link is identified, the next step is to define the clinical cut-off value for the assay, which is the assay value that splits the patient population

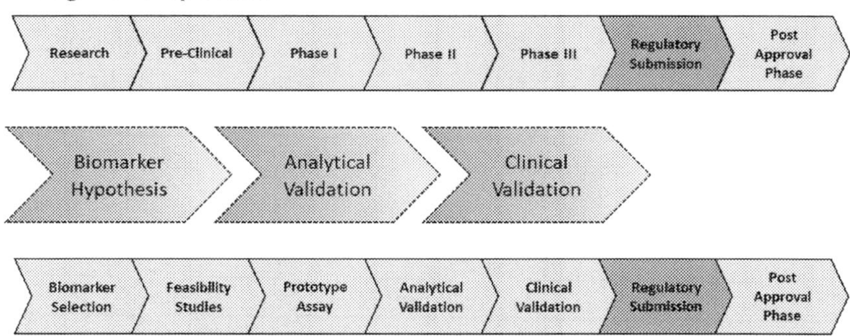

FIG. 2 The traditional drug-diagnostic codevelopment model with an aligned regulatory submission at the end of phase III. In relation to the CDx assays development, the three critical elements are: (1) generation of the biomarker hypothesis; (2) the analytical validation; and (3) the clinical validity [18]. *(Reprinted with permission from J.T. Jørgensen, The drug-diagnostic codevelopment model, in: J.T. Jørgensen (Ed.), Complementary Diagnostics: From Biomarker Discovery to Clinical Implementation, Academic Press, London, 2019, Elsevier.)*

into likely responders or non-responders. This can be a tricky exercise as the amount of clinical outcome data available at this early stage of the drug development project is often limited. When the cut-off value for the assay has be established the analytical validation can be finalized [16, 21].

During the analytical validation of the CDx or CoDx assay, it must be demonstrated that the measurements of the biomarker(s) in question are accurate and reliable [21]. When an assay is going to be used prospectively to select patients into a clinical trial, it is important that it has demonstrated to be sufficiently analytically robust, especially around the selected cut-off value. In general, the analytical validation studies are performed in order to evaluate any critical performance characteristics of the assay. Normally, the analytical validation has to be completed before the assay is used in a pivotal clinical trial, which is going to support claims in relation to a drug-diagnostic combination [21, 22]. With the recent appearance of tumor agnostic drugs, where a biomarker solely defines the indication, the analytical validation of the CDx assay before start of a clinical trial has become even more important. Examples of this type of drugs are the tropomyosin receptor kinase (TRK) inhibitors larotrectinib (Vitrakvi, Loxo Oncology/Bayer) and entrectinib (Rozlytrek, Roche/Genentech), which have been approved very recently for patients with solid tumors who harbored a neurotrophic receptor tyrosine kinase (*NTRK*) gene fusion [23–25].

The aim of the clinical validation is to demonstrate that the CDx or CoDx assay is able to detect the biomarkers of interest in the intended use population and to identify the patients who are expected to benefit from the investigational drug [10, 26]. The clinical validation of the assay should not be initiated before the analytical validation has been concluded, as shown in Fig. 2. It is important that the pivotal clinical trials are conducted with the final analytical validated version of the CDx or CoDx assay as the analytical performance can greatly affect the patient selection process [26]. It is crucial that the right patients receive the investigational drug in order to achieve the best possible outcome of the treatment. The clinical validation of a CDx or CoDx assay is normally performed at the same time as the safety and efficacy of the investigational drug is demonstrated in a phase II/III pivotal clinical trial. In relation to this validation, it is important to remember the "one rule" that says, only to use <u>one version of the assay</u>, namely the final analytically validated version, and furthermore, only to use <u>one testing site</u> in order to reduce possible inter-laboratory variability [21]. If a non-validated assay or different versions of the CDx assay are used in a pivotal clinical trial a bridging study will normally be required before the assay can obtain regulatory approval [16, 26]. In a bridging study, it must be documented that the final validated CDx assay has performance characteristic similar to assay(s) used in the pivotal trial. Furthermore, a reanalysis of the primary clinical outcome data should be performed, in order to assure that the use of the final validated version of the CDx assay does not alter the conclusion on the safety and efficacy of the drug [16, 22]. Bridging studies is both costly and time-consuming and should be avoided if possible. After regulatory approval of the CDx assay, it will normally be distributed to a number of local laboratories and in order secure the quality of the assay results, proficiency testing is important.

4. Regulatory requirements

The changes in drug development over the last decades, especially within oncology, are also reflected in the regulatory requirements. Drug-diagnostic codevelopment has played an increasing role and more and more drugs are now being developed together with a CDx or CoDx assay. This has resulted in a growing number of discussion papers, draft guidelines and guidance documents related to CDx assays issued by the regulatory drug agencies. Especially the US FDA has been at the forefront in relation to the implementation of regulatory strategies for CDx assays and drug-diagnostic codevelopment [21].

4.1 US FDA

CDx assays take up a central role as an important treatment decision tools for a number of oncology drugs, that is also reflected in the way in which the US FDA risk classifies these assays. In the US, CDx assays are classified as IVD Class III products, which mean high risk and consequently a high level of regulatory control [27]. IVD Class III products will normally require a pre-market approval (PMA) application in order to secure a high standard for both the analytical and clinical performance of the assays. Compared to other types of IVD submissions, the PMA requires the most comprehensive documentation level, which further underlines the critical role of CDx assays [16].

Over the years, the US FDA has issued several important guidance documents describing the requirements for development of CDx assays and specifically three documents should attract our attention. In 2014, the guidance document on In Vitro *Companion Diagnostic Devices* was issued in which a CDx assay was regulatory defined for the first time [12]. Furthermore, this guidance document outlined a number of administrative procedures in relation to the development of a therapeutic product for which a CDx assay is evaluated essential. A couple of years later, in 2016, an important draft guidance on *Principles for Codevelopment of an In Vitro Companion Diagnostic Device With a Therapeutic Product* was issued [28]. This guidance document aims in a practical manner to provide information on the design and implementation of a successful drug-diagnostic codevelopment program. Over the last decade, we have seen that cancer drugs have been developed for smaller and smaller molecular defined subsets of patients. This is in fact often the situation in tumor agnostic drug development, where a drug can be effective in multiple tumor types that harbors a specific molecular aberration, often at low frequency. By the end of 2018, a guidance document on *Developing Targeted Therapies in Low-Frequency Molecular Subsets of a Disease* was issued by US FDA, describing the approach to evaluation of targeted therapies in these groups of patients [29]. As drug-diagnostic codevelopment is an integrate approach, involving different disciplines, the above-mentioned guidance documents have all been issued based on a cooperation between Center for Biologics Evaluation and Research (CBER), Center for Drug Evaluation and Research (CDER) and Center for Devices and Radiological Health (CDRH) at the US FDA.

4.2 The European Union and other countries

The current regulatory framework for CDx assays in the European Union (EU) is the IVD Directive 98/79/EC issued in 1998 [30]. However, this directive does not mention CDx assays in the definition of an IVD, and likewise the classification system does not consider this type of assay at all. Currently, any CDx assay developed for the EU market is classified as general IVD, which means low risk devices. Through the so-called self-certification procedure, the manufacturer has to perform a conformity assessment where it is stated that the essential requirements of the IVD directive are fulfilled and subsequently the assay can be CE-marked and marketed. Compared with the PMA approval process in the US, the EU self-certification procedure must be considered a very different regulatory path, which does not take into consideration the critical role of the CDx assay as an important decision tool [16].

However, changes are under way. In April 2017, the European Commission and European Parliament adopted new regulations on IVD medical devices, which is more up-to-date with the current thinking of the role of CDx assays [20]. This regulation will enter into force in 2022 after which all CDx assays used in EU must comply with the requirements in the new regulation. This new IVD regulation will introduce a number of substantial changes and CDx assays will no longer be considered low risk devices. According to the new regulation, CDx assays will be classified as class C, which means a high individual risk or moderate public health risk. Another major change is that self-certification will no longer be a possibility, the requirement will be that the conformity assessment must be performed by an independent notified body and that this notified body also has to consult the European Medicines Agency (EMA) or one of the medical product national competent authorities. Furthermore, the general requirements for both the technical and clinical performance documentation as well as the post-market surveillance activities have been strengthened [16, 20]. The EU is not the only region that has strengthened and updated the requirements for CDx development. In a number of other countries such as Japan, Australia, and Canada new regulatory guidance documents have been issued recently [31].

5. Concluding remark

CDx and CoDx should be seen as an important element in the realization of precision medicine. The number of drugs that have such an assay linked to their use is rapidly increasing and within oncology, they have already had major impact on drug development and on how these drugs are used in the clinic. Most pharmaceutical companies have recognized that the full therapeutic potential of their targeted cancer drugs can only be realized if CDx assays are developed concomitantly. As pointed out in the introduction, the progress in autoimmunity is limited, but will likely be accelerated by success in oncology.

Looking at the *List of Cleared or Approved Companion Diagnostic Devices*, issued by the US FDA, all the listed assays are more or less linked to different type of targeted oncological and hematological drugs [9]. A likely explanation for this is that most malignant diseases are driven by somatic gene mutations in the respective tumors, whereas it for most other relevant diseases are germline mutations. However, as a kind of "supplement" to the list of approved CDx assays the US FDA have issued a *Table of Pharmacogenomic Biomarkers in Drug Labeling*, which covers a large number of drugs used for different disease indications [32]. This table lists a large number of therapeutic products indicated for a variety of diseases and where pharmacogenomic information is found in the drug labeling. This information can appear in different sections of the labeling depending on the actions and covers both genomic and protein biomarkers.

With the increased molecular understanding of the pathophysiology and disease heterogeneity, the CDx assays based on single biomarker detection will need to be supplemented by assays with a more complex signature. These assays will be a kind of "composite biomarker" consisting of multiple biomarkers of different nature, which is combined using a specific algorithm depending on the type of drug they are meant to guide. These algorithms will be developed based on the principles of system biology and pharmacology and will require extensive verification and validation studies before they will be ready for clinical implementation [17].

References

[1] J.T. Jørgensen, Twenty years with personalized medicine: past, present, and future of individualized pharmacotherapy, Oncologist 24 (2019) e432–e440.

[2] C.J. Klett, E.C. Moseley, The right drug for the right patient, J. Consult. Psychol. 29 (1965) 546–551.

[3] C.R. Galbrecht, C.J. Klett, Predicting response to phenothiazines: the right drug for the right patient, J. Nerv. Ment. Dis. 147 (1968) 173–183.

[4] L.J. Lesko, S. Schmidt, Individualization of drug therapy: history, present state, and opportunities for the future, Clin. Pharmacol. Ther. 92 (2012) 458–466.

[5] D.J. Slamon, G.M. Clark, S.G. Wong, et al., Human breast cancer: correlation of relapse and survival with amplification of the HER-2/neu oncogene, Science 235 (1987) 177–182.

[6] D.J. Slamon, B. Leyland-Jones, S. Shak, et al., Use of chemotherapy plus a monoclonal antibody against HER2 for metastatic breast cancer that overexpresses HER2, N. Engl. J. Med. 344 (2001) 783–792.

[7] J.T. Jørgensen, H. Winther, The development of the HercepTest—from bench to bedside, in: J.T. Jørgensen, H. Winther (Eds.), The Key Driver of Personalized Cancer Medicine, Pan Stanford, Singapore, 2010, pp. 43–60.

[8] D.F. Hayes, HER2 and breast cancer—a phenomenal success story, N. Engl. J. Med. 381 (2019) 1284–1286.
[9] US Food and Drug Administration, List of Cleared or Approved Companion Diagnostic Devices (In Vitro and Imaging Tools) (Update: August 2, 2019) http://www.fda.gov/MedicalDevices/ProductsandMedicalProcedures/InVitroDiagnostics/ucm301431.htm. (Accessed 24 September 2019).
[10] J.T. Jørgensen, M. Hersom, Companion diagnostics—a tool to improve pharmacotherapy, Ann. Transl. Med. 4 (2016) 482.
[11] N. Papadopoulos, K.W. Kinzler, B. Vogelstein, The role of companion diagnostics in the development and use of mutation-targeted cancer therapies, Nat. Biotechnol. 24 (2006) 985–995.
[12] US Food and Drug Administration, Guidance for Industry and Food and Drug Administration Staff. In Vitro Companion Diagnostic Devices (Issued: 09/06/14) http://www.fda.gov/downloads/MedicalDevices/DeviceRegulationandGuidance/GuidanceDocuments/UCM262327.pdf. (Accessed 26 September 2019).
[13] J.T. Jørgensen, M. Hersom, An introduction to companion and complementary diagnostics, in: J.T. Jørgensen (Ed.), Companion and Complementary Diagnostics: From Biomarker Discovery to Clinical Implementation, Academic Press/Elsevier, London, 2019, pp. 1–10.
[14] European Union, Regulation of the European Parliament and of the Council on In Vitro Diagnostic Medical Devices, 5 April 2017, 2012/0267 (COD) http://eur-lex.europa.eu/legal-content/EN/TXT/PDF/?uri=CONSIL:PE_15_2017_INIT&from=EN. (Accessed 26 September 2019).
[15] J.T. Jørgensen, Companion diagnostic assays for PD-1/PD-L1 checkpoint inhibitors in NSCLC, Expert. Rev. Mol. Diagn. 16 (2016) 131–133.
[16] J.T. Jørgensen, M. Hersom, Clinical and regulatory aspects of companion diagnostic development in oncology, Clin. Pharmacol. Ther. 103 (2018) 999–1008.
[17] J.A. Beaver, et al., An FDA perspective on the regulatory implications of complex signatures to predict response to targeted therapies, Clin. Cancer Res. 23 (2017) 1368–1372.
[18] Y.F. Hu, Development of companion diagnostics—an FDA perspective, in: Presentation at the CBA Workshop, Rockville, MD, August 11, 2018. https://cba-usa.org/sites/default/files/workshop/workshop201808/Development%20of%20companion%20diagnostics%20-%20an%20FDA%20Perspective%20Fu.pdf. (Accessed 8 February 2020).
[19] J.T. Jørgensen, Companion and complementary diagnostics—clinical and regulatory perspectives, Trends Cancer 2 (2016) 706–712.
[20] Regulation (EU) 2017/746 of the European Parliament and of the Council of 5 April 2017 on In Vitro Diagnostic Medical Devices and Repealing Directive 98/79/EC and Commission Decision 2010/227/EU. http://eur-lex.europa.eu/legal-content/EN/TXT/PDF/?uri=CELEX:32017R0746&from=EN. (Accessed 1 October 2019).
[21] J.T. Jørgensen, The drug-diagnostic codevelopment model, in: J.T. Jørgensen (Ed.), Companion and Complementary Diagnostics: From Biomarker Discovery to Clinical Implementation, Academic Press/Elsevier, London, 2019, pp. 11–25.
[22] US FDA, Principles for Codevelopment of an In Vitro Companion Diagnostic Device With a Therapeutic Product. Draft Guidance, 15 July 2016. https://www.fda.gov/downloads/MedicalDevices/DeviceRegulationandGuidance/GuidanceDocuments/UCM510824.pdf. (Accessed 28 September 2019).
[23] S. Lemery, P. Keegan, R. Pazdur, First FDA approval agnostic of cancer site—when a biomarker defines the indication, N. Engl. J. Med. 377 (2017) 1409–1412.
[24] A. Drilon, T.W. Laetsch, S. Kummar, et al., Efficacy of larotrectinib in TRK fusion-positive cancers in adults and children, N. Engl. J. Med. 378 (2018) 731–739.

[25] J.T. Jørgensen, A paradigm shift in biomarker guided oncology drug development, Ann. Transl. Med. 7 (2019) 148.
[26] D.M. Roscoe, Y.F. Hu, R. Philip, Companion diagnostics: a regulatory perspective from the last 5 years of molecular companion diagnostic approvals, Expert. Rev. Mol. Diagn. 15 (7) (2015) 869–880.
[27] US FDA, Overview of IVD Regulation, Updated: 09/16/2019 https://www.fda.gov/MedicalDevices/DeviceRegulationandGuidance/IVDRegulatoryAssistance/ucm123682.htm. (Accessed 30 September 2019).
[28] US FDA, Draft Guidance on Principles for Codevelopment of an In Vitro Companion Diagnostic Device With a Therapeutic Product, Issued: 07/15/2016 https://www.fda.gov/media/99030/download. (Accessed 30 September 2019).
[29] US FDA, Developing Targeted Therapies in Low-Frequency Molecular Subsets of a Disease, 10/15/18 https://www.fda.gov/media/117173/download. (Accessed 1 October 2019).
[30] Directive 98/79/EC of the European Parliament and of the Council of 27 October 1998 on In Vitro Diagnostic Medical Devices. http://eur-lex.europa.eu/legal-content/EN/TXT/PDF/?uri=CELEX:31998L0079&from=EN. (Accessed 1 October 2019).
[31] A. Craig, Personalised medicine with companion diagnostics: the intercept of medicines and medical devices in the regulatory landscape, EMJ Innov. 1 (2017) 47–53.
[32] US FDA, Table of Pharmacogenomic Biomarkers in Drug Labeling, Last Updated: December 2019 https://www.fda.gov/media/124784/download. (Accessed 8 February 2020).

Chapter 5

Checkpoint inhibitors: Interface of cancer and autoimmunity: Opportunity for second level precision medicine

Savino Sciascia[a], Marie Hudson[b], Marvin J. Fritzler[c], Minoru Satoh[d], and Michael Mahler[e]

[a]*Department of Clinical and Biological Sciences, and SCDU Nephrology and Dialysis, Center for Research of Immunopathology and Rare Diseases-Coordinating Center of Piemonte and Valle d'Aosta Network for Rare Diseases, S. Giovanni Bosco Hospital, Turin, Italy,* [b]*Department of Medicine, Division of Rheumatology, McGill University, Jewish General Hospital, and Lady Davis Institute for Medical Research, Montreal, QC, Canada,* [c]*Cumming School of Medicine, University of Calgary, Calgary, AB, Canada,* [d]*Department of Clinical Nursing, University of Occupational and Environmental Health, Kitakyushu, Japan,* [e]*Inova Diagnostics, Inc., San Diego, CA, United States*

Abbreviations

AI	artificial intelligence
AIM	autoimmune myopathies
ANA	antinuclear antibodies
ATA	anti-tumor autoantibodies
CTLA-4	cytotoxic T lymphocyte-associated antigen 4
DMARDs	disease modifying antirheumatic drugs
HCQ	hydroxychloroquine
IA	inflammatory arthritis
ICI	immune checkpoint inhibitor
IMNM	immune-mediated necrotizing myopathy
IrAEs	immune-related adverse events
mAb	monoclonal antibody
NSAIDs	nonsteroidal anti-inflammatory drugs
PD-1	programmed cell death protein 1
PD-L1	PD-1 ligand
PM	precision medicine
RF	rheumatoid factor
SLE	systemic lupus erythematosus

Precision Medicine and Artificial Intelligence. https://doi.org/10.1016/B978-0-12-820239-5.00011-5

TIF1 transcription intermediary factor 1
TNF tumor necrosis factor

1 Introduction

Immune checkpoint inhibitors (ICI) have dramatically improved the treatment of certain types of cancer and represent an example of applied precision medicine (PM). However, some patients treated with ICI can develop immune-related adverse events (irAE) that can progress to overt life-threatening diseases [1–8]. In addition, treatment decisions for patients with pre-existing autoimmune diseases who develop cancer and are candidates for ICI are particularly challenging [9]. A multi-disciplinary care team composed of oncologists and specialists with expertise in managing autoimmune and inflammatory conditions is recommended to improve the management of irAE and the outcomes of patients with pre-existing autoimmune diseases [3]. A recent European League Against Rheumatism (EULAR) task force has provided recommendations for the diagnosis and management of rheumatic irAEs with four overarching principles and 10 recommendations [10]. This chapter summarizes the history of ICI and provides perspective on irAEs, with special attention to rheumatic irAE and patients with pre-existing autoimmune diseases.

2 History of immune checkpoint inhibitors

The history of ICI dates back to 1910 when two Austrian physicians, Freund and Kaminer, noticed that serum derived from healthy individuals could "dissolve" cancer cells whereas serum of cancer patients could not. Later on in 1924 the same group identified a substance in the intestines of cancer patients which, when added to normal serum, reduced its ability to dissolve cancer cells (reviewed in [11]). In 1966, a Swedish couple (Karl and Ingegerd Hellstrom) observed that serum taken from mice with chemically induced tumors suppressed the reaction of lymphocytes. They attributed this to some sort of blocking factor.

It would take until 1987 to elucidate the blocking mechanism as a result of a discovery made by a French group, led by Jean-François Brunet, who identified "cytotoxic T lymphocyte-associated antigen 4" (CTLA-4) [12–14]. The exact role CTLA-4 remained unknown until 1995 when two teams lead by Tak Wah Mak and Allison and Bluestone showed that CTLA-4 could inhibit the activity of T-cells. Allison was the first to realize the same mechanism could provide a mechanism of treating cancer and consequently developed a monoclonal antibody (mAb) to block CTLA-4. Encouragingly, the mAb inhibited the growth of tumors in mice.

Based on these results, Allison began looking for a commercial partner to develop his idea further. After significant pushback by many pharmaceutical and biotechnology companies, in 1999 Allison licensed his technology to Medarex, a small biotechnology company, which shortly after (in 2000) launched its first clinical trial with a human mAb that bound to CTLA-4. This paved the way to the Food and Drug Administration's (United States: FDA) approval of ipilimumab for the treatment of metastatic melanoma in 2011 (the first ICI to reach market) [15].

Three years later, the FDA approved nivolumab, a mAb blocking programmed cell death protein 1 (PD-1: identified by Honjo et al. in 1992) developed by Medarex. Subsequent studies contributed to the elucidation of the mechanisms related to the activity of CTLA-4. Soon after, among others, Freeman and colleagues demonstrated that cancer cells were capable of hijacking the PD-1 protein to evade attack by the immune system. Since the success of ipilimumab, several other ICI pathways became heavily studied for the treatment of cancer [16]. This has been aided by ongoing research into the regulation of immune responses. Today, thousands of clinical trials on ICI are either underway or have already concluded. Such trials are exploring not only new ICI pathways but also different combinations of ICI together with radiation, chemotherapy and targeted therapies [16]. Milestones of the development of ICI are shown in Fig. 1.

3 Mechanisms of immune checkpoint inhibitors

An important component of the immune system is its ability to differentiate between "self" and "foreign" cells, allowing the immune system to attack the foreign cells while "tolerating" normal cells. To do this, it uses "checkpoints"—molecules on immune cells that need to be activated or inactivated to start or stop an immune response (summarized in Fig. 2). Cancer cells sometimes use these checkpoints to escape attack by the immune system. Drugs that target these checkpoints hold significant promise as cancer treatments [17–23]. The list of currently commercially available ICI is shown in Table 1 and Fig. 3.

3.1 PD-1 or PD-L1 inhibitors

Programmed cell death protein 1 (PD-1) is a checkpoint protein on T-cells which acts as a switch to keep T-cells from attacking normal cells in the body. This mechanism gets activated when PD-1 binds to its ligand (PD-L1), a protein located on normal cells. When PD-1 binds to PD-L1, it basically tells the T-cell to tolerate these cells. Some cancer cells express high levels of PD-L1 on their surface, which helps them evade immune attack. mAb can be titrated to target and block either PD-1 or PD-L1 thereby enhancing natural anti-tumor immune responses. PD-1 or PD-L1 inhibitors have been shown to be helpful in treating several types of cancer, including cutaneous melanoma, non-small cell lung cancer, renal cell carcinoma, bladder cancer, head and neck cancers as well as Hodgkin's lymphoma. In some countries, such as Japan, indications are expanded to include gastric cancer and mesothelioma. PD-1 or PD-L1 inhibitors are also being studied for use against many other types of cancer [17]. Examples of drugs that target PD-1 include: nivolumab (Opdivo), pembrolizumab (Keytruda), cemiplimab (Libtayo). PD-L1 inhibitors include: atezolizumab (Tecentriq), avelumab (Bavencio), and durvalumab (Imfinzi).

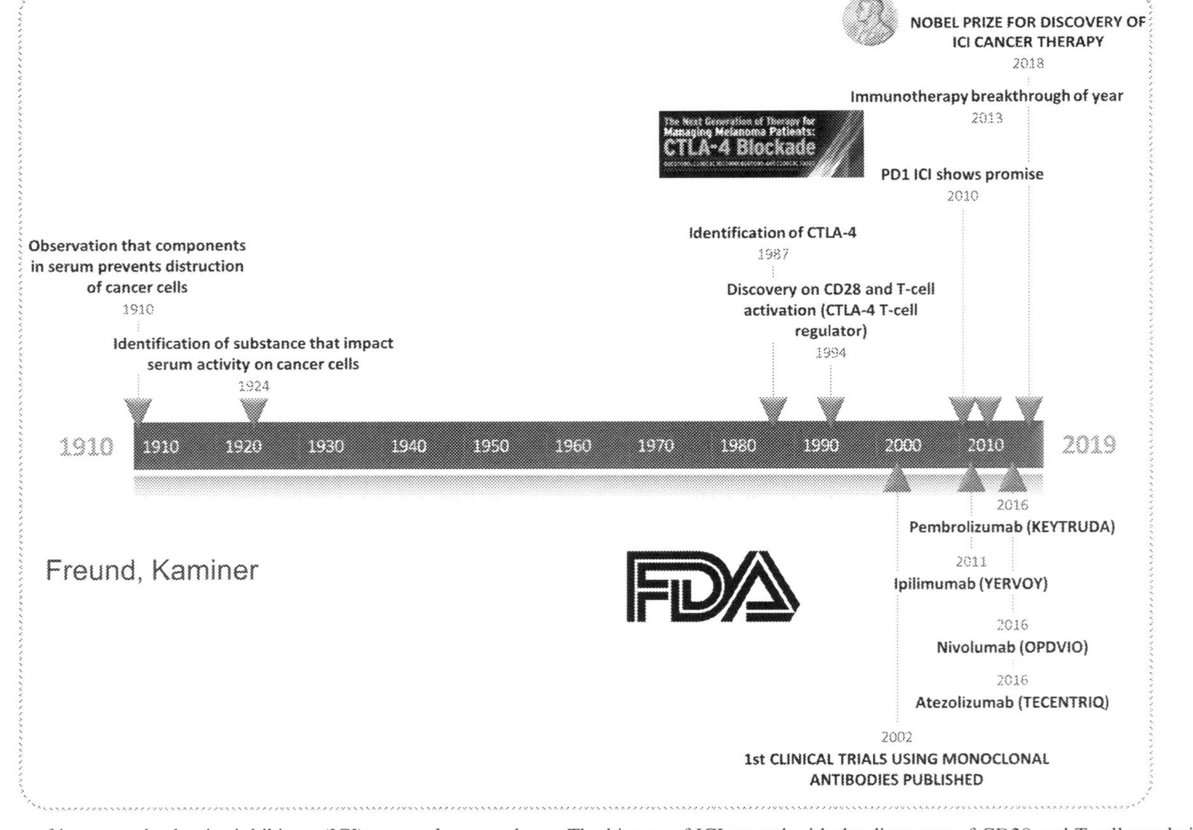

FIG. 1 History of immune checkpoint inhibitors (ICI) as novel cancer drugs. The history of ICI started with the discovery of CD28 and T-cell regulation in 1994 followed by the first clinical trials between 2000 and 2003. Finally, PD1 ICI showed promise in clinical trials around 2009. Later on, several antibodies including ipilimumab (2010), pembrolizumab, atezolizumab and nivolumab were cleared for treatment of different forms of cancer. The full recognition of the impact of ICI on cancer was formally recognized in 2018 when James Allison and Tasuku Honjo received the Nobel Prize in Physiology or Medicine for the discovery of cancer therapy by inhibition of negative immune regulation.

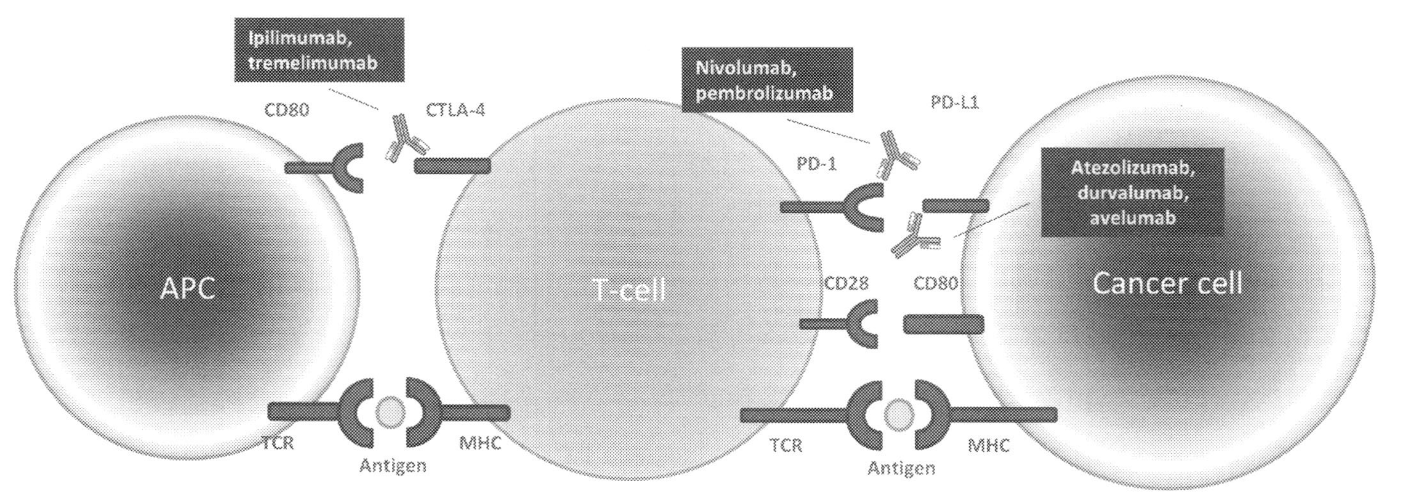

FIG. 2 Simplified overview and mechanism of immune checkpoint inhibitors (ICI). The two main mechanisms PD-1 and CTLA-4 are illustrated. In addition, some of the most relevant ICI are listed next to their respective target. Several new targets for ICI are being studied in various stages including LAG-3, BTLA4, CD80/86, CD200, CD276, KIR, LAG-3, TIM-1, 3, TIGIT, VISTA (not shown). *APC*, antigen presenting cell; *CD*, cell differentiation; *TCR*, T-cell receptor; *CTLA-4*, cytotoxic T-lymphocyte-associated protein 4; *PD-1*, programmed cell death protein 1; *PD-L1*; PD-1 ligand; *MHC*, major histocompatibility complex.

TABLE 1 Overview of commercially available immune checkpoint inhibitors (ICI) and those in phase III trials.

Drug name	Generic name	Target	Company	FDA approved indications
Yervoy	Ipilimumab	CTLA-4	Bristol-Myers Squibb	Metastatic melanoma
Opdivo	Nivolumab	PD-1	Bristol-Myers Squibb	Metastatic melanoma
Keytruda	Pembrolizumab	PD-1	Merck & Co.	Metastatic melanoma, PD-L1 positive non-small cell lung cancer, head and neck cancer
Tecentriq	Atezolizumab	PD-L1	Roche	Urothelial cell carcinoma, non-small cell lung cancer
NA	Nivolumab + Ipilimumab	PD-1, CTLA-4	NA	Metastatic melanoma
Bavencio	Avelumab	PD-L1	Merch KGaA/Pfizer	Metastatic Merkel cell carcinoma
Libtayo	Cemiplimab	PD-1	Sanofi	Metastatic cutaneous squamous cell carcinoma (CSCC) or locally advanced CSCC
Imfinzi	Durvalumab	PD-L1	AstraZeneca	Unresectable stage III non-small cell lung cancer
INCB-24360	Epacadostat	IDO	Incyte	Phase III
MEDI-1123	Tremelimumab	CTLA-4	Pfizer/AstraZeneca	Phase III

Abbreviations: *CTLA-4*, cytotoxic T-lymphocyte-associated protein 4; NA, not applicable; *PD-1*, programmed cell death protein 1; *IDO*, indoleamine 2,3-dioxygenase.

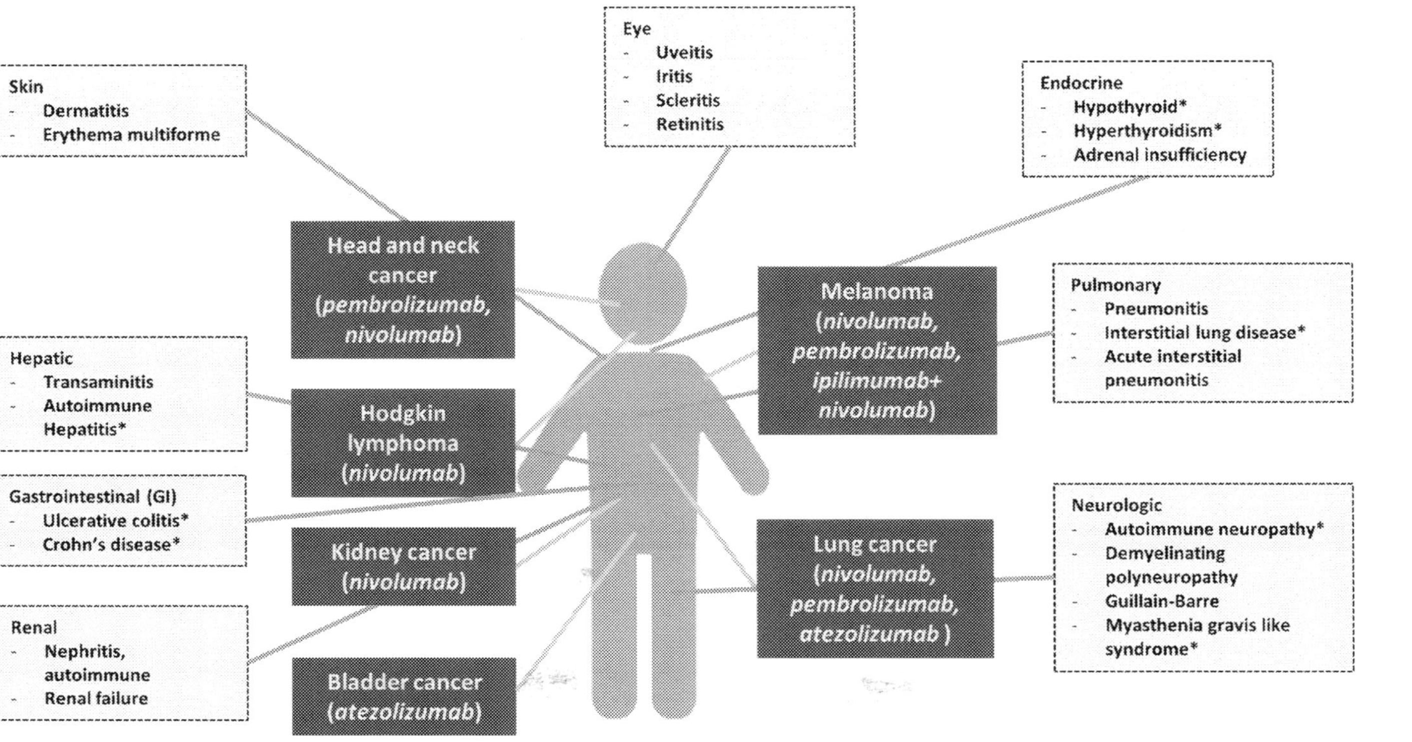

FIG. 3 Overview of clinical indications, available drugs and potential adverse events of immune checkpoint inhibitors (ICI). Indication and drug names are indicated in the *dark gray boxes*. Most common immune-related adverse effects (irAE) are presented in the *red boxes* (*gray* in print version) in labeled with *.

One concern with ICI is that they also can facilitate an attack of some normal cells and organs in the body, which can lead to serious side effects in some patients. Common side effects can include fatigue, cough, nausea, loss of appetite, skin rash, and itching. Less often they can cause more serious problems in the lungs, intestines, liver, kidneys, endocrine glands, or other organs. Some of these clinical features might or might not be related to irAE which will be discussed in more detail later.

3.2 CTLA-4 inhibitors

CTLA-4 is another protein expressed on some T-cells that acts as an "off switch" to keep the immune system in check. Ipilimumab (Yervoy), a mAb that attaches to and inactivates CTLA-4, can enhance the body's immune response against cancer cells. This class of ICI are most commonly used to treat cutaneous melanoma and some other types of cancers. CTLA-4 inhibitor side effects are more common and severe, and the spectrum of side effects, although similar, is not identical to that of PD-1/PD-L1 inhibitors [24, 25]. This highlights the distinct roles of various checkpoints in immune regulation [26].

4 The interface between cancer and autoimmunity

Cancer and autoimmunity are connected in different and complex ways [27, 28]. Many autoimmune diseases or subtypes of those conditions are associated with a higher risk of cancer. For example, subsets of patients with autoimmune myopathies (AIM) have a higher risk of cancer [29, 30]. Interestingly, this association is strongly linked to the presence of certain autoantibodies, such as anti-TIF1-y [27, 31, 32]. In addition, patients with or without signs of autoimmune diseases, but with autoantibodies to centromere protein F have a higher risk of breast and other cancers [33]. Interestingly, there are also autoantibodies that seemingly exhibit a negative association with the presence or development of cancer. Along those lines, anti-centromere (CENP-A, B, and C) positive patients with systemic sclerosis (SSc) have been reported to bear a lower risk for cancer [33]. The occurrence of irAE in patients treated with ICI might provide more insights on the ethology of autoimmune conditions [34, 35].

Lastly, certain autoantibodies, especially when tested in arrays, might have value in the detection and diagnosis of cancer. For example, anti-p53 antibodies were approved in Japan in 2007 as a laboratory screening test for cancer of the esophagus, colon, and breast. However, such systems have not been systematically introduced into clinical practice due to the lack of regulatory clearance (in other geographies). Similarly, while the presence of circulating anti-tumor autoantibodies (ATA) was initially proposed as a potential tool for diagnostic and monitoring purposes in different types of carcinoma, they were never systematically adopted [36]. However, ATA might have the potential to aid in the early diagnosis of different forms of cancer which should be explored in further studies [37–52].

5 Immune-related adverse events

There is tremendous interest in irAE in patients treated with ICI and the body of literature is growing quickly [53–62] (Table 2). Such irAE might shed more light on the pathophysiology of autoimmunity. In addition to the more common irAE such as hypothyroidism [73] and colitis [76], irAE include rheumatic-irAE such as inflammatory arthritis [63–65, 75], Sjögren's syndrome [72], systemic

TABLE 2 Diseases reported in patients treated with immune checkpoint inhibitors.

Disease group	Diseases	References
Systemic autoimmune rheumatic diseases	Systemic lupus erythematosus Systemic sclerosis Polymyositis/dermatomyositis Immune-mediated necrotizing myopathy, rhabdomyolysis Rheumatoid arthritis Sjögren's syndrome/Sicca syndrome	[8, 63–71]
Other rheumatic diseases	Polymyalgia rheumatica Psoriatic arthritis Crystal-induced oligoarthritis Remitting seronegative symmetrical synovitis with pitting edema (RS3PE) syndrome Eosinophilic granulomatosis with polyangiitis Autoimmune fasciitis Anti-phospholipid antibody syndrome	[8, 72]
Hematologic	Autoimmune thrombocytopenia Autoimmune hemolytic anemia Autoimmune neutropenia Bone marrow failure	[73]
Cardiac	Myocarditis Sick sinus syndrome	[74]
Pulmonary disease	Interstitial lung disease	[75]
Liver disease	Autoimmune hepatitis	
Gastrointestinal disease	Ulcerative colitis Crohn's disease Esophagitis Gastritis	[76]
Thyroid disease	Hyperthyroidism Hypothyroidism Chronic thyroiditis (Hashimoto's disease)	[53, 73]

Continued

TABLE 2 Diseases reported in patients treated with immune checkpoint inhibitors—cont'd

Disease group	Diseases	References
Endocrine disease	Type I diabetes Pituitary dysfunction Hypophysitis Isolated ACTH deficiency Polyendocrinopathy Hypoparathyroidism Diabetes insipidus	[77–81]
Skin disease	Vitiligo Lichenoid dermatitis Psoriasis Bullous pemphigoid	[82, 83]
Neuromuscular disease	Myasthenia gravis Guillain-Barré syndrome Demyelinating polyradiculoneuropathy Autoimmune encephalitis Limbic encephalitis Neuromyelitis optica Multifocal choroiditis Acute cerebellar ataxia Necrotizing encephalopathy Cerebral vasculitis	[84–87]
Renal	Glomerulonephritis Interstitial nephritis	[88]
Ophthalmologic	Vogt-Koyanagi-Harada disease-like posterior uveitis	[75]
Otolaryngologic	Autoimmune inner ear disease Sudden hearing loss	[153]

sclerosis [66], systemic lupus erythematosus (SLE) [8, 67, 68, 89, 90], inflammatory myopathy [69, 70, 74, 91–94], vasculitis [88, 95–97], and others.

Interestingly, recent case studies have reported the development of immune-mediated necrotizing myopathy (IMNM) [69, 70, 74] which has generally been associated with anti-SRP antibodies or in patients on statin therapy with anti-HMGCR antibodies [69, 70, 74]. The reported patients were negative for anti-HMGCR or anti-SRP antibodies. A characteristic clinic-pathological pattern, including a myasthenia gravis-like syndrome plus myositis was found in patients receiving PD-1 and PD-L1 inhibitors. A large component of macrophages resembling granulomas seems to be the pathological hallmark of this novel syndrome [98]. Thus, rheumatic-irAE seem to have similarities and differences with primary autoimmune diseases. In addition, the spectrum of ICI-related myositis extends beyond IMNM, with cases of dermatomyositis [99, 100], granulomatous myositis [101] and inflammatory myopathy with tertiary lymphoid organs [102], also having been reported. This highlights the fact that a common trigger has the potential to recapitulate the spectrum of autoimmune diseases.

Similarly, unlike primary autoimmune diseases which are usually chronic, most irAE resolve after treatment. Nevertheless, some irAE, and in particular rheumatic-irAE, have been reported to persist despite the cessation of ICI (selected cases presented in Table 3). The reasons for this remain largely unknown.

6 Patients with pre-existing autoimmunity

Treatment of cancer patients with pre-existing autoimmune disease, or at high risk thereof (e.g., strong family history of autoimmunity) with ICI requires careful consideration since these patients were excluded from clinical trials (Fig. 4) and potentially at higher risk of irAE. However, emerging data suggests that the benefits of treatment outweigh the risks in this patient population [89, 116–119]. Thus, in general, experts agree that pre-existing autoimmunity is not a contra-indication for ICI [9, 117, 120–122].

7 Opportunities for precision medicine

Biomarkers that could predict IrAE in cancer patients scheduled for ICI treatment would be highly valuable [123–127]. They would facilitate early detection and management of irAE, and even help identify preventive interventions [128–130]. Indeed, the increasing use of ICI and irAE might provide impetus for a paradigm shift in the use and application of autoantibodies and biomarkers. Precision health is based on the model of "intent to PREVENT" [131], and in the near future, diagnostics should include an early symptom/risk-based approach, as opposed to a disease-based approach. Novel panels of autoantibodies eventually combined with other multi-analyte "omic" profiles should be designed to guide the choice of ICI, aiming to provide for effective and personalized interventions for the malignancy, and, at the same time, reduce the risk of irAEs.

However, the efficiency of currently available biomarkers might be limited by what is referred to as the "serological gap." For example, although the antinuclear antibodies by indirect immunofluorescence test is usually regarded as the screening test of choice for some rheumatic diseases, to date there is insufficient evidence to prescribe its use when planning ICI therapy. Fig. 5 shows the proposal of a second level precision medicine approach for patients treated with ICI.

Newer diagnostic platforms that utilize emerging megatrends such as deep learning and augmented/artificial intelligence (AI) have the potential to close the gaps in identifying patients at higher risk to develop irAEs before ICI treatment is initiated. The challenge for the near future is to improve sensitivity, specificity and predictive value of autoantibody testing by integrating AI

TABLE 3 Reported cases of persistent irAEs after immune checkpoint inhibitor (ICI) treatment.

Author	Study design	Agent	Malignancy	Persistent irAE/ (n. of patient)	Antibodies profile
Hematological malignancy					
Saini and Chua [103]	CR	Nivolumab (+ azacitidine)	AML	AIM (1)	
Lung malignancy					
Kudo et al. [99]	CR	Nivolumab	Lung adenocarcinoma	AIM (1)	
John et al. [104]	CR	Tremelimumab + Durvalumab	Lung adenocarcinoma	AIM (1)	Anti-striated muscle (ASM) IgG was detected at a low titer of 1:40; it was not detected in archived pre-treatment serum
Thapa et al. [105]	CR	Nivolumab	NSCLC	GBS (1)	
Fukumoto et al. [106]	CR	Nivolumab	NSCLC	GBS (1)	
Chen et al. [107]	CR	Nivolumab	Squamous cell carcinoma	MG (1)	
Skin malignancy					
Zitouni et al. [90]	CR (series)	Nivolumab	Melanoma	SCLE (2)	Anti-Ro/SS-A, anti-La/SS-B, ANA (1/640)
Lidar et al. [108]	OS	Pembrolizumab	Melanoma	IA (4)	ACPA in one patient
Lidar et al. [108]	OS	Nivolumab	Melanoma	IA (2)	
Shah et al. [109]	CR	Pembrolizumab	Melanoma	Vitiligo (1)	

Ogawa et al. [110]	CR	Nivolumab	Melanoma	PM (1)	
Gupta et al. [111]	CR	Nivolumab + Ipilimumab	Melanoma	APS (1)	aPL (anti-B2GP1 IgM), confirmed 12 weeks a part supporting the diagnosis of APS
Renal malignancy					
Kuswanto et al. [112]	CRs	Pembrolizumab	Renal cell carcinoma	IA and PMR-like symptoms (1)	
Kuswanto et al. [112]	CRs	Nivolumab	Renal cell carcinoma	PMR (1)	
Kuswanto et al. [112]	CRs	Nivolumab	Renal cell carcinoma	Polyarticular arthralgias and synovitis (1)	
Yildirim et al. [113]	CR	Nivolumab	Renal cell carcinoma	GBS (1)	
Miscellaneous					
Kuswanto et al. [112]	CRs	Atezolizumab (+ bevacizumab)	Collecting duct carcinoma	IA (1)	
Lidar et al. [108]	OS	Pembrolizumab	Endometrial carcinoma	Eosinophilic fasciitis (1)	
Kang et al. [114]	CR	Nivolumab	Head and neck squamous cell carcinoma	MG (1)	Anti-AChR, AChR blocking antibodies, anti-striated muscle antibodies
Yuen et al. [115]	CR	Avelumab	Ovarian adenocarcinoma	MG (1)	

IA, inflammatory arthritis; *CR*, case report; *CRs*, case report series; *OS*, observational study; *ANA*, anti-nuclear antibodies; *ACPA*, anticitrullinated protein antibodies; *SCLE*, subacute cutaneous lupus erythematosus; *AIM*, autoimmune myositis; *PMR*, polymyalgia rheumatica; *GBS*, Guillain-Barré syndrome; *MG*, myasthenia gravis; *aPL*, anti-phospholipid antibodies; *APS*, anti-phospholipid syndrome.

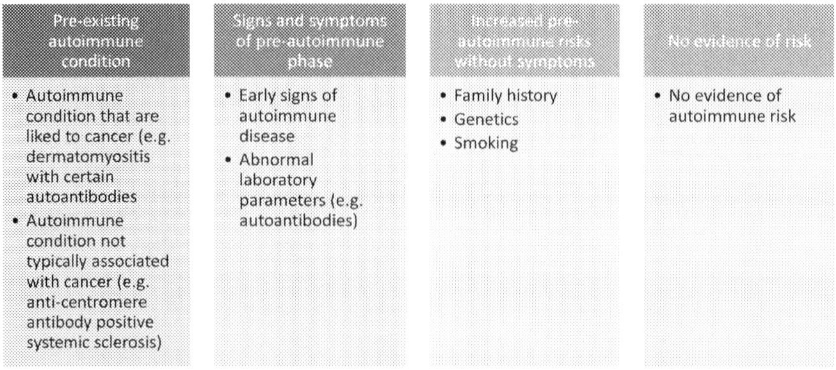

Pre-existing autoimmune condition	Signs and symptoms of pre-autoimmune phase	Increased pre-autoimmune risks without symptoms	No evidence of risk
• Autoimmune condition that are liked to cancer (e.g. dermatomyositis with certain autoantibodies • Autoimmune condition not typically associated with cancer (e.g. anti-centromere antibody positive systemic sclerosis)	• Early signs of autoimmune disease • Abnormal laboratory parameters (e.g. autoantibodies)	• Family history • Genetics • Smoking	• No evidence of autoimmune risk

FIG. 4 Immune checkpoint inhibitors (ICI) and autoimmune features. Four main clinical scenarios for ICI can be defined and should be considered when managing patients treated with ICI.

approaches to multi-analyte solid phase assays (so called multi-analyte arrays with algorithmic analyses) [131, 132].

In terms of prevention, novel strategies in high-risk patients could be explored. For example, hydroxychloroquine (HCQ) is a widely used drug in patients with systemic autoimmune diseases with a relatively good safety profile. Noteworthy, HCQ has shown potential to delay or prevent the development of SLE [67, 133] and was also used in a prevention trial for RA [134]. Consequently, one potential strategy to prevent or reduce irAE in patients treated with ICI could be the combination therapy of ICI with HCQ in a risk-based approach. Similarly, sulfasalazine has been reported to have anti-neoplastic effects and could also represent potential preventive option [135, 136]. The identification of such predictive biomarkers [123] will require biobanking of specimens from ICI trials and observational cohorts obtained before and after treatment [53].

8 Management of immune-related adverse events

Once a patient develops an irAE, several questions arise for the multidisciplinary team [3], including: proper diagnosis, continuation, discontinuation and/or switch to a different ICI, and treatment of the irAE.

(1) Discriminating between a concomitant new onset/flare of a rheumatic disease in a patient treated with ICI and rheumatic-irAEs.

While it sounds logical to assume that new inflammatory symptoms or flares of pre-existing autoimmune disease after recent exposure to ICI should be related to the treatment and considered as irAEs, discriminating between a concomitant new onset/flare of a rheumatic disease in some patients treated with ICI and irAEs with rheumatologic symptoms can be challenging [137]. Table 4 highlights the main similarities and differences when comparing primary RA, polymyalgia rheumatica (with or without giant cell arteritis) and AIM with

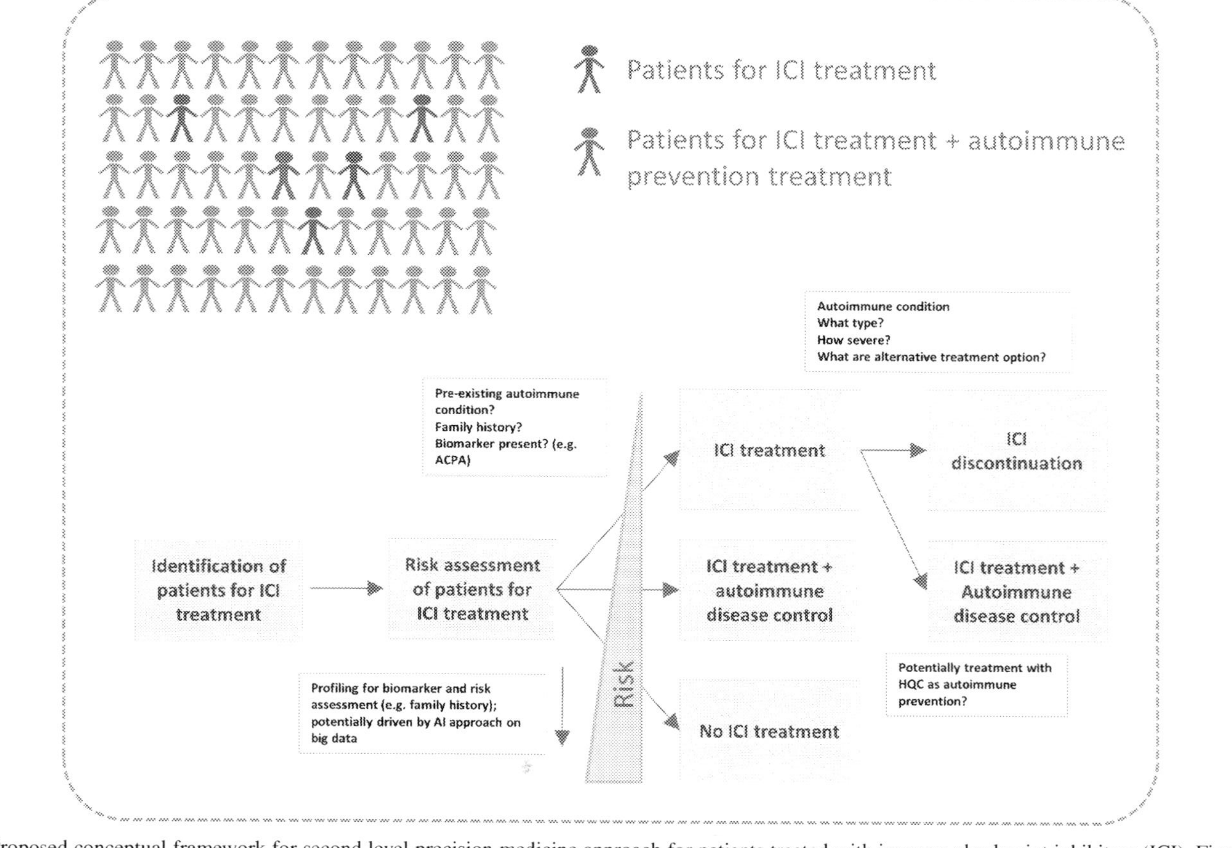

FIG. 5 Proposed conceptual framework for second level precision medicine approach for patients treated with immune checkpoint inhibitors (ICI). First, patients with malignancies that would benefit from ICI treatment have to be identified. Subsequently, a risk profile should be established for each patient based on family history, signs for pre-clinical autoimmunity or biomarker profiles. This might be combined with artificial intelligence (AI) solutions or risk scores. In a potential second step, patients that might develop autoimmune related adverse events might require different treatment approach (potentially combined treatment). Abbreviations: *ACPA*, anti-citrullinated peptide/protein antibodies; *HQC*, hydroxichloroquin.

TABLE 4 Comparison between characteristic of traditional form of rheumatic diseases and irAE with rheumatic symptoms.

Rheumatic diseases comparator	Similarities	Differences
Polymyalgia rheumatica and/or giant cell arteritis	- Age of onset (> 50 years) - Histological features at biopsy	Heterogeneity in the elevation of inflammatory markers and response to glucocorticoids
Rheumatoid arthritis	Similar joint distribution	- Higher rate of erosion and tendon involvement in ICI treated patients - RF and CCP often negative - Similar incidence in male and female (albeit the different prevalence of cancer in men and female can represent a bias)
Ankylosing spondylitis and psoriatic arthritis	Rate of enthesitis, inflammatory back pain, and dactylitis	- Higher rate of erosion and lower rate of psoriasis in ICI treated patients - Likely not associated with HLA-B27
Dermatomyositis, polymyositis and immune-mediated necrotizing myopathy	Histological features at muscle biopsy	Skin involvement rarely described

Abbreviations: *ICI*, immune checkpoint inhibitor; *HLA-B27*, Human Leukocyte Antigen B27; *RF*, rheumatoid factor; *CCP*, cyclic citrullinated peptide.

rheumatic-irAE. Immunological laboratory profiling [including rheumatoid factor, RF; anti-CCP, antinuclear antibodies (ANA); anti-extractable nuclear antigens (ENA); and anti-dsDNA, as appropriate according to the clinical presentation] might help to discriminate between the two conditions, with serologies being generally negative in rheumatic-irAE.

(2) In case of irAEs with rheumatologic signs and symptoms, should the patient continue the ICI therapy, and what is the future oncology treatment plan?

When a patient has an early irAE or partial tumor response to immunotherapy, the oncologist might wish to continue immunotherapy as long as possible. On the other hand, if ICIs are continued, this may lead to a chronic autoimmune condition. Patient and oncologist's preferences regarding treatment continuation, discontinuation or switch are important. In addition, several professional associations have published useful recommendations on the management of irAE [138, 139]. Understanding the expectations and prognosis for cancer therapy will facilitate treatment decision-making. As additional irAEs might occur long after initiation or even after cessation of ICI therapy, it is crucial to have serial monitoring by the rheumatologist and oncologist regarding this problem, or regarding loss of ICI efficacy.

(3) Management of irAEs.

The optimal management of irAE is currently largely unknown and based on expert opinion [138]. One of the main challenges is that many treatments used to manage irAE are meant to suppress the immune system, which might in theory inhibit the anti-tumor effect of the ICI [140–142]. Of note, many clinical trials investigating ICI did not allow the use of glucocorticoids at > 10 mg daily dose of prednisolone (or equivalent) or DMARDs alongside ICI treatment.

From a practical point of view, treatment decisions are made on a case-by-case basis and tailored according to the severity of the rheumatic-irAE. In general, steroid exposure is minimized as much as possible. After ruling out infections, nonsteroidal anti-inflammatory drugs (NSAIDs) and intra-articular steroids can be useful to manage mild arthritis or arthritis when only one joint is involved. For patients with severe arthritis, DMARDs can be considered [46]. Anecdotal experience with TNF inhibitors to control immune-related colitis or inflammatory arthritis in melanoma patients exposed to ICI reported no difference in tumor response. In fact, there might be a potentiation of ICI with TNF inhibitors [143, 144] and there is at least one randomized controlled trial exploring the combined use (NCT03293784). Patients with myositis and immune-mediated thrombocytopenia might benefit from intravenous immunoglobulin administration, albeit its systemic use and efficacy is still based on anecdotal experience. Of note, ir-myositis is associated with significant risk of mortality and needs to be treated aggressively [69, 145]. Sicca symptoms can often be treated with topical, oral or ocular therapies or sialagogues, all of which would not be expected to have any effect on tumor response.

9 Autoantibodies and immune-related adverse events

Evidence for the hypothesis that irAE may result from unmasking a predisposition for autoimmune disease comes from studies in which patients who developed RA had pre-ICI treatment samples containing anti-CCP antibodies (1), which are generally present many years prior to onset of clinical disease and are highly specific for RA [146–148]. In addition, pre-existing autoantibodies may predict irAE in as yet unexplained ways. In a series of eight patients with thymoma, four had pre-existing autoantibodies to acetylcholine receptor and all four developed inflammatory myositis [149]. Acetylcholine receptor antibodies are usually associated with myasthenia gravis, which only rarely overlaps with AIM [74]. However, the overlap in the setting of irAE is much more common [69], and may point to as yet unknown mechanisms. Nevertheless, most cases of ICI-associated RA, AIM (5) and other conditions usually associated with autoantibodies such as type 1 diabetes [77] and thyroid dysfunction [150] have been seronegative, suggesting that unmasking immune predisposition is one of possibly many mechanisms of irAE.

ICI may contribute to irAE through a break in tolerance and production of *de novo* autoantibodies. In a study of ipilimumab-treated melanoma patients, 19/99 (19.2%) who were autoantibody negative before treatment developed autoantibodies to 23 common specificities post-treatment [53]. Those with anti-thyroid antibodies after ipilimumab treatment had significantly more thyroid dysfunction than those who did not [7/11 (54.6%) vs 7/49 (14.3%), odds ratio 10.0 (95% CI, 2.0–51.1)]. However, overall, although those who developed autoantibodies had more irAE [14/19 (73.7%) vs 37/80 (46.3%), odds ratio 3.6 (95% CI 1.1–11.8)], the new antibodies did not associate with an irAE in the organ system that would have been predicted for that autoantibody. Similarly, two patients with metastatic melanoma treated with high-dose intra-lesional Bacillus Calmette-Guérin followed by ipilimumab developed autoantibodies against a broad panel of autoantigens prior to developing high-grade irAE [126]. The breadth of the autoimmune reactivity was suggestive of a systemic B-cell dysregulation, rather than a focused immune response.

There has been considerable interest in identifying biomarkers of irAE. Toi et al. recently showed that patients with subclinical rheumatoid factor, antinuclear antibodies and anti-thyroid antibodies were more likely to develop irAE (OR 3.3, 95% CI 1.6–6.7), but were also more likely to have better tumor outcomes (OR for disease progression or death 0.5, 95% CI 0.4–0.8) [151]. Tahir et al. performed large-scale screening of autoantibodies against cDNA expression libraries of target organs and identified autoantibodies that correlated with the development of hypophysitis (anti-GNAL and anti-ITM2B autoantibodies) and pneumonitis (anti-CD74 autoantibody) [152]. These results, although of considerable interest, were obtained in relatively small samples and require validation.

10 Conclusions

A variety of irAEs with rheumatic and musculoskeletal features have been described during treatment with ICI. Arthralgia and myalgia and even new onset full-blown autoimmune conditions have been reported.

The increasing use of ICI is likely to affect the prevalence of autoimmune clinical conditions. Similarly, some patients may develop multiple autoimmune complications. The precise pathogenesis remains unknown and further research is therefore needed. Future studies should focus on knowledge gaps in epidemiology, evaluation, and treatment of irAE. A major challenge is the identification of patients at higher risk of developing irAEs. Designing novel integrated laboratory approaches will help to guide a tailored choice of ICI when balancing safety and efficacy aspects.

Acknowledgments

This work was supported by JSPS KAKENHI (Grants-in-Aid for Scientific Research) grant number 19K08617 (for MS).

Competing interests

M. Mahler, Y. is employed at Inova Diagnostics selling autoantibody assays. M.J. Fritzler has been paid honoraria by Inova Diagnostics.

References

[1] C. Calabrese, et al., Rheumatic immune-related adverse events of checkpoint therapy for cancer: case series of a new nosological entity, RMD Open 3 (1) (2017) e000412.
[2] L.H. Calabrese, C. Calabrese, L.C. Cappelli, Rheumatic immune-related adverse events from cancer immunotherapy, Nat. Rev. Rheumatol. 14 (10) (2018) 569–579.
[3] L. Calabrese, X. Mariette, The evolving role of the rheumatologist in the management of immune-related adverse events (irAEs) caused by cancer immunotherapy, Ann. Rheum. Dis. 77 (2) (2018) 162–164.
[4] M.C. Ornstein, et al., Myalgia and arthralgia immune-related adverse events (irAEs) in patients with genitourinary malignancies treated with immune checkpoint inhibitors, Clin. Genitourin. Cancer 17 (3) (2019) 177–182.
[5] M. Kostine, et al., Addressing immune-related adverse events of cancer immunotherapy: how prepared are rheumatologists? Ann. Rheum. Dis. 78 (6) (2019) 860–862.
[6] G.D. Sebastiani, C. Scirocco, M. Galeazzi, Rheumatic immune related adverse events in patients treated with checkpoint inhibitors for immunotherapy of cancer, Autoimmun. Rev. 18 (8) (2019) 805–813.
[7] I.G. Motofei, Malignant melanoma: autoimmunity and supracellular messaging as new therapeutic approaches, Curr. Treat. Options in Oncol. 20 (6) (2019) 45.
[8] F. Gediz, S. Kobak, Immune checkpoint inhibitors-related rheumatic diseases: what rheumatologist should know? Curr. Rheumatol. Rev. 15 (3) (2019) 201–208.
[9] M. Pantuck, D. McDermott, A. Drakaki, To treat or not to treat: patient exclusion in immune oncology clinical trials due to preexisting autoimmune disease, Cancer 125 (20) (2019) 3506–3513.
[10] M. Kostine, et al., EULAR recommendations for the diagnosis and the management of rheumatic immune-related adverse events due to cancer immunotherapy, Ann. Rheum. Dis. 78 (Suppl. 2) (2019) 158.1–158.
[11] S. Singh, et al., Immune checkpoint inhibitors: a promising anticancer therapy, Drug Discov. Today 25 (1) (2020) 223–229.
[12] J.F. Brunet, et al., The inducible cytotoxic T-lymphocyte-associated gene transcript CTLA-1 sequence and gene localization to mouse chromosome 14, Nature 322 (6076) (1986) 268–271.
[13] J.F. Brunet, et al., A new member of the immunoglobulin superfamily—CTLA-4, Nature 328 (6127) (1987) 267–270.
[14] J.F. Brunet, et al., CTLA-1 and CTLA-3 serine esterase transcripts are detected mostly in cytotoxic T cells, but not only and not always, J. Immunol. 138 (12) (1987) 4102–4105.
[15] D.M. Pardoll, The blockade of immune checkpoints in cancer immunotherapy, Nat. Rev. Cancer 12 (4) (2012) 252–264.

[16] H. Babey, et al., Immune-checkpoint inhibitors to treat cancers in specific immunocompromised populations: a critical review, Expert. Rev. Anticancer. Ther. 18 (10) (2018) 981–989.
[17] A. Hoos, Development of immuno-oncology drugs—from CTLA4 to PD1 to the next generations, Nat. Rev. Drug Discov. 15 (4) (2016) 235–247.
[18] M.A. Postow, M.K. Callahan, J.D. Wolchok, Immune checkpoint blockade in cancer therapy, J. Clin. Oncol. 33 (17) (2015) 1974–1982.
[19] The Lancet Oncology, Calling time on the immunotherapy gold rush, Lancet Oncol. 18 (8) (2017) 981.
[20] P. Sharma, J.P. Allison, The future of immune checkpoint therapy, Science 348 (6230) (2015) 56–61.
[21] P. Sharma, From the guest editor: immune checkpoint therapy as a weapon against cancer, Cancer J. 22 (2) (2016) 67.
[22] M.J. Flynn, et al., Challenges and opportunities in the clinical development of immune checkpoint inhibitors for hepatocellular carcinoma, Hepatology 69 (5) (2019) 2258–2270.
[23] P. Sharma, J.P. Allison, Dissecting the mechanisms of immune checkpoint therapy, Nat. Rev. Immunol. 20 (2) (2020) 75–76.
[24] M.A. Postow, R. Sidlow, M.D. Hellmann, Immune-related adverse events associated with immune checkpoint blockade, N. Engl. J. Med. 378 (2) (2018) 158–168.
[25] C. Boutros, et al., Safety profiles of anti-CTLA-4 and anti-PD-1 antibodies alone and in combination, Nat. Rev. Clin. Oncol. 13 (8) (2016) 473–486.
[26] S.C. Wei, et al., Distinct cellular mechanisms underlie anti-CTLA-4 and anti-PD-1 checkpoint blockade, Cell 170 (6) (2017) 1120–1133.e17.
[27] L.C. Cappelli, A.A. Shah, The relationships between cancer and autoimmune rheumatic diseases, Best Pract. Res. Clin. Rheumatol. 34 (2020) 101472.
[28] C.H. June, J.T. Warshauer, J.A. Bluestone, Is autoimmunity the Achilles' heel of cancer immunotherapy? Nat. Med. 23 (5) (2017) 540–547.
[29] M. Mahler, M.J. Fritzler, Detection of myositis-specific antibodies: additional notes, Ann. Rheum. Dis. 78 (5) (2019) e45.
[30] M. Mahler, et al., Standardisation of myositis-specific antibodies: where are we today? Ann. Rheum. Dis. (2019), https://doi.org/10.1136/annrheumdis-2019-216003.
[31] Z.E. Betteridge, et al., Investigation of myositis and scleroderma specific autoantibodies in patients with lung cancer, Arthritis Res. Ther. 20 (1) (2018) 176.
[32] M. Best, et al., Use of anti-transcriptional intermediary factor-1 gamma autoantibody in identifying adult dermatomyositis patients with cancer: a systematic review and meta-analysis, Acta Derm. Venereol. 99 (3) (2019) 256–262.
[33] M.J. Fritzler, et al., Historical perspectives on the discovery and elucidation of autoantibodies to centromere proteins (CENP) and the emerging importance of antibodies to CENP-F, Autoimmun. Rev. 10 (4) (2011) 194–200.
[34] M. van der Vlist, et al., Immune checkpoints and rheumatic diseases: what can cancer immunotherapy teach us? Nat. Rev. Rheumatol. 12 (10) (2016) 593–604.
[35] M.J. Olde Nordkamp, B.P. Koeleman, L. Meyaard, Do inhibitory immune receptors play a role in the etiology of autoimmune disease? Clin. Immunol. 150 (1) (2014) 31–42.
[36] Y.T. Lee, et al., Circulating anti-tumor and autoantibodies in breast carcinoma: relationship to stage and prognosis, Breast Cancer Res. Treat. 6 (1) (1985) 57–65.
[37] Y. Jin, et al., Use of autoantibodies against tumor-associated antigens as serum biomarkers for primary screening of cervical cancer, Oncotarget 8 (62) (2017) 105425–105439.
[38] D. Djureinovic, et al., Detection of autoantibodies against cancer-testis antigens in non-small cell lung cancer, Lung Cancer 125 (2018) 157–163.

[39] H. Zhao, et al., Circulating anti-p16a IgG autoantibodies as a potential prognostic biomarker for non-small cell lung cancer, FEBS Open Bio. 8 (11) (2018) 1875–1881.
[40] J. Qiu, et al., Autoantibodies as potential biomarkers in breast cancer, Biosensors (Basel) 8 (3) (2018) 67.
[41] M. Ushigome, et al., Multi-panel assay of serum autoantibodies in colorectal cancer, Int. J. Clin. Oncol. 23 (5) (2018) 917–923.
[42] N. Karin, Autoantibodies to chemokines and cytokines participate in the regulation of cancer and autoimmunity, Front. Immunol. 9 (2018) 623.
[43] D. Jiang, et al., A panel of autoantibodies against tumor-associated antigens in the early immunodiagnosis of lung cancer, Immunobiology 225 (1) (2019) 151848.
[44] L. Pei, et al., Discovering novel lung cancer associated antigens and the utilization of their autoantibodies in detection of lung cancer, Immunobiology 225 (1) (2019) 151891.
[45] H. Huang, et al., The diagnostic efficiency of seven autoantibodies in lung cancer, Eur. J. Cancer Prev. 29 (4) (2019) 315–320.
[46] H. Wang, et al., Autoantibodies as biomarkers for colorectal cancer: a systematic review, meta-analysis, and bioinformatics analysis, Int. J. Biol. Markers 34 (4) (2019) 334–347.
[47] L. Xu, et al., Improved detection of prostate cancer using a magneto-nanosensor assay for serum circulating autoantibodies, PLoS One 14 (8) (2019) e0221051.
[48] S. Yadav, et al., Autoantibodies as diagnostic and prognostic cancer biomarker: detection techniques and approaches, Biosens. Bioelectron. 139 (2019) 111315.
[49] M. Yin-Yu, et al., Performance evaluation of an enzyme-linked immunosorbent assay for seven autoantibodies in lung cancer, Clin. Lab. 65 (4) (2019).
[50] R. Zhang, et al., Diagnostic value of multiple tumor-associated autoantibodies in lung cancer, Onco. Targets. Ther. 12 (2019) 457–469.
[51] C. Qiu, et al., Establishment and validation of an immunodiagnostic model for prediction of breast cancer, Oncoimmunology 9 (1) (2020) 1682382.
[52] T. Wang, et al., Screening of tumor-associated antigens based on Oncomine database and evaluation of diagnostic value of autoantibodies in lung cancer, Clin. Immunol. 210 (2020) 108262.
[53] E.C. de Moel, et al., Autoantibody development under treatment with immune-checkpoint inhibitors, Cancer Immunol. Res. 7 (1) (2019) 6–11.
[54] F. Graus, J. Dalmau, Paraneoplastic neurological syndromes in the era of immune-checkpoint inhibitors, Nat. Rev. Clin. Oncol. 16 (9) (2019) 535–548.
[55] T.J. Williams, et al., Association of autoimmune encephalitis with combined immune checkpoint inhibitor treatment for metastatic cancer, JAMA Neurol. 73 (8) (2016) 928–933.
[56] S. Khan, D.E. Gerber, Autoimmunity, checkpoint inhibitor therapy and immune-related adverse events: a review, Semin. Cancer Biol. 64 (2020) 93–101.
[57] L.C. Cappelli, et al., Rheumatic and musculoskeletal immune-related adverse events due to immune checkpoint inhibitors: a systematic review of the literature, Arthritis Care Res. (Hoboken) 69 (11) (2017) 1751–1763.
[58] A.S. Tocheva, A. Mor, Checkpoint inhibitors: applications for autoimmunity, Curr. Allergy Asthma Rep. 17 (10) (2017) 72.
[59] L. Calabrese, V. Velcheti, Checkpoint immunotherapy: good for cancer therapy, bad for rheumatic diseases, Ann. Rheum. Dis. 76 (1) (2017) 1–3.
[60] A. Bertrand, et al., Immune related adverse events associated with anti-CTLA-4 antibodies: systematic review and meta-analysis, BMC Med. 13 (2015) 211.
[61] J.M. Michot, et al., Immune-related adverse events with immune checkpoint blockade: a comprehensive review, Eur. J. Cancer 54 (2016) 139–148.

[62] A. Tarhini, Immune-mediated adverse events associated with ipilimumab CTLA-4 blockade therapy: the underlying mechanisms and clinical management, Scientifica (Cairo) 2013 (2013) 857519.
[63] T.J. Braaten, et al., Immune checkpoint inhibitor-induced inflammatory arthritis persists after immunotherapy cessation, Ann. Rheum. Dis. 79 (3) (2020) 332–338.
[64] L.C. Cappelli, et al., Immune checkpoint inhibitor-induced inflammatory arthritis as a model of autoimmune arthritis, Immunol. Rev. 294 (1) (2020) 106–123.
[65] L. Calabrese, X. Mariette, Chronic inflammatory arthritis following checkpoint inhibitor therapy for cancer: game changing implications, Ann. Rheum. Dis. 79 (3) (2020) 309–311.
[66] N.S. Barbosa, et al., Scleroderma induced by pembrolizumab: a case series, Mayo Clin. Proc. 92 (7) (2017) 1158–1163.
[67] C. Kosche, J.L. Owen, J.N. Choi, Widespread subacute cutaneous lupus erythematosus in a patient receiving checkpoint inhibitor immunotherapy with ipilimumab and nivolumab, Dermatol. Online J. 25 (10) (2019). 13030/qt4md713j8.
[68] F. Fadel, K. El Karoui, B. Knebelmann, Anti-CTLA4 antibody-induced lupus nephritis, N. Engl. J. Med. 361 (2) (2009) 211–212.
[69] C. Anquetil, et al., Immune checkpoint inhibitor-associated myositis, Circulation 138 (7) (2018) 743–745.
[70] M. Tauber, et al., Severe necrotizing myositis associated with long term anti-neoplastic efficacy following nivolumab plus ipilimumab combination therapy, Clin. Rheumatol. 38 (2) (2019) 601–602.
[71] H. Kamo, et al., Pembrolizumab-related systemic myositis involving ocular and hindneck muscles resembling myasthenic gravis: a case report, BMC Neurol. 19 (1) (2019) 184.
[72] J. Ghosn, et al., A severe case of neuro-Sjogren's syndrome induced by pembrolizumab, J. Immunother. Cancer 6 (1) (2018) 110.
[73] T. Jotatsu, et al., Immune-mediated thrombocytopenia and hypothyroidism in a lung cancer patient treated with nivolumab, Immunotherapy 10 (2) (2018) 85–91.
[74] S. Suzuki, et al., Nivolumab-related myasthenia gravis with myositis and myocarditis in Japan, Neurology 89 (11) (2017) 1127–1134.
[75] M. Nguyen, et al., Pembrolizumab induced ocular hypotony with near complete vision loss, interstitial pulmonary fibrosis and arthritis, Front. Oncol. 9 (2019) 944.
[76] K. Yoshino, et al., Severe colitis after PD-1 blockade with nivolumab in advanced melanoma patients: potential role of Th1-dominant immune response in immune-related adverse events: two case reports, BMC Cancer 19 (1) (2019) 1019.
[77] V.H.M. Tsang, et al., Checkpoint inhibitor-associated autoimmune diabetes is distinct from type 1 diabetes, J. Clin. Endocrinol. Metab. 104 (11) (2019) 5499–5506.
[78] M. Zezza, et al., Combined immune checkpoint inhibitor therapy with nivolumab and ipilimumab causing acute-onset type 1 diabetes mellitus following a single administration: two case reports, BMC Endocr. Disord. 19 (1) (2019) 144.
[79] L. Spiers, N. Coupe, M. Payne, Toxicities associated with checkpoint inhibitors—an overview, Rheumatology (Oxford) 58 (Supplement_7) (2019) vii7–vii16.
[80] H.M.A. Abdullah, et al., Rapid onset type-1 diabetes and diabetic ketoacidosis secondary to nivolumab immunotherapy: a review of existing literature, BMJ Case Rep. 12 (8) (2019) e229568.
[81] K.A. Farina, M.P. Kane, Programmed cell death-1 monoclonal antibody therapy and type 1 diabetes mellitus: a review of the literature, J. Pharm. Pract. (2019), https://doi.org/10.1177/0897190019850929.

[82] C.A. Nelson, et al., Bullous pemphigoid after anti-PD-1 therapy: a retrospective case-control study evaluating impact on tumor response and survival outcomes, J. Am. Acad. Dermatol. S0190-9622 (20) (2020), https://doi.org/10.1016/j.jaad.2019.12.068. 30048-7.
[83] C.D. Sadik, et al., Checkpoint inhibition may trigger the rare variant of anti-LAD-1 IgG-positive, anti-BP180 NC16A IgG-negative bullous pemphigoid, Front. Immunol. 10 (2019) 1934.
[84] A. Zaremba, et al., Metastatic Merkel cell carcinoma and myasthenia gravis: contraindication for therapy with immune checkpoint inhibitors? J. Immunother. Cancer 7 (1) (2019) 141.
[85] R. Thummalapalli, et al., Checkpoint inhibitor-induced autoimmune encephalitis reversed by rituximab after allogeneic bone marrow transplant in a patient with Hodgkin lymphoma, Leuk. Lymphoma 61 (1) (2020) 228–230.
[86] A. Zekeridou, V.A. Lennon, Neurologic autoimmunity in the era of checkpoint inhibitor cancer immunotherapy, Mayo Clin. Proc. 94 (9) (2019) 1865–1878.
[87] S. Fujiwara, et al., Elevated adenosine deaminase levels in the cerebrospinal fluid in immune checkpoint inhibitor-induced autoimmune encephalitis, Intern. Med. 58 (19) (2019) 2871–2874.
[88] P. Boland, J. Heath, S. Sandigursky, Immune checkpoint inhibitors and vasculitis, Curr. Opin. Rheumatol. 32 (1) (2020) 53–56.
[89] M.D. Richter, et al., Brief report: cancer immunotherapy in patients with preexisting rheumatic disease: the Mayo Clinic experience, Arthritis Rheumatol. 70 (3) (2018) 356–360.
[90] N.B. Zitouni, et al., Subacute cutaneous lupus erythematosus induced by nivolumab: two case reports and a literature review, Melanoma Res. 29 (2) (2019) 212–215.
[91] G. Hunter, C. Voll, C.A. Robinson, Autoimmune inflammatory myopathy after treatment with ipilimumab, Can. J. Neurol. Sci. 36 (4) (2009) 518–520.
[92] M. Yoshioka, et al., Case of respiratory discomfort due to myositis after administration of nivolumab, J. Dermatol. 42 (10) (2015) 1008–1009.
[93] T. Kobayashi, et al., Relationship between clinical course of nivolumab-related myositis and immune status in a patient with Hodgkin's lymphoma after allogeneic hematopoietic stem cell transplantation, Int. J. Hematol. 109 (3) (2019) 356–360.
[94] T. Kimura, et al., Myasthenic crisis and polymyositis induced by one dose of nivolumab, Cancer Sci. 107 (7) (2016) 1055–1058.
[95] C.S. Nabel, et al., Anti-PD-1 immunotherapy-induced flare of a known underlying relapsing vasculitis mimicking recurrent cancer, Oncologist 24 (8) (2019) 1013–1021.
[96] A. Rutgers, et al., Systemic vasculitis developed after immune checkpoint inhibition: comment on the article by Cappelli et al, Arthritis Care Res. (Hoboken) 70 (8) (2018) 1275.
[97] A. Daxini, K. Cronin, A.G. Sreih, Vasculitis associated with immune checkpoint inhibitors—a systematic review, Clin. Rheumatol. 37 (9) (2018) 2579–2584.
[98] A. Matas-Garcia, et al., Emerging PD-1 and PD-1L inhibitors-associated myopathy with a characteristic histopathological pattern, Autoimmun. Rev. 19 (2) (2020) 102455.
[99] F. Kudo, et al., Advanced lung adenocarcinoma with nivolumab-associated dermatomyositis, Intern. Med. 57 (15) (2018) 2217–2221.
[100] T.P. Nguyen, et al., Dermatomyositis-associated sensory neuropathy: a unifying pathogenic hypothesis, J. Clin. Neuromuscul. Dis. 16 (1) (2014) 7–11.
[101] N. Uchio, et al., Inflammatory myopathy with myasthenia gravis: thymoma association and polymyositis pathology, Neurol. Neuroimmunol. Neuroinflamm. 6 (2) (2019) e535.
[102] S. Matsubara, et al., Tertiary lymphoid organs in the inflammatory myopathy associated with PD-1 inhibitors, J. Immunother. Cancer 7 (1) (2019) 256.

[103] L. Saini, N. Chua, Severe inflammatory myositis in a patient receiving concurrent nivolumab and azacitidine, Leuk. Lymphoma 58 (8) (2017) 2011–2013.

[104] S. John, et al., Progressive hypoventilation due to mixed CD8(+) and CD4(+) lymphocytic polymyositis following tremelimumab-durvalumab treatment, J. Immunother. Cancer 5 (1) (2017) 54.

[105] B. Thapa, et al., Nivolumab-associated Guillain-Barre syndrome in a patient with non-small-cell lung cancer, Am. J. Ther. 25 (6) (2018) e761–e763.

[106] Y. Fukumoto, et al., Acute demyelinating polyneuropathy induced by nivolumab, J. Neurol. Neurosurg. Psychiatry 89 (4) (2018) 435–437.

[107] W.X. Chen, et al., Significant response to anti-PD-1 based immunotherapy plus lenvatinib for recurrent intrahepatic cholangiocarcinoma with bone metastasis: a case report and literature review, Medicine (Baltimore) 98 (45) (2019) e17832.

[108] M. Lidar, et al., Rheumatic manifestations among cancer patients treated with immune checkpoint inhibitors, Autoimmun. Rev. 17 (3) (2018) 284–289.

[109] N.J. Shah, et al., Safety and efficacy of immune checkpoint inhibitors (ICIs) in cancer patients with HIV, hepatitis B, or hepatitis C viral infection, J. Immunother. Cancer 7 (1) (2019) 353.

[110] T. Ogawa, et al., Polymyositis induced by PD-1 blockade in a patient in hepatitis B remission, J. Neurol. Sci. 381 (2017) 22–24.

[111] A. Gupta, et al., Antiphospholipid syndrome associated with combined immune checkpoint inhibitor therapy, Melanoma Res. 27 (2) (2017) 171–173.

[112] W.F. Kuswanto, et al., Rheumatologic symptoms in oncologic patients on PD-1 inhibitors, Semin. Arthritis Rheum. 47 (6) (2018) 907–910.

[113] N. Yildirim, et al., Fatal acute motor axonal neuropathy induced by nivolumab: a case report and literature review, Clin. Genitourin. Cancer 17 (6) (2019) e1104–e1107.

[114] K.H. Kang, et al., Immune checkpoint-mediated myositis and myasthenia gravis: a case report and review of evaluation and management, Am. J. Otolaryngol. 39 (5) (2018) 642–645.

[115] C. Yuen, et al., Myasthenia gravis induced by avelumab, Immunotherapy 11 (14) (2019) 1181–1185.

[116] D.B. Johnson, R.J. Sullivan, A.M. Menzies, Immune checkpoint inhibitors in challenging populations, Cancer 123 (11) (2017) 1904–1911.

[117] A.M. Menzies, et al., Anti-PD-1 therapy in patients with advanced melanoma and preexisting autoimmune disorders or major toxicity with ipilimumab, Ann. Oncol. 28 (2) (2017) 368–376.

[118] N. Abdel-Wahab, et al., Use of immune checkpoint inhibitors in the treatment of patients with cancer and preexisting autoimmune disease: a systematic review, Ann. Intern. Med. 168 (2) (2018) 121–130.

[119] N. Abdel-Wahab, et al., Use of immune checkpoint inhibitors in the treatment of patients with cancer and preexisting autoimmune disease, Ann. Intern. Med. 169 (2) (2018) 133–134.

[120] C. Kyi, et al., Ipilimumab in patients with melanoma and autoimmune disease, J. Immunother. Cancer 2 (1) (2014) 35.

[121] D.B. Johnson, et al., Ipilimumab therapy in patients with advanced melanoma and preexisting autoimmune disorders, JAMA Oncol. 2 (2) (2016) 234–240.

[122] F.X. Danlos, et al., Safety and efficacy of anti-programmed death 1 antibodies in patients with cancer and pre-existing autoimmune or inflammatory disease, Eur. J. Cancer 91 (2018) 21–29.

[123] P. Sharma, Immune checkpoint therapy and the search for predictive biomarkers, Cancer J. 22 (2) (2016) 68–72.

[124] G. Manson, et al., Biomarkers associated with checkpoint inhibitors, Ann. Oncol. 27 (7) (2016) 1199–1206.
[125] R. Das, et al., Early B cell changes predict autoimmunity following combination immune checkpoint blockade, J. Clin. Invest. 128 (2) (2018) 715–720.
[126] J. Da Gama Duarte, et al., Autoantibodies may predict immune-related toxicity: results from a phase I study of intralesional Bacillus Calmette-Guerin followed by ipilimumab in patients with advanced metastatic melanoma, Front. Immunol. 9 (2018) 411.
[127] M.F. Gowen, et al., Baseline antibody profiles predict toxicity in melanoma patients treated with immune checkpoint inhibitors, J. Transl. Med. 16 (1) (2018) 82.
[128] J.T. Jorgensen, K.B. Nielsen, Companion and complementary diagnostics for first-line immune checkpoint inhibitor treatment in non-small cell lung cancer, Transl. Lung Cancer Res. 7 (Suppl. 2) (2018) S95–S99.
[129] M. Hersom, J.T. Jorgensen, Companion and complementary diagnostics-focus on PD-L1 expression assays for PD-1/PD-L1 checkpoint inhibitors in non-small cell lung cancer, Ther. Drug Monit. 40 (1) (2018) 9–16.
[130] S. Lu, et al., Comparison of biomarker modalities for predicting response to PD-1/PD-L1 checkpoint blockade: a systematic review and meta-analysis, JAMA Oncol. 5 (8) (2019) 1195–1204.
[131] M.J. Fritzler, et al., The utilization of autoantibodies in approaches to precision health, Front. Immunol. 9 (2018) 2682.
[132] M.J. Fritzler, M. Mahler, Redefining systemic lupus erythematosus—SMAARTT proteomics, Nat. Rev. Rheumatol. 14 (8) (2018) 451–452.
[133] M.Y. Choi, et al., Preventing the development of SLE: identifying risk factors and proposing pathways for clinical care, Lupus 25 (8) (2016) 838–849.
[134] N.M. Steven, B.A. Fisher, Management of rheumatic complications of immune checkpoint inhibitor therapy—an oncological perspective, Rheumatology (Oxford) 58 (Supplement_7) (2019) vii29–vii39.
[135] L.P. Yco, et al., Effect of sulfasalazine on human neuroblastoma: analysis of sepiapterin reductase (SPR) as a new therapeutic target, BMC Cancer 15 (2015) 477.
[136] M.R. Mooney, et al., Anti-tumor effect of sulfasalazine in neuroblastoma, Biochem. Pharmacol. 162 (2019) 237–249.
[137] L.H. Calabrese, Sorting out the complexities of autoimmunity and checkpoint inhibitors: not so easy, Ann. Intern. Med. 168 (2) (2018) 149–150.
[138] J.R. Brahmer, et al., Management of immune-related adverse events in patients treated with immune checkpoint inhibitor therapy: American Society of Clinical Oncology clinical practice guideline, J. Clin. Oncol. 36 (17) (2018) 1714–1768.
[139] H. Arima, et al., Management of immune-related adverse events in endocrine organs induced by immune checkpoint inhibitors: clinical guidelines of the Japan Endocrine Society, Endocr. J. 66 (7) (2019) 581–586.
[140] G. Fuca, et al., Modulation of peripheral blood immune cells by early use of steroids and its association with clinical outcomes in patients with metastatic non-small cell lung cancer treated with immune checkpoint inhibitors, ESMO Open 4 (1) (2019) e000457.
[141] F. Martins, et al., Adverse effects of immune-checkpoint inhibitors: epidemiology, management and surveillance, Nat. Rev. Clin. Oncol. 16 (9) (2019) 563–580.
[142] F. Martins, et al., New therapeutic perspectives to manage refractory immune checkpoint-related toxicities, Lancet Oncol. 20 (1) (2019) e54–e64.
[143] E. Perez-Ruiz, et al., Prophylactic TNF blockade uncouples efficacy and toxicity in dual CTLA-4 and PD-1 immunotherapy, Nature 569 (7756) (2019) 428–432.

[144] M. Alvarez, et al., Impact of prophylactic TNF blockade in the dual PD-1 and CTLA-4 immunotherapy efficacy and toxicity, Cell Stress 3 (7) (2019) 236–239.
[145] M. Touat, et al., Immune checkpoint inhibitor-related myositis and myocarditis in patients with cancer, Neurology 91 (10) (2018) e985–e994.
[146] J.R. Kolfenbach, et al., Autoimmunity to peptidyl arginine deiminase type 4 precedes clinical onset of rheumatoid arthritis, Arthritis Rheum. 62 (9) (2010) 2633–2639.
[147] K.D. Deane, Preclinical rheumatoid arthritis and rheumatoid arthritis prevention, Curr. Rheumatol. Rep. 20 (8) (2018) 50.
[148] K.D. Deane, V.M. Holers, The natural history of rheumatoid arthritis, Clin. Ther. 41 (7) (2019) 1256–1269.
[149] A.L. Mammen, et al., Pre-existing antiacetylcholine receptor autoantibodies and B cell lymphopaenia are associated with the development of myositis in patients with thymoma treated with avelumab, an immune checkpoint inhibitor targeting programmed death-ligand 1, Ann. Rheum. Dis. 78 (1) (2019) 150–152.
[150] I. Mazarico, et al., Low frequency of positive antithyroid antibodies is observed in patients with thyroid dysfunction related to immune check point inhibitors, J. Endocrinol. Investig. 42 (12) (2019) 1443–1450.
[151] Y. Toi, et al., Profiling preexisting antibodies in patients treated with anti-PD-1 therapy for advanced non-small cell lung cancer, JAMA Oncol. 5 (3) (2019) 376–383.
[152] S.A. Tahir, et al., Autoimmune antibodies correlate with immune checkpoint therapy-induced toxicities, Proc. Natl. Acad. Sci. USA 116 (44) (2019) 22246–22251.
[153] M. Zibelman, N. Pollak, A.J. Olszanski, Autoimmune inner ear disease in a melanoma patient treated with pembrolizumab, J. Immunother. Cancer 4 (8) (2016) s40425–16–0114–4.

Chapter 6

The evolving potential of precision medicine in the management of autoimmune liver disease

Gary L. Norman[a], Nicola Bizzaro[b], Danilo Villalta[c], Diego Vergani[d], Giorgina Mieli-Vergani[e], Gideon M. Hirschfield[f], and Michael Mahler[a]

[a]*Inova Diagnostics, Inc., San Diego, CA, United States,* [b]*Laboratory of Clinical Pathology, San Antonio Hospital, Tolmezzo, Italy,* [c]*Immunology and Allergy Unit, Santa Maria degli Angeli Hospital, Pordenone, Italy,* [d]*Institute of Liver Studies, MowatLab, King's College London Faculty of Life Sciences & Medicine at King's College Hospital, London, United Kingdom,* [e]*Paediatric Liver, GI & Nutrition Centre, MowatLab, King's College London Faculty of Life Sciences & Medicine at King's College Hospital, London, United Kingdom,* [f]*Division of Gastroenterology and Hepatology, Toronto Centre for Liver Disease, University of Toronto, Toronto, ON, Canada*

Abbreviations

AI-ALF autoimmune acute liver failure
AIH-1 autoimmune hepatitis, type 1
AIH-2 autoimmune hepatitis, type 2
ALD autoimmune liver disease
AMA anti-mitochondrial antibody
ANA anti-nuclear antibody
ANCA anti-neutrophil cytoplasmic antigens
ASCA anti-Saccharomyces cerevisiae
ASGPR asialoglycoprotein receptor
β2GPI beta2 glycoprotein
CCA cholangiocarcinoma
CCP cyclic citrullinated peptide
CIA chemiluminescent immunoassay
DILI drug-induced liver injury
F-actin filamentous actin
FGF19 fibroblast growth factor 19
FXR farnesoid x receptor
GGT gamma glutamyl transferase
GPA gastric parietal antibody

Precision Medicine and Artificial Intelligence. https://doi.org/10.1016/B978-0-12-820239-5.00012-7

GWAS	genome wide association study
HBV	hepatitis B virus
HCC	hepatocellular carcinoma
HCV	hepatitis C virus
HIP1R	Huntingtin-interacting protein 1-related protein
HK-1	hexokinase 1
HLA	human leukocyte antigen
IF	intrinsic factor
IIF	indirect immunofluorescence
KLHL12	Kelch-like 12
KL-p	Kelch-like 12 peptide
LC1	liver cytosol 1
LKM-1	liver-kidney microsome-1
LT	liver transplant
MZGP2	major zymogen glycoprotein 2
NAFLD	nonalcoholic fatty liver disease
NASH	nonalcoholic steatohepatitis
OCA	obeticholic acid
PBC	primary biliary cholangitis
PH	precision health
PM	precision medicine
PR3	serine proteinase 3
PSC	primary sclerosing cholangitis
SLA	soluble liver antigen
SMA	smooth muscle antibodies
tTG	tissue transglutaminase
UC	ulcerative colitis
UDCA	ursodeoxycholic acid

1 Introduction

It is increasingly recognized that many autoimmune diseases result from the convergence and interactions of multiple factors including, but likely not limited to genetic susceptibility, gender, age, environmental factors, infectious agents, and epigenetics [1]. In autoimmune liver disease (ALD), as in many other diseases, there may be a spectrum of disease presentations and courses. Whether this spectrum truly reflects different "flavors" of one disease or may reflect distinct subtypes is now becoming clearer as vast amounts of data on affected patients is assembled and analyzed. Using artificial intelligence (AI) and machine learning (ML) methodologies to analyze detailed data on the clinical history, demographics, laboratory measures, clinical features, genetic variations, disease progress and treatment response of large number of patients with similar disease characteristics, precision medicine (PM) seeks to more precisely guide the accurate diagnosis, management, and optimum outcome for subsets of patients. The ultimate goal is the often cited concept of the "right treatment, to right patient, at right time" [2].

ALD is a prime target for PM since the initiating etiology is usually not clear, the spectrum of disease presentation is wide, early identification and treatment is ideal, and importantly, diagnosis is often delayed as a result of its relatively rarity and overlapping of symptoms with other more common conditions [3]. In this chapter we provide an overview of ALD, its history, laboratory diagnosis, spectrum of disease, challenges in differential diagnosis and patient management, and thoughts on the evolution and application of PM for enhancing care of patients with ALD.

2 History of autoimmune liver disease

The first mention of a liver disease resembling primary biliary cholangitis (PBC) dates back to 1851, when Addison and Gull [4] described a patient with obstructive jaundice not involving major bile ducts. Based on elevated cholesterol levels and the presence of cutaneous xanthelasmas, they called this disease "xanthomatous biliary cirrhosis." However, it was not until almost a century later that a series of seminal studies paved the way for a better definition of what would be eventually classified as an ALD.

In 1949, Dauphinee [5] and Ahrens [6] described in detail the clinical picture of a disease of the biliary tree and named it PBC. One year later, Waldenstrom described a young woman with chronic hepatitis and hypergammaglobulinemia who responded well to corticosteroids [7]. Since this form of chronic hepatitis was found to be associated with a positive lupus erythematosus (LE) cell test and antinuclear antibodies (ANA) [8], the term lupoid hepatitis was initially coined by Mackay [9]. However, it was soon clear that this disorder was not always associated with lupus, but constituted instead a characteristic type of ANA-associated chronic hepatitis.

A milestone for the definition of ALD was the discovery of autoantibodies as serological hallmark. For PBC, anti-mitochondrial antibodies (AMA) were first described by Mackay et al. [10]. The authors of this landmark study found high-titers of circulating complement-fixing autoantibodies directed against liver, kidney, and other human tissue antigens. In 1965, a team of researchers led by Sheila Sherlock at the Royal Free Hospital in London developed the first serological test for AMA detection using an indirect immunofluorescence (IIF) assay with rat liver and kidney tissues [11].

In 1987, the molecular target of AMA was identified by Gershwin et al. as the 2-oxo-acid dehydrogenase mitochondrial enzyme complex [12]. In the same year, Szostecki identified sp100 protein as an autoantigenic target [13] and almost simultaneously, antibodies reactive to nuclear lamins [14] and to a transmembrane glycoprotein of the nuclear envelope [15] were described. Eventually, Nickowitz identified the glycoprotein autoantigen as the gp210 nuclear pore membrane protein [16] and Courvalin demonstrated that anti-gp210 antibodies were present in up to 25% of patients with PBC, even in the absence of AMA [17]. Recently, two novel PBC-specific autoantibodies directed toward hexokinase 1 (HK-1) and to the Kelch-like 12 (KLHL12) protein have expanded the number of autoantibodies associated with PBC [18].

In autoimmune hepatitis (AIH), smooth muscle antibodies (SMA) were first observed in 1965 by IIF on unfixed sections of rat stomach in the sera of patients with chronic active hepatitis [19]. Farrow then demonstrated in 1971 that the primary autoantibody target of SMA were the actin filaments in liver cells [20] and actin was later confirmed as the major autoantigen by Gabbiani et al. [21] and Kurki [22].

Several years later in 1973, Rizzetto [23] described anti-microsomal antibodies, later named anti-liver-kidney microsome-1 (LKM-1) antibodies by Homberg [24, 25], as a serological marker of AIH type II (AIH-2) targeting cytochrome P450IID6 [26]. This finally led to a new classification for AIH: Type I (AIH-1) characterized by the presence of SMA and ANA, and type II (AIH-2) marked by anti-LKM-1 antibodies.

At about the same time, additional AIH-associated autoantibodies were discovered, namely antibodies against soluble liver antigen/liver pancreas (SLA/LP) [27], antibodies to the asialoglycoprotein-receptor (ASGPR) [28], and antibodies directed against the liver cytosol 1 antigen (LC1) [29]. In 1993, the International Autoimmune Hepatitis Group (IAIHG) introduced the term AIH to harmonize the various terms used to describe the disease and also proposed diagnostic criteria [30], which were later updated to include anti-SLA/LP (now usually referred to as only anti-SLA) as a criteria biomarker and also to simplify and to make the criteria more applicable for routine clinical use [31]. Fig. 1 presents a timeline of milestones in the history of ALD.

3 Spectrum of autoimmune liver disease

The three major forms of ALD are AIH, PBC, and primary sclerosing cholangitis (PSC) and reflect where different components of the liver become targets of an auto-aggressive immune attack. AIH, PBC, and PSC are complex disorders that result from interaction between genetic and environmental factors. All three have a progressive course that, if untreated, leads to liver failure requiring liver transplantation [35]. Treatment aims at abolishing or reducing inflammation, cholestasis, and progression of fibrosis. Some of the basic features of the various forms of autoimmune liver disease are summarized in Table 1.

3.1 Autoimmune hepatitis

AIH is characterized by active inflammation, liver cell necrosis, and fibrosis, which may then lead to hepatic failure, cirrhosis, and ultimately death. It is diagnosed based on a histological picture of interface hepatitis in the context of elevated transaminase and serum immunoglobulin G levels and in the presence of autoantibodies [47, 48].

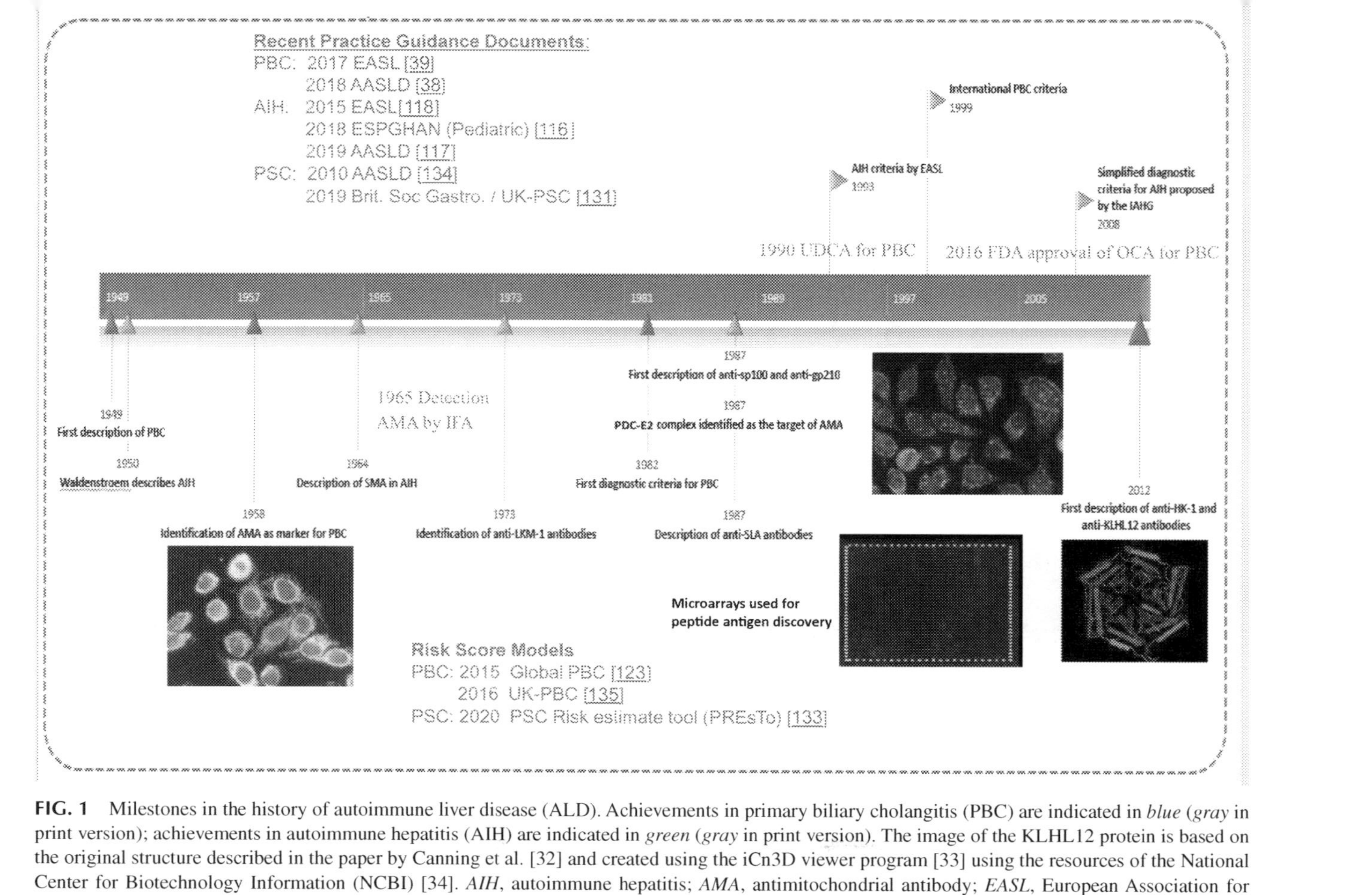

FIG. 1 Milestones in the history of autoimmune liver disease (ALD). Achievements in primary biliary cholangitis (PBC) are indicated in *blue* (*gray* in print version); achievements in autoimmune hepatitis (AIH) are indicated in *green* (*gray* in print version). The image of the KLHL12 protein is based on the original structure described in the paper by Canning et al. [32] and created using the iCn3D viewer program [33] using the resources of the National Center for Biotechnology Information (NCBI) [34]. *AIH*, autoimmune hepatitis; *AMA*, antimitochondrial antibody; *EASL*, European Association for Study of Liver Disease; *HK1*, hexokinase 1; *IAIHG*, International Autoimmune Hepatitis Group; *KLHL12*, Kelch-like protein 12; *LC1*, liver cytosol 1; *LKM-1*, liver kidney microsome 1; *PBC*, primary biliary cholangitis; *PDC-E2*, pyruvate dehydrogenase complex-E2 subunit; *SLA*, soluble liver antigen; *SMA*, smooth muscle antibody.

TABLE 1 Overview of autoimmune liver diseases.

Disease	Clinical characteristics	
AIH-1	Chronic progressive, inflammatory liver disease, unknown etiology. More common form of AIH, without treatment usually progresses to end-stage liver disease (ESLD)	[36]
AIH-2	5–10% of all AIH, more aggressive, mostly commonly affecting children and young adults (females)	[37]
PBC	Chronic necroinflammatory injury of small, interlobular bile ducts and their branches, progressive cholestasis and fibrosis, can lead to cirrhosis and ESLD requiring liver transplantation	[38, 39]
PSC	Chronic progressive, necroinflammatory injury of intrahepatic and extrahepatic large bile ducts, unknown cause, death or liver transplantation 15–20 years, 60–70% concomitant IBD, risk cholangiocarcinoma	[40]
AIH/PBC	PBC-specific antibodies can sometimes be detected in AIH, persistent elevation cholestatic enzymes, Variant syndrome, 3–20% of PBC patients may have features of AIH, criteria for diagnosis of AIH in PBC have been established, treatment in these usually toward predominant disease	[39, 41–43]
AIH/PSC	1–53% overlap, diagnostic criteria inconsistent	[44]
ASC	Overlap clinical features PSC and AIH, children-young adults, ANA and SMA positivity, high IgG levels, interface hepatitis, usually treated as AIH	[45, 46]

AIH, autoimmune hepatitis; *ANA*, antinuclear antibodies; *ASC*, autoimmune sclerosing cholangitis; *ESLD*, end-stage liver disease; *IBD*, inflammatory bowel disease; *PBC*, primary biliary cholangitis; *PSC*, primary sclerosing cholangitis; *SMA*, smooth muscle antibodies.

AIH is divided into two main subgroups according to the type of autoantibody detected. AIH-1 is associated with SMA and/or ANA and affects both adults and children, while AIH-2, associated with anti-LKM-1 and/or anti-LC1 antibodies, affects mainly children and adolescents. Patients with AIH-2 more frequently have an acute or even fulminant presentation and often present with IgA deficiency. Concurrent autoimmune disorders occur in approximately 40% of patients (thyroiditis, diabetes type 1, inflammatory bowel disease (IBD), vitiligo, celiac disease) and a family history of autoimmunity is frequent.

There is a strong genetic influence upon the development of AIH. Patients are usually female and frequently possess the haplotype HLA A1-B8-MICA*008-TNFA*2-DRB3*0101-DRB1*0301-DQB1*0201 and null (i.e., non-productive) C4 complement genes. Susceptibility to AIH-1 is conferred by the possession of either DRB1*0301 or DRB1*0401 genes, while predisposition to AIH-2 is imparted by DRB1*0701 or DRB1*0301. The key role of the HLA genes in predisposing to AIH has been confirmed in a genome wide association study (GWAS) [49].

A liver biopsy is required to establish the diagnosis of AIH. The typical histological picture is interface hepatitis, which is characterized by a portal mononuclear cell infiltrate that erodes the limiting plate and spills over into the parenchyma, where inflammatory cells surround dying hepatocytes. Histological signs of cirrhosis can be found in up to 30% of patients at diagnosis. The onset of AIH tends to be insidious in most cases, although the presentation may be that of an acute, even fulminant, hepatitis particularly in young patients. The course is variable, but the disease is usually controlled by a lifelong treatment with immunosuppressive drugs. Alterations of liver function tests do not necessarily reflect the extent of the histological lesion.

Standard therapy consists of a combination of steroids and azathioprine, which is effective in about 80% of patients; however progression may occur despite effective treatment in about 10%. Other immunosuppressive agents such as mycophenolate mofetil, calcineurin inhibitors, anti-B-cell, and anti-tumor necrosis factor (TNF) alpha monoclonal antibodies have been tried with variable results in refractory cases [47, 48]. Cytokine and cellular treatments to boost defective immunoregulation are being investigated.

3.2 Primary biliary cholangitis

PBC is a chronic liver disease of unknown etiology characterized by slowly progressive intrahepatic cholestasis caused by an inflammatory destruction of small intrahepatic bile ducts [50]. PBC is 10 times more common in women vs. men and is most common over 50 years of age. Concomitant autoimmune disorders are present in 30–50% of cases (autoimmune thyroiditis, systemic lupus erythematosus, systemic sclerosis) and autoimmune conditions in 14–20% of their family members, but, unlike AIH, PBC does not respond to immunosuppressive treatment [35].

The clinical course of PBC is variable. While in the past it was one the most common indications for liver transplantation, in more recent years, probably due to effective treatment, patients tend to have a less severe form with normal life-expectancy. PBC is characterized by the presence of AMA (present in over 90% of cases), specific ANAs (multiple nuclear dots and rim-like), cholestatic biochemistry, and elevated serum IgM. AMAs are strong disease predictors, since their accidental detection is virtually always followed by the development of the disease. This, however, may take up to 20 years to manifest itself. Recent GWAS and Immunochip (Ichip)-association studies confirmed a key role for HLA genes and a link with a number of genes involved in innate or acquired immune responses [35].

The histological picture of PBC is characterized by lymphocyte infiltration and destruction of small and medium size bile ducts, ductular proliferation, periductular granulomas, fibrosis, and ultimately cirrhosis. These histological alterations are frequently patchy and for this reason a liver biopsy is considered consistent with, rather than diagnostic of, PBC.

The earliest symptoms of PBC are fatigue and pruritus, though the disease can be asymptomatic and diagnosed incidentally in the presence of AMA or cholestatic biochemistry performed at check-up or for different medical reasons. Jaundice, darkening of the skin in exposed areas and manifestations resulting from impaired bile excretion follow. The latter include steatorrhea and impaired absorption of lipid-soluble vitamins, leading to osteomalacia, bruising, and occasionally night blindness, deriving from vitamin D, vitamin K, and vitamin A deficiency, respectively. While physical examination can be rather unremarkable, laboratory investigations are informative: alkaline phosphatase (ALP) levels are elevated early in the course of the disease, and AMA at a titer > 1:40 is present in over 90% of patients. Serum bilirubin rises over time, an increase in aspartate aminotransferase (AST) levels is frequently seen, but it is never dramatic. Hyperlipidaemia is common. There is significant heterogeneity in disease severity. Some patients have disease that remains more mild, whereas others have disease that progresses faster through to end-stage liver disease.

Ursodeoxycholic acid (UDCA) is the first-line licensed therapy for PBC. UDCA has a favorable effect on long-term survival, slowing progression of fibrosis, with a 10-year survival rate of > 95% [51]. Patients who do not respond to UDCA have a 10-year survival rate < 80%. For poor responders, other drugs, including fibrates, methotrexate, and colchicine have been used in combination with UDCA, but their benefit is debated. A promising novel agent is obeticholic acid (OCA), a bile acid agonist of the Farsenoid X receptor (FXR), which is involved in bile acids metabolism and enterohepatic circulation [35]. OCA was approved by US Food and Drug Administration (FDA) for the treatment of PBC in combination with UDCA in adults with an inadequate response to UDCA, or as monotherapy in adults unable to tolerate UDCA [38].

3.3 Primary sclerosing cholangitis

PSC is a chronic cholestatic disorder of unknown etiology, characterized by inflammation and progressive obliterative fibrosis of the intrahepatic and/or extrahepatic bile ducts. It progresses slowly and often asymptomatically to biliary cirrhosis, liver failure, and/or cholangiocarcinoma (CCA) [52]. PSC is atypical among autoimmune diseases since it has a male predominance (M:F ratio, 2:1), no disease-specific autoantibodies, and responds poorly to immunosuppression. However, predisposition to PSC is imparted by genes within the HLA region, in common with the other ALDs. Moreover, presence of non-specific autoantibodies, including ANA, SMA, atypical anti-neutrophil cytoplasmic antibodies and anti-serine proteinase 3 (PR3) [53], as well as a strong association with other autoimmune or immune-mediated disorders (in particular IBD) in about 70% of patients, do suggest an immune-mediated pathogenesis.

UDCA improves serum liver tests in PSC, but its effect on long-term survival is debated, and there is no overwhelming evidence for true benefit. While high UDCA doses (17–23 mg/kg/day) are associated with serious adverse effects [54], lower doses have been in some settings reported to improve transplant-free survival [35].

3.4 Overlap syndromes

AIH can be associated with a variety of liver and extrahepatic autoimmune diseases. Some patients with AIH may have clinical, histological, and serological features of PBC or PSC. These overlap syndromes are probably clinical descriptions rather than distinct pathological entities, the dominant component of the disease determining management. In the presence of AIH features, immunosuppression can be of benefit also in PBC and PSC [55].

In children and young adults, a form of sclerosing cholangitis, named autoimmune sclerosing cholangitis (ASC), and characterized by ANA and SMA positivity, high levels of IgG, and interface hepatitis is recognized [45, 46]. As ALP and gamma glutamyl transpeptidase (GGT) levels are often normal at presentation, a cholangiography is needed for diagnosis. In the absence of bile duct imaging, these patients are diagnosed and treated as AIH. ASC affects equally males and females. If treatment is started early, the parenchymal liver damage responds well to the same immunosuppressive treatment used for AIH, with good medium to long-term survival [45, 46, 56], but ultimate prognosis is worse than AIH as the bile duct disease can progress despite treatment [45]. Also in an adult series, patients with overlapping features between PSC and AIH, treated with immunosuppression, appear to have a better prognosis than those with classical PSC, suggesting that immunosuppression might have a positive effect on disease progression [57].

4 Laboratory biomarkers of autoimmune liver disease

4.1 Serological markers for autoimmune hepatitis

As previously discussed, AIH is divided into two types: AIH-1 and AIH-2. ANA and anti-SMA are associated with AIH-1, whereas anti-LKM-1 and anti-LC-1 antibodies are the hallmarks of AIH-2. These autoantibodies can be detected by IIF testing: ANA on HEp-2 cells and the other autoantibodies on rodent triple tissue (kidney, stomach, liver). When these markers are absent, it is useful to test for autoantibodies to anti-SLA and anti-neutrophil cytoplasmic antigens (ANCA). All these antibodies are included in the International Autoimmune Hepatitis Group (IAIHG) diagnostic scoring system [30, 58].

Although ANA can be present in 50–80% of AIH-1 patients [59], they are not disease-specific, since they may also be detected in connective tissues diseases, as well as in non-ALD [60]. ANA in AIH are directed against a variety of antigens, including histones, chromatin, double- and single-stranded DNA, centromere, laminin, small nuclear-ribonucleoproteins, cyclin A, and other not yet identified nuclear antigens. The most frequent encountered patterns are homogeneous and speckled.

SMA can be detected in about 50–80% of AIH-1 patients [61], mainly associated with ANA (~70% of cases). However, like ANA, they are not a disease-specific marker. SMA represent a heterogeneous group of autoantibodies directed to various antigens including actin and other components of the cytoskeleton (tubulin, vimentin, desmin) [62]. On kidney tissue, the vascular/glomerular (VG) and the vascular/glomerular/tubular (VGT) IIF staining patterns are more specific for AIH than the vascular (V) pattern [63]. Indeed, VG and VGT patterns are more strictly associated with the presence of autoantibodies against filamentous actin (F-actin), which seems to be the primary specific target of SMA. The presence of anti-F-actin autoantibodies may be confirmed by ELISA assays or by IIF using the VSM47 or rat intestinal epithelial cell line substrates [64–66].

Anti-LKM-1 antibodies are a marker of AIH-2 (up to 90% of cases). By IIF on rodent tissues they stain the cytoplasm of the hepatocytes and of the cells of the proximal, but not the distal renal tubules, or the gastric parietal cells; the last feature may help differentiation from AMA, which in contrast stain all rodent tissues. Following the identification of cytochrome P450 (CYP) 2D6 as the target antigen of anti-LKM-1 antibodies [67], solid-phase assays using native or recombinant proteins have been developed for their detection and confirmation of IIF results [36]. Since anti-LKM-1 can also be found in some patients with hepatitis C virus infection (HCV), HCV should be ruled out whenever anti-LKM-1 reactivity is detected.

Anti-LC-1 antibodies, alone or in combination with anti-LKM-1, are present in about 30% of AIH-2 [65]. By IIF, they stain hepatocytes, but spare the centrolobular areas of the liver. However, when anti-LKM-1 antibodies are coexistent, they can completely mask the presence of anti-LC-1. Since identification of formiminotranferase cyclodeamidase as the antigenic target of anti-LC-1 antibodies [68], ELISA, chemiluminescent immunoassays (CIA), and immunoblot assays have been developed for their detection [69]. Like anti-LKM-1 antibodies, anti-LC-1 antibodies are not entirely specific for AIH, as they may also be found in some patients with HCV [70].

In contrast, anti-SLA antibodies are highly specific for AIH [63, 65], but present in only 10–20% of patients. They cannot be detected by IIF, but only with immunoassays, where the autoantigen Sep-*O*-phosphoseryl-tRNA:selenocysteine-tRNA synthetase (SepSecS) is coated on the solid phase. Anti-SLA antibodies are present in AIH-1 and using a high sensitivity ligand assay have been reported in patients with AIH-2. They are associated with a more severe clinical course and worse prognosis [71–75]. Since anti-SLA antibodies may be the only antibody present, they should be searched in all cases of suspected AIH [73, 76].

When the conventional autoantibodies described above are absent, positivity for ANCA (atypical p-ANCA) and anti-ASGPR antibodies may aid the diagnosis. However, these autoantibodies are not-disease specific. In particular, atypical p-ANCA can also detected in IBD and PSC, while anti-ASGPR can be detected in many non-ALD. Anti-ASGPR antibodies have been reported to correlate with disease activity [77]. Other autoantibodies found in patients with AIH, include anti-Ro52 (which is almost always present in anti-SLA-positive patients), anti-chromatin, ASCA, and anti-cyclic citrullinated peptide (CCP) [74, 78–80]. With the exception of ASCA, the presence of anti-Ro52, anti-chromatin, and anti-CCP were associated with more severe disease, but the additive clinical significance of these biomarkers remains to be established.

4.2 Serological markers of primary biliary cholangitis

AMA are considered a hallmark for the diagnosis of PBC and are present in more than 90% of patients with PBC [63]. AMA are mainly directed to the E2 subunits of the 2-oxo-acid dehydrogenase complex (pyruvate dehydrogenase complex [PDC-E2], the 2-oxoglutarate dehydrogenase complex [OGD-E2], and the branched-chain 2-oxoacid dehydrogenase complex [BCOAD-E2]) found on the inner mitochondrial membrane and are classically detected by IIF on rodent tissues. Following the identification of the mitochondrial antigens targeted by AMA, a variety of solid-phase assays were developed. These methods typically use either a mixture of recombinant E2 subunits [81] or a fusion protein combining all the three immunodominant E2 subunits (MIT3) [82, 83]. These assays provide higher diagnostic sensitivity compared to IIF, since they enable detection of additional, though yet not all, AMA IIF-negative patients [82, 84]. Although AMA or specific anti-E2 subunit positivity determined by solid phase assays are a strong indicator of PBC in patients with otherwise unexplained abnormal liver biochemistry, only 1/6 with AMA positivity and normal ALP or GGT were reported to develop PBC within 5 years [85].

In addition to AMA, ANA displaying a multiple nuclear dot (MND) pattern on HEp-2 cells (AC-6 according to the International Consensus of ANA patterns (ICAP)) [86] or a rim-like/membranous punctate nuclear envelope staining pattern on HEp-2 cells (AC-12, according to ICAP) are detected in 25–30% of PBC patients and in a portion of AMA-negative subjects, increasing the sensitivity of serological tests for diagnosing PBC. Autoantibodies responsible for the MND pattern are mainly directed against the sp100 or p62 proteins, while those causing the rim-like/membranous pattern are mainly directed against the nuclear pore protein gp210. Anti-gp210 positivity has been associated with a poor outcome and a higher mortality rate in PBC patients [87, 88], while the prognostic value of anti-sp100 antibodies is still controversial [50, 87]. Anti-centromere antibodies, typically found in patients with systemic sclerosis, have been found in 20–30% of patients with PBC and associated with portal hypertension type progression [87].

Despite the utilization of new sensitive and specific assays for AMA and PBC-specific ANA, some patients remain serologically negative. In the effort to close the serological gap and to improve the diagnostic detection rate, novel specific markers for PBC, such as anti-KLHL12, anti-KLHL12 peptide (KL-p), and anti-HK-1 antibodies were recently identified. It has been shown that detection of anti-sp100 and anti-gp210, together with the additional detection of anti-HK-1 and anti-KLHL12 antibodies, resulted 68.5% of AMA-negative patients positive for at least one serological marker [18, 89, 69]. A recent study suggests that in addition to their diagnostic value, antibodies to HK-1 may be associated with worse prognosis and shorter transplant-free survival time [90].

4.3 Serological markers of primary sclerosing cholangitis

Immunoserology in PSC is generally considered unspecific and therefore of limited value to establish the diagnosis [39]. The most prevalent serum antibodies in PSC are atypical pANCA, which can be found in up 90% of patients [91]. However, these autoantibodies are also found in patients with IBD without PSC, in patients with AIH, and to a lesser extent in PBC, so that they can support the diagnosis of PSC only in selected patients, mainly those lacking associated IBD. Recently several groups have shown that anti-serine proteinase 3 (PR3), typically associated with vasculitis, can be found in approximately 1/3 of PSC patients [53]. Anti-Glycoprotein 2 (GP2) antibodies, also referred to as Major Zymogen Glycoprotein 2 (MZGP2), have also been identified in patients with PSC and initial results suggest these antibodies may help identify patients with worse outcome [92–94]. Both anti-PR3 IgG and anti-GP2 IgA have been shown to be increased in patients with cholangiocarcinoma (CCA), but additional validation of these observations is necessary [95,94]. Anti-F-actin IgA and anti-gliadin IgA (whole molecule, not deamidated gliadin peptide) were recently shown to be associated with poor disease outcome in PSC, and thus are additional potential parameters to consider if they can be validated in additional studies [96].

The search for new biomarkers to improve the detection and management of patients with autoimmune disease continues to advance. Not surprisingly, the immunological alterations in patients with ALD and the linkages with other autoimmune disease lead to the detection of antibodies to a variety of targets. Table 2 summarizes many of these observations. It should be understood however, that many of these observations have not been externally validated and their clinical significance remains to be established.

TABLE 2 Overview of biomarkers associated with autoimmune liver diseases.

Autoantibody	Clinical association	Prevalence estimate	Comments	References
AMA (IIF)	PBC	> 90%	Predominant and classic marker of PBC	[63]
MIT3	PBC	96% AMA +, 25% AMA −	AMA target, fusion antigen immunodominant portions of the PDC-E2 complex	[76, 83, 97]
sp100	PBC	30–50%	Nuclear dot pattern, AMA-negative PBC	[98]
gp210	PBC	15–20% AMA +, 30–50% AMA −	Nuclear rim pattern, association with more severe disease, higher HCC?, AMA-negative PBC	[17, 87]
HK-1	PBC	45.7% MIT3 +, 24.7% MIT3 −	Note MIT3 ELISA used for reference detection, associated poor prognosis, AMA-negative PBC	[18, 89, 90]
KLHL12 (KLp)	PBC	24.9% MIT3 +, 19.2% MIT3 −	Note MIT3 ELISA used for reference detection AMA-negative PBC	[18, 89]
HK-1 and/or KLp	PBC	58.0% MIT3 +, 38.4% MIT3 −	Note MIT3 ELISA used for reference detection, AMA-negative PBC	[18, 89]
GPA/IF	PBC		Increased in PBC, associated lower B12, association recurrent PBC after OLT	[99, 100]
Centromere	PBC	30%	Portal hypertension, poor prognosis, may suggest overlap with systemic sclerosis	[87, 101–106]
dsDNA	PBC/AIH overlap	37.5% by Crithidia luciliae IIF	Not supported by CIA using synthetic DNA substrate	[106]
Autotaxin[a]	PBC		Levels associated with disease progression and pruritis	[98, 107]
ANA	AIH	75%	Also positive in PBC, PSC, HBV, HCV, DILI, NAFLD	[37, 98]
F-actin IgG	AIH-1	75%	Not AIH-2	[66, 98]
SMA	AIH-1	95%	Multiple targets-F-actin, tubulin, micro/intermediate filaments	[62, 98]

Continued

TABLE 2 Overview of biomarkers associated with autoimmune liver diseases—cont'd

Autoantibody	Clinical association	Prevalence estimate	Comments	References
SLA	AIH-1	20–50%	May be only autoantibody detected, associated relapse after corticosteriod withdrawal, detected in AIH-2 by radioligand assay	[71, 73, 98]
LKM1	AIH-2	> 60%	Absent AIH-1, some HCV	[98]
LC1	AIH-2	2/3	Very rare AIH-1, some HCV	[98]
Chromatin	AIH	36%	Associated relapse after corticosteriod withdrawal	[74, 108]
A-pANCA (ANNA)	AIH/PSC (UC)	50–90%	Overlap with UC, not specific to PSC	[108]
HIP1R	AIH	71–85%	Reported higher specificity vs. ANA and SMA	[109]
B2GPI IgA	AIH, less PBC	51% AIH, 34% PBC	Associated severity, biochemical markers of severity, clinical significance unclear	[110]
aCL IgA	AIH, less PBC	33% AIH	Associated severity, biochemical markers of severity, clinical significance unclear	[110]
ASCA	AIH, PBC	20–30%	Common but no clinical relevance	[74, 111]
MZGP2/GP2 IgA	PSC (UC)	31–50%	Overlap with UC, increased patients with CCA, associated with severity	[92,93,95,94]
PR3-ANCA	PSC (UC)	30–40%	Overlap with UC, increased patients with CCA, associated with severity	[53,94]
F-actin IgA	PSC	28%	Poor outcome in PSC (also marker mucosal damage in celiac disease)	[96, 112]
ASGPR	AIH, PBC, PSC		Potential association with disease activity	[77]
AMA/tTG	AI-ALF	15%/8%	May assist discrimination of AI-ALF from indeterminate ALF	[113, 114]

aCL, anti-cardiolipin antibody; *AI-ALF*, autoimmune acute liver failure; *AIH*, autoimmune hepatitis; *AMA*, anti-mitochondrial antibody; *ANA*, anti-nuclear antibody; *A-pANCA (ANNA)*, atypical perinuclear neutrophil cytoplasmic antibody; *ASCA*, anti-Saccharomyces cerevisiae; *ASGPR*, asialoglycoprotein receptor; *B2GPI IgA*, beta2 glycoprotein I; *CCA*, cholangiocarcinoma; *CCP*, cyclic citrullinated protein; *CIA*, chemiluminescent immunoassay; *DILI*, drug-induced liver injury; *F-actin*, filamentous actin; *GPA*, gastric parietal antibody; *HBV*, hepatitis B virus; *HCC*, hepatocellular carcinoma; *HCV*, hepatitis C virus; *HIP1R*, Huntingtin-interacting protein 1-related protein; *HK1*, hexokinase 1; *IF*, intrinsic factor; *IIF*, indirect immunofluorescence; *KLHL12*, Kelch-like 12; *LC1*, liver cytosol; *LKM1*, liver kidney microsome 1; *MZGP2*, major zymogen glycoprotein 2; *NAFLD*, nonalcoholic fatty liver disease; *PBC*, primary biliary cholangitis; *PSC*, primary sclerosing cholangitis; *PR3*, serine proteinase 3; *SLA*, soluble liver antigen; *SMA*, smooth muscle antibody; *tTG*, tissue transglutaminase; *UC*, ulcerative colitis.

[a] *Not an autoantibody.*

5 Challenges in differential diagnosis of autoimmune liver disease

As outlined above, autoimmune liver diseases are largely defined into three distinct clinical groupings: PBC, PSC and AIH. In reaching a diagnosis for an individual patient it is essential to carefully evaluate the clinical presentation and follow-up with serum liver tests, including critical appraisal of treatment effects and understanding what constitutes response. Clearly in the absence of true diagnostic markers of disease, there will always be challenges in a proportion of patients to reach a secure diagnosis (Fig. 2). Frequently this challenge is greatest in patients presenting with more subtle features and milder liver disease. This should be borne in mind because therapy need not be started in everyone immediately, not least because of side-effects, but equally for lack of evidence of efficacy in some settings.

In PBC, a large part of the diagnosis is secured by the combination of cholestatic serum liver chemistry with accompanying AMA reactivity. The most common diagnostic pitfalls are: (a) very mild changes in liver biochemistry in patients with low titer AMA reactivity, but risk factors for non-alcoholic fatty liver disease (NAFLD); (b) more marked transaminase activity than expected for the degree of cholestasis as evident by the peak ALP; and (c) patients with cholestasis, but AMA-negative serology. In any clinical setting, nothing can mitigate for a failure to grasp pre-test probabilities and to contextualize information regarding why testing was performed, the local test context of testing/ test performance, as well as local demographics. In addition, other information known about a patient, e.g., clinical evidence of obesity with ultrasound features of fatty liver and clinical features to support metabolic syndrome needs to be considered. Subsequent to this knowledge is the ability to further investigate by selective use of liver biopsy, opinioned by an expert liver pathologist, alongside the appropriate use of extended liver serologic testing: ANA by IIF, specific assays for anti-gp210, anti-sp100, and anti-centromere reactivity, as well as review of the pattern of IgG, IgM, and IgA elevations. With this additional information, variants such as AMA-negative PBC, small duct PSC, overt AIH-overlap can be considered. Additional alternate diagnoses such as NAFLD can also be reached. It is critical to treat a patient as an individual. Errors often repeated include failing to give UDCA sufficient time to work, rushing to the

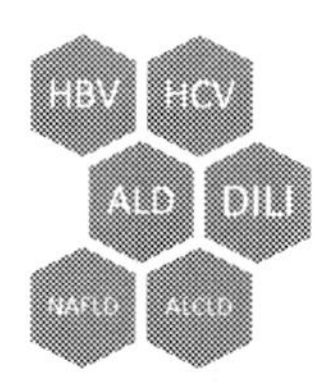

FIG. 2 Autoimmune liver disease and relevant diseases in the differential diagnosis. *AILD*, autoimmune liver disease; *HBV*, hepatitis B virus infection; *HCV*, hepatitis C virus infection; *DILI*, drug-induced liver disease; *NAFLD*, non-alcoholic fatty liver disease; *ALCLD*, alcoholic liver disease.

addition of corticosteroids because of over-diagnosis of AIH-overlap, and treating before clinically indicated. PSC is of course much easier to diagnose, but clinicians are very reliant on the quality of their radiologist reporting the biliary tree magnetic resonance imaging (MRI). Serology has not been proven to facilitate diagnosis at this point, and it is a misunderstanding of the literature to use ANCA diagnostically, given the high rate of positive results in AIH and PBC, as well as high positive findings in PSC. Liver histology is relevant where patients don't present classically (IBD; biliary features on MRI or ultrasound), but if turning to invasive endoscopic cholangiography, clarity is needed what the information will be used for. Much is made of IgG4 disease and how it can mimic PSC. Firstly, while this is true, IgG4 disease is rarer than PSC. Secondly most patients with IgG4 disease present in other ways. This leaves only a few patients where PSC and IgG4-cholangitis can be hard to distinguish; efforts to do so can rest on empiric treatment with corticosteroids, or alterations of IgG1:IgG4 ratios.

In contrast, AIH is a diagnosis of exclusion, so when considering the diagnosis, the importance of clinical history, drug history, viral exposure and testing, as well as exclusion of rare metabolic disease is essential. In this context, liver biopsy remains a key test alongside immunologic testing. For most adult patients ANA and SMA testing provides sufficient information, given the usual elevation in IgG alongside a compatible liver biopsy. Pediatric patients may present with AIH-2 more commonly than adults, in which case anti-LKM-1 assays confirm disease, or if needed anti-LC1 testing. There is some controversy about the use of anti-F-actin testing, primarily concerning low titer F-actin reactivity in healthy patients or those with low titer SMAs. While some advocate IIF pattern of SMA, the expertise to perform and interpret the assay is not readily available; it is not helpful to clinicians to recommend investigations that are not validated and accessed by a wide body of treating clinicians. Anti-SLA testing is quite useful, given its tight association with AIH in the 20% of patients with AIH and seroreactivity, as well as some evidence that it is a marker of more severe disease [72, 73]. Certainly, in ANA-negative/SMA-negative atypical patients where AIH remains a possibility, extended liver autoantibody profiling can help clarify a fair proportion of cases.

Treatments for ALD are slow to evolve in the context of PM. Clinicians focus on the use of UDCA, corticosteroids, azathioprine and mycophenolate mofetil for the large part. Other immunosuppressants can include tacrolimus and rituximab. Biliary disease, most notably PBC, is positively impacted by UDCA. PSC is not, but clinician familiarity means many clinicians still use UDCA for PSC patients despite no benefit. Improved cholestatic therapy in PBC relates to the licensed use of obeticholic acid (first in-class semi-synthetic FXR agonist) and the off-label use of bezafibrate (evidence based in small randomized datasets) [115]. Enthusiasm for new treatments across all ALD remains, but is hampered by slow natural history, the rare nature of disease and challenges in defining meaningful trial endpoints. Most efforts are now turning toward PSC with a variety of approaches, biliary and immune-modulating therapies, as well as microbiome manipulation. AIH is a disease worthy of precision immunomodulation, but development in this area has been slow because primary therapy is quite effective, and clinical trial recruitment of homogenous at-risk patients is very hard.

6 Development of precision medicine in autoimmune liver disease

As the preceding sections have established, ALD is complex, with considerable interrelationships and overlap among the various subforms (AIH and PBC, PSC and PBC), as well as with other autoimmune conditions. Furthermore, ALD symptoms can vary from mild to severe and progression can be rapid or extend over many decades. This spectrum of features adds to the challenges of ALD management.

One potential first step to develop a PM model is to generate detailed profiles of known ALD patients, as well as those suspected of ALD due to symptoms, risk-factors, or incidental findings. While patients presenting with "classic" symptoms and clinical features may have a relatively clear management path, patients with early or mild disease may be the most difficult to diagnose and to predict their prognosis. The clinician may be faced with uncertainty as to the actual diagnosis and how they should be managed. Should they be treated or should the patient be followed and treatment delayed? Without additional input, the clinician may make a "working" diagnosis of the patient and may begin an accepted, but "non-precision" medicine guided, empirical treatment. PM seeks to improve this process by first assembling a detailed profile of individual patients using demographics, family history, and clinical symptoms.

The second step toward an ALD PM model is to collect results of clinical observations, records of treatment and outcome, and data on variety of biomarkers including conventional biochemistry markers, inflammatory biomarkers, autoantibodies, cytokines, genetic susceptibility, HLA haplotype, and co-morbidities for each patient via electronic health records (EHR). A huge number of potential variables can be considered for inclusion in order to generate a comprehensive profile of the individual (Table 3). The more data that can be included in the profile, the more detailed stratification of patients might be possible. Longitudinal data is particularly valuable to correlate changes in biomarkers with the disease natural history and outcome. Viewed together, this data matrix provides a unique signature for each individual patient. While there may be a concern that "too much" data could decrease specificity by identifying clinically non-significant variations in the population or generating statistical artifacts, as the datasets increase in size they will allow better understanding of the baseline values of the population.

TABLE 3 Potential parameters to be considered in development of Precision Medicine models for autoimmune liver disease.

Parameters to consider for inclusion in Precision Medicine Model development
Demographics
Age
Gender
Age at diagnosis
Duration of disease
Co-morbidities
Autoimmune disease
Other
Disease stage
Disease severity-mild, moderate, severe
Disease progression-slow, moderate, rapid
Fibrosis stage
Cirrhosis
Decompensation
Histological stage
Treatment-current
Treatment-response?
Varices
Pruritis
Viral hepatitis
Variceal bleeding
Laboratory biomarkers
Alanine transaminase
Aspartate transaminase
Albumin
Bilirubin
Platelets
Autoantibodies (see Table 2)
GGT

TABLE 3 Potential parameters to be considered in development of Precision Medicine models for autoimmune liver disease—cont'd

Parameters to consider for inclusion in Precision Medicine Model development
Markers of fibrosis
HLA and genetic markers associated with the disease
Current and potentially useful biomarkers
All autoantibodies—liver, gastrointestinal, connective tissue, phospholipid
Bile biomarkers—FGF19
Glycosylated, methylated markers
Microbiome
microRNA
Intestinal permeability markers
Metabolomics
IgG4
Cancer markers
Lipidomics

FGF19, fibroblast growth factor 19; *GGT*, gamma glutamyl transferase; *HLA*, human leukocyte antigen.

PM then aggregates all the unique individual profiles and clinical histories into a combined database which can be analyzed using AI and ML techniques to confirm expected correlations and importantly, to identify unexpected and non-obvious patterns that can improve the clinician's ability to make more accurate and precision decisions for optimal diagnosis, management, and treatment of patients. The PM model likely will start as a "frozen" model, but is expected to be dynamic (self-learning). Continual updating will strengthen and increase its precision as more patients and their data are integrated into the model for analysis.

PM ideally will address many questions that confront clinicians. Clarifying diagnosis and management, especially in the case of mild disease or overlapping disease. Timing of treatment initiation? What is the risk they will progress rapidly, slowly, or not at all? What is the best treatment for a patient with a particular profile? Can PM identify patients who are unlikely to respond to particular treatment and therefore save wasted time, effort, and financial resources? How does the age at diagnosis, the duration of disease, and the severity of disease impact outcome? What are the key clinical and laboratory factors and specific combinations of parameters that influence progression? The questions go on and on.

The formulation of "guidelines," to help other clinicians to diagnose and treat patients, is essentially an early form of PM. Guidelines for AIH, PBC, PSC, and overlap conditions have been developed by multiple societies and have become more formalized and transparent in the process leading to their recommendations. The International Autoimmune Hepatitis Group (IAIHG) proposed a scoring system for AIH in 1993, revised in 1999, consisting of approximately 13 variables subdivided into a total of 30 items each with an accompanying score. Features included demographic, histological, and serological parameters. This scoring system was later simplified for easier routine use and updated to included antibodies to soluble liver antigen as an additional criteria parameter, while the more detailed system was retained for more difficult patients [31]. Several detailed practice guidelines on the diagnosis of AIH have recently been developed [116, 117]. Drug-induced liver injury (DILI) may resemble AIH and its differential diagnosis can be challenging [117–120]. Up to 30–40% of patients with suspected DILI may have detectable autoantibodies such as ANA and or SMA adding to diagnostic uncertainty [121]. This appears to be an area where PM can help provide clarity for diagnosis and management.

For PBC, current diagnostic guidelines list three criteria, namely "biochemical evidence of cholestasis based on elevated ALP, presence of AMA or other PBC-specific autoantibodies including anti-sp100 or anti-gp210 if AMA-negative, histologic evidence of non-suppurative destructive cholangitis, and destruction of interlobular bile ducts." The American Association for the Study of Liver disease (AASLD) recently suggested new PBC-specific biomarkers such as anti-HK-1 and anti-KLHL12 may assist diagnosis, especially in cases of suspected AMA-negative PBC [38]. One fundamental difference between traditional diagnostic guidelines and PM concepts is the approach used to develop and evolve recommendations. In diagnostic guidelines, experience and evidence-based concepts are utilized, often supported by data collection and analyses tools (in supervised fashion). In contrast, PM leverages AI and ML to select features and is data-driven. Eventually, it is very likely that this concept will be used to form diagnostic guidelines, potentially starting with a hybrid of the two approaches.

While basic guidelines are necessary for the practical recognition and treatment of disease, increasing emphasis is now placed on how and when to treat patients. In the case of ALD, where initial disease can be mild and progression can be highly variable, guidance on who will most likely progress is critical to choosing optimal management of the patient. Earlier models of risk progression, for example the Barcelona, Paris I and II, Rotterdam, and Toronto used dichotomous prognostic criteria where patients were stratified into responders or non-responders at high or low risk of progression to end-stage liver disease (ESLD) as described by Carbone et al. [3]. Since disease progression is rarely dichotomous, the development of continuous prognostic scores are preferable [3]. Using information on 1916 patients, an online calculator (UK-PBC risk calculator) was developed to calculate the risk of patients progressing to

ESLD over specified time intervals by assessing values for alanine transaminase (ALT), aspartate transaminase (AST), ALP, bilirubin, albumin, and platelets after 12 months of UDCA therapy [122]. The Globe PBC Score uses the same basic parameters as the UK-PBC score, with the addition of age, to stratify patients and predict liver transplantation (LT) or liver-related death at 3, 5, 10, and 15 years [123]. Since these scoring systems were developed using Western populations, Yang et al. assessed a Chinese PBC population and validated that the systems also worked in this population [124].

The UK-PBC and Globe scores currently do not include autoantibodies. The Yang group showed that the presence of anti-gp210 antibodies, which have been associated with more severe disease and outcome, especially in Asian studies, improved the prognostic value of the UK-PBC and Globe scores and suggested they could be added to the scores [124]. Recent studies have suggested that anti-HK-1 antibodies were associated with shorter time to LT or liver-related death and might also be added to scoring systems [90]. The addition of new biomarkers into the scoring systems will accelerate as their significance is validated in new studies. However, the more features that are built into scoring systems or PM models, the more patients are needed to validate the system.

PSC, a devastating disease with a variable, but mostly progressive and poor outcome, is a prime candidate for PM models. There is a strong relationship between PSC and IBD, especially ulcerative colitis (UC). Between 60% and 80% of patients with PSC have IBD, while 2.5–7.5% of patients with IBD will develop PSC (which is significantly higher vs. general population). The diagnosis of PSC includes assessment of cholestatic liver biochemistry, normal ultrasound, and liver screening including autoimmune serological biomarkers, and magnetic resonance cholangiopancreatography (MRCP). Once diagnosed however, patient median survival is only 13.2–21.3 years without LT [125]. While there are no specific biomarkers for PSC, it is possible that combined analysis of a profile of factors including family history, HLA haplotypes, genotypic variations, and biomarkers together with clinical features could identify individuals at a higher risk of PSC for increased targeted surveillance. Better understanding the reason and impact of co-existing IBD in PSC patients might shed more light on pathogenesis and provide strategies for improved management. IBD patients with PSC have a sevenfold increased mortality and a fivefold increased risk of cancer than those with IBD alone [126]. There is growing recognition that PSC patients have both bacterial and fungal dybiosis and that the microbiome of patients with PSC, with or without IBD, is different than that of patients with IBD alone or healthy individuals [127–129]. Assessment of individual's microbiome may therefore deliver important biomarkers for detecting and monitoring for changes that may indicate disease progression from health to disease. In addition to general trends such as reduced diversity of particular species, it is likely that at least some alterations may be individual-specific. A PM targeted therapy to restore a more normal microbiome might offer one treatment (prevention) option for PSC [130].

Several risk models for patients with PSC have been developed [131]. The UK-PSC risk score is one of the most recently developed and is focused on prediction of LT or death over both the short- and long-term based on age, bilirubin, ALP, albumin, platelets, presence of extrahepatic liver disease, and variceal hemorrhage [132]. Using some of the same variables (age, bilirubin, alkaline phosphatase, albumin, platelets) and additional commonly measured parameters, including AST, hemoglobin, sodium, and time since diagnosis of PSC, an improved PSC Risk Estimate Tool (PREsTo) was recently derived using ML methods [133]. In addition to demonstrating the value of ML using gradient boosting modeling, a key feature of this new prognostic tool is that the clinical end-point was hepatic decompensation as opposed to LT. These risk assessment tools can help guide clinicians in their management of patients with PSC and will be undoubtably refined and augmented with additional parameters over time using AI methodologies. Table 4 summarizes recent practice guidelines and risk assessment models that have been discussed.

Patients with PSC have a 400-fold increased risk of developing CCA, an epithelial cancer with extremely poor prognosis and which represents the second

TABLE 4 Recent practice guidance and risk assessment models.

			References
Recent practice guidance documents			
PBC	2017	EASL practice guidelines for diagnosis and management of PBC	[39]
	2018	AASLD PBC practice guidance	[38]
AIH	2015	EASL practice guidelines AIH	[118]
	2018	ESPGHAN diagnosis and management of pediatric ALD	[116]
	2019	AASLD practice guidance and guidelines for AIH in adults and children	[117]
PSC	2010	AASLD diagnosis and management of PSC	[134]
	2019	British Society of Gastroenterology and UK-PBC guidelines for diagnosis and management of PSC	[131]
Risk score models			
PBC	2015	Global PBC score	[123]
	2016	UK-PBC score	[135]
PSC	2020	PSC risk estimate tool (PREsTo)	[133]

Abbreviations: *EASL*, European Association for the Study of the Liver; *AASLD*, American Society for the Study of Liver Disease; *ESPGHAN*, European Society for Pediatric Gastroenterology, Hepatology, and Nutrition.

most common primary liver cancer. Incidence appears to be increasing. While only a small number of PSC patients will develop CCA, there is no effective approach to determine which patients will develop CCA, necessitating the ongoing stringent surveillance of all patients. PM models to better identify PSC patients at higher risk for CCA are needed. In this context, anti-glycoprotein 2 (GP2) IgA antibodies have been associated with poor outcome and CCA [95,94]. In addition, a recent study reported that of 13 PSC patients with CCA, 12/13 were positive by a high sensitivity chemiluminescent assay for anti-PR3 IgG, 1/13 for anti-GP2 IgA, and 13/13 for at least one of the two biomarkers [136,137,94]. If these results can be replicated in additional studies, it would add to biomarkers with potential value for managing patients with PSC.

Fibrosis and cirrhosis are critical indicators of disease progression in ALD, as well as liver disease in general, and therefore inclusion of fibrosis markers are important parameters for inclusion in PM models. Multiple scoring systems to non-invasively assess fibrosis have been developed and are continually being improved, but they tend to be best at assessing fibrosis at the low or high ends of the fibrosis spectrum and are most predictive for those with risk factors rather than the general population [138]. Selection of biomarkers in these scoring systems may not necessarily be based on the optimum combination of biomarkers, but more on the practical realities of availability, complexity, instrumentation, laboratory constraints, intellectual property, and reimbursement.

The development of accurate PM models for assessing and monitoring development of fibrosis and cirrhosis extends far beyond patients with ALD. The rapidly growing problem of NAFLD, which is thought to affect 30% of the world's population, is a critical area in need of better diagnostics and predictive modeling. Like ALD, NAFLD can develop slowly and be clinically unrecognized for years and therefore is an important target for PM. Another target of PM is the need to recognize and manage individuals developing the more serious and progressive form of NAFLD, non-alcoholic steatohepatitis (NASH).

Microbial dysbios, regardless of whether it is a cause or effect, indicates possible disease progression. Both bacterial and fungal dysbiosis have been shown in various gastrointestinal and liver diseases, including IBD, PBC, AIH, PSC, alcoholic liver disease, NAFLD, fibrosis, and cirrhosis [129, 139, 140]. While measurement of the microbiome is complex and not routinely available, the continued accumulation of evidence showing changes in microbial signatures with disease status suggest this will be an important additional parameter in evolving PM models.

New studies identifying specific measurable factors that are associated with progression and outcomes of early disease, as was recently described for early PBC, will assist in the refinement and effectiveness of PM models [141]. A particularly exciting area that offers significant potential for improved diagnostics, prognosis, and stratification of patients is protein glycosylation. Glycosylated proteins may serve as specific biomarkers of disease [142]. Changes in cell surface receptors by glycosylation could affect disease progression. Changes in glycosylation patterns have been shown as the healthy liver progresses to fibrosis, cirrhosis, and hepatocellular carcinoma (HCC) as well as from the health liver to steatosis and NASH [143].

Current scoring systems for risk prediction usually focus on easily obtained laboratory parameters to allow for practical implementation. New areas such as protein glycosylation offer significant potential, but will need simplification of the often complex instrumentation and analysis to be realistically adopted by clinical laboratories. Technology for accurately and simultaneously measuring multiple analytes, antigens, antibodies, cytokines, metabolites, microbiome, and other biomarkers continues to evolve, improve, and move from the research to the routine clinical laboratory (multi-omics, see Chapter 8). Together with increased availability of high-power computing, this will allow comprehensive profiles incorporating multiple biomarkers and clinical parameters to assess an individual's current health as well as predicting low or high risk of future clinical progression. Accurate risk prediction can then be used to guide the aggressiveness of treatment for that specific patient.

There is continuous outpouring of studies highlighting the presence or association of particular biomarkers with a specific disease. The great majority never reach the clinical laboratory. A fundamental question is determining which of these biomarkers actually add value to diagnostic or prognostic models and which are redundant. For example, in PBC the predominant biomarker is AMA. In addition to being present in some AMA-negative PBC patients, additional markers such as anti-gp210 and anti-HK1 antibodies may be associated with poorer prognosis, while anti-centromere may suggest portal hypertension type progression and overlapping systemic sclerosis [87]. We know however, that multiple other biomarkers are present in patients with autoimmune disease. Determining whether their presence at levels greater than that seen in a healthy population is clinically significant to the clinical outcome of patients with PBC or is a result of a non-specific increase in autoimmune activity is often unclear. However, since it is estimated that 14–44% of patients with AIH have concurrent autoimmune disease [117], detection of non-AIH specific biomarkers such as anti-tissue transglutaminase for celiac disease or markers of autoimmune thyroid disease are clinically important for management. A large variety of autoantibodies found in patients with other autoimmune diseases is included in Table 3 and may or may not have impact the future disease course of the patient. By collecting and analyzing detailed data with AI and ML/DL on a broad cohort of patients with detailed clinical history, PM will assist clinicians in their management decisions. Intuitively if one subset of patients develops a specific antibody, while another does not, it suggests there is some sort of difference between the groups. But, is it clinically significant? Does the appearance or disappearance of a specific antibody or pattern of antibodies indicate a change in the disease course? In addition, increased understanding of the significance of the presence of multiple antibodies in a patient is needed. Do they reflect a different subtype with a different risk profile? Does the presence of multiple antibodies reflect a stage in the disease progression or an increased risk of a clinical event. Along those lines, for antiphospholipid syndrome, it has been convincingly demonstrated that triple biomarker positivity (lupus anticoagulant, anti-cardiolipin, and anti-β2GPI antibodies) is associated with a higher risk of thrombosis [144].

There may be multiple pathways leading to disease initiation and progression in patients. The detection of one biomarker may be significant for one patient, while for another patient following a slightly different clinical pathway, it may be a different biomarker. In the past, patients were bundled together by general symptomatic and clinical features. An illustrative example is the case of patients with myositis. Rather than being one homogenous group, the multiple autoantibodies seen in myositis are now thought to correlate with specific subtypes with different clinical features [145]. As the scientific community discovers more biomarkers and accumulates more information on clinical features, progression, and therapeutic responses by the analysis of large cohorts of well-characterized patients, it is likely that PM will stratify many ALD diseases into more precise subgroups with more homogeneous clinical characteristics and perhaps increasing effective targets of therapeutic approaches.

An important goal of PM for those with chronic liver conditions is encouraging the paradigm shift from intent to treat, to intent to prevent disease. This concept relies on an accurate predictor of disease onset as well as treatment options that are cost-effective and exhibit favorable safety profiles [146–148]. For ALD we still don't know the why some people develop disease and why some don't. While it is well known that autoantibodies, including those in ALD can precede the onset of disease for many years [149], it is unclear if the disease can be stopped or only slowed down once these antibodies have appeared. It is reasonable to expect however, that early recognition and monitoring might better enable future therapeutic intervention (see also example of rheumatoid arthritis in Chapter 7). Guiding use of PM guided therapeutics, predicting who will respond or not respond to specific drugs will be a key goal of PM, and with continual recruitment of patients and expansion of the database with outcome results, the performance of PM will continue to improve.

7 Conclusions

PM offers an exciting pathway for dealing with the complexities of ALD. New biomarkers, new technology, and new advancements in AI and ML systems will continue to improve the care of patients with ALD. These advances will allow the increasingly more accurate and finer stratification of patients into smaller and smaller subgroups, ultimately to a subgroup of one, thus allowing optimal precision medical diagnosis, treatment, and management.

Competing interests

G.L. Norman and M. Mahler are employees of Inova Diagnostics.

References

[1] T. Hardy, D.A. Mann, Epigenetics in liver disease: from biology to therapeutics, Gut 65 (11) (2016) 1895–1905.

[2] E.M. Haldorsen, The right treatment to the right patient at the right time, Occup. Environ. Med. 60 (4) (2003) 235–236.

[3] M. Carbone, et al., Toward precision medicine in primary biliary cholangitis, Dig. Liver Dis. 48 (8) (2016) 843–850.

[4] T. Addison, W. Gull, On a certain affection of the skin, vitiligoidea plana and vitiliogidea tuberose, with remarks, Guy's Hosp. Rep. 7 (1851) 256–276.

[5] J.A. Dauphinee, J.C. Sinclair, Primary biliary cirrhosis, Can. Med. Assoc. J. 61 (1) (1949) 1–6.

[6] E.H. Ahrens Jr., et al., Primary biliary cirrhosis, Medicine (Baltimore) 29 (4) (1950) 299–364.

[7] J.W. Liber, Blutporteine und Nahrungseiweiss, Dtsch. Z. Verdau. Stoffwechselkr. 2 (1950) 113–119.

[8] H.G. Kunkel, Extreme hypergammaglobinemia in young women with liver disease of unknown etiology, J. Clin. Invest. 30 (1951) 654.

[9] D.C. Cowling, I.R. Mackay, L.I. Taft, Lupoid hepatitis, Lancet 271 (6957) (1956) 1323–1326.

[10] I.R. Mackay, Primary biliary cirrhosis showing a high titer of autoantibody; report of a case, N. Engl. J. Med. 258 (4) (1958) 185–188.

[11] J.G. Walker, et al., Serological tests in diagnosis of primary biliary cirrhosis, Lancet 1 (7390) (1965) 827–831.

[12] M.E. Gershwin, et al., Identification and specificity of a cDNA encoding the 70 kd mitochondrial antigen recognized in primary biliary cirrhosis, J. Immunol. 138 (10) (1987) 3525–3531.

[13] C. Szostecki, et al., Autoimmune sera recognize a 100 kD nuclear protein antigen (sp-100), Clin. Exp. Immunol. 68 (1) (1987) 108–116.

[14] J. Wesierska-Gadek, et al., Antibodies to nuclear lamin proteins in liver disease, Immunol. Investig. 18 (1–4) (1989) 365–372.

[15] U.F. Greber, A. Senior, L. Gerace, A major glycoprotein of the nuclear pore complex is a membrane-spanning polypeptide with a large lumenal domain and a small cytoplasmic tail, EMBO J. 9 (5) (1990) 1495–1502.

[16] R.E. Nickowitz, H.J. Worman, Autoantibodies from patients with primary biliary cirrhosis recognize a restricted region within the cytoplasmic tail of nuclear pore membrane glycoprotein Gp210, J. Exp. Med. 178 (6) (1993) 2237–2242.

[17] J.C. Courvalin, et al., The 210-kD nuclear envelope polypeptide recognized by human autoantibodies in primary biliary cirrhosis is the major glycoprotein of the nuclear pore, J. Clin. Invest. 86 (1) (1990) 279–285.

[18] G.L. Norman, et al., Anti-kelch-like 12 and anti-hexokinase 1: novel autoantibodies in primary biliary cirrhosis, Liver Int. 35 (2) (2015) 642–651.

[19] G.D. Johnson, E.J. Holborow, L.E. Glynn, Antibody to smooth muscle in patients with liver disease, Lancet 2 (7418) (1965) 878–879.

[20] L.J. Farrow, E.J. Holborow, W.D. Brighton, Reaction of human smooth muscle antibody with liver cells, Nat. New Biol. 232 (2) (1971) 186–187.

[21] G. Gabbiani, et al., Human smooth muscle autoantibody. Its identification as antiactin antibody and a study of its binding to "nonmuscular" cells, Am. J. Pathol. 72 (3) (1973) 473–488.

[22] P. Kurki, et al., Smooth muscle antibodies of actin and "non-actin" specificity, Clin. Immunol. Immunopathol. 9 (4) (1978) 443–453.
[23] M. Rizzetto, G. Swana, D. Doniach, Microsomal antibodies in active chronic hepatitis and other disorders, Clin. Exp. Immunol. 15 (3) (1973) 331–344.
[24] J.C. Homberg, C. Andre, N. Abuaf, A new anti-liver-kidney microsome antibody (anti-LKM2) in tienilic acid-induced hepatitis, Clin. Exp. Immunol. 55 (3) (1984) 561–570.
[25] J.C. Homberg, et al., Chronic active hepatitis associated with antiliver/kidney microsome antibody type 1: a second type of "autoimmune" hepatitis, Hepatology 7 (6) (1987) 1333–1339.
[26] U.M. Zanger, et al., Antibodies against human cytochrome P-450db1 in autoimmune hepatitis type II, Proc. Natl. Acad. Sci. USA 85 (21) (1988) 8256–8260.
[27] M. Manns, et al., Characterisation of a new subgroup of autoimmune chronic active hepatitis by autoantibodies against a soluble liver antigen, Lancet 1 (8528) (1987) 292–294.
[28] B.M. McFarlane, et al., Serum autoantibodies reacting with the hepatic asialoglycoprotein receptor protein (hepatic lectin) in acute and chronic liver disorders, J. Hepatol. 3 (2) (1986) 196–205.
[29] E. Martini, et al., Antibody to liver cytosol (anti-LC1) in patients with autoimmune chronic active hepatitis type 2, Hepatology 8 (6) (1988) 1662–1666.
[30] P.J. Johnson, I.G. McFarlane, Meeting report: international autoimmune hepatitis group, Hepatology 18 (4) (1993) 998–1005.
[31] E.M. Hennes, et al., Simplified criteria for the diagnosis of autoimmune hepatitis, Hepatology 48 (1) (2008) 169–176.
[32] P. Canning, C.D. Cooper, T. Krojer, J.W. Murray, A. Chaikuad, T. Keates, et al., Structural basis for Cul3 protein assembly with the BTB-Kelch family of E3 ubiquitin ligases, J. Biol. Chem. 288 (2013) 7803–7814.
[33] J. Wang, P. Youkharibache, D. Zhang, C.J. Lanczycki, R.C. Geer, J. Madej, L. Phan, et al., iCn3D, a web-based 3D viewer for sharing 1D/2D/3D representations of biomolecular structures, Bioinformatics 36 (2020) 131–135.
[34] E.W. Sayers, J. Beck, E.E. Bolton, D. Bourexis, J.R. Brister, K. Canese, D.C. Comeau, et al., Database resources of the National Center for Biotechnology Information, Nucleic Acids Res. (2020).
[35] M. Carbone, J.M. Neuberger, Autoimmune liver disease, autoimmunity and liver transplantation, J. Hepatol. 60 (1) (2014) 210–223.
[36] M.P. Manns, et al., Diagnosis and management of autoimmune hepatitis, Hepatology 51 (6) (2010) 2193–2213.
[37] M. Sebode, et al., Autoantibodies in autoimmune liver disease-clinical and diagnostic relevance, Front. Immunol. 9 (2018) 609.
[38] K.D. Lindor, et al., Primary biliary cholangitis: 2018 practice guidance from the American Association for the Study of Liver Diseases, Hepatology 69 (1) (2019) 394–419.
[39] European Association for the Study of the Liver, Electronic address, easloffice@easloffice.eu and European Association for the Study of the Liver, EASL clinical practice guidelines: the diagnosis and management of patients with primary biliary cholangitis, J. Hepatol. 67 (1) (2017) 145–172.
[40] J.J. Tischendorf, et al., Characterization, outcome, and prognosis in 273 patients with primary sclerosing cholangitis: a single center study, Am. J. Gastroenterol. 102 (1) (2007) 107–114.
[41] M. Sebode, et al., Autoimmune hepatitis: from current knowledge and clinical practice to future research agenda, Liver Int. 38 (1) (2018) 15–22.
[42] N.K. Gatselis, et al., Autoimmune hepatitis, one disease with many faces: etiopathogenetic, clinico-laboratory and histological characteristics, World J. Gastroenterol. 21 (1) (2015) 60–83.

[43] P.J. Trivedi, et al., Risk stratification in autoimmune cholestatic liver diseases: opportunities for clinicians and trialists, Hepatology 63 (2) (2016) 644–659.
[44] A.C. Cheung, et al., Effects of age and sex of response to ursodeoxycholic acid and transplant-free survival in patients with primary biliary cholangitis, Clin. Gastroenterol. Hepatol. 17 (10) (2019) 2076–2084.e2.
[45] G.V. Gregorio, et al., Autoimmune hepatitis/sclerosing cholangitis overlap syndrome in childhood: a 16-year prospective study, Hepatology 33 (3) (2001) 544–553.
[46] B. Terziroli Beretta-Piccoli, D. Vergani, G. Mieli-Vergani, Autoimmune sclerosing cholangitis: evidence and open questions, J. Autoimmun. 95 (2018) 15–25.
[47] G. Mieli-Vergani, et al., Autoimmune hepatitis, Nat. Rev. Dis. Primers 4 (2018) 18017.
[48] M.P. Manns, A.W. Lohse, D. Vergani, Autoimmune hepatitis—update 2015, J. Hepatol. 62 (1 Suppl) (2015) S100–S111.
[49] Y.S. de Boer, et al., Genome-wide association study identifies variants associated with autoimmune hepatitis type 1, Gastroenterology 147 (2) (2014) 443–452.e5.
[50] B. Terziroli Beretta-Piccoli, et al., The challenges of primary biliary cholangitis: what is new and what needs to be done, J. Autoimmun. 105 (2019) 102328.
[51] European Association for the Study of the Liver, EASL clinical practice guidelines: management of cholestatic liver diseases, J. Hepatol. 51 (2) (2009) 237–267.
[52] T.H. Karlsen, et al., Primary sclerosing cholangitis—a comprehensive review, J. Hepatol. 67 (6) (2017) 1298–1323.
[53] L.M. Stinton, et al., PR3-ANCA: a promising biomarker in primary sclerosing cholangitis (PSC), PLoS One 9 (11) (2014) e112877.
[54] K.D. Lindor, et al., High-dose ursodeoxycholic acid for the treatment of primary sclerosing cholangitis, Hepatology 50 (3) (2009) 808–814.
[55] J.M. Vierling, Autoimmune hepatitis and overlap syndromes: diagnosis and management, Clin. Gastroenterol. Hepatol. 13 (12) (2015) 2088–2108.
[56] A. Di Giorgio, et al., Seamless management of juvenile autoimmune liver disease: long-term medical and social outcome, J. Pediatr. 218 (2020) 121–129.e3.
[57] R. Zenouzi, A.W. Lohse, Long-term outcome in PSC/AIH "overlap syndrome": does immunosuppression also treat the PSC component? J. Hepatol. 61 (5) (2014) 1189–1191.
[58] F. Alvarez, et al., International autoimmune hepatitis group report: review of criteria for diagnosis of autoimmune hepatitis, J. Hepatol. 31 (5) (1999) 929–938.
[59] F. Meda, et al., Serum autoantibodies: a road map for the clinical hepatologist, Autoimmunity 41 (1) (2008) 27–34.
[60] D.P. Bogdanos, et al., Autoimmune liver serology: current diagnostic and clinical challenges, World J. Gastroenterol. 14 (21) (2008) 3374–3387.
[61] P. Obermayer-Straub, C.P. Strassburg, M.P. Manns, Autoimmune hepatitis, J. Hepatol. 32 (1 Suppl) (2000) 181–197.
[62] D.P. Bogdanos, G. Mieli-Vergani, D. Vergani, Autoantibodies and their antigens in autoimmune hepatitis, Semin. Liver Dis. 29 (3) (2009) 241–253.
[63] D. Vergani, et al., Liver autoimmune serology: a consensus statement from the committee for autoimmune serology of the International Autoimmune Hepatitis Group, J. Hepatol. 41 (4) (2004) 677–683.
[64] D. Villalta, et al., Diagnostic accuracy of four different immunological methods for the detection of anti-F-actin autoantibodies in type 1 autoimmune hepatitis and other liver-related disorders, Autoimmunity 41 (1) (2008) 105–110.
[65] D. Villalta, et al., Autoantibody profiling in a cohort of pediatric and adult patients with autoimmune hepatitis, J. Clin. Lab. Anal. 30 (1) (2016) 41–46.

[66] C. Frenzel, et al., Evaluation of F-actin ELISA for the diagnosis of autoimmune hepatitis, Am. J. Gastroenterol. 101 (12) (2006) 2731–2736.
[67] M. Gueguen, et al., Anti-liver kidney microsome antibody recognizes a cytochrome P450 from the IID subfamily, J. Exp. Med. 168 (2) (1988) 801–806.
[68] P. Lapierre, et al., Formiminotransferase cyclodeaminase is an organ-specific autoantigen recognized by sera of patients with autoimmune hepatitis, Gastroenterology 116 (3) (1999) 643–649.
[69] D. Villalta, et al., Evaluation of a novel extended automated particle-based multi-analyte assay for the detection of autoantibodies in the diagnosis of primary biliary cholangitis, Clin. Chem. Lab. Med. 58 (9) (2020) 1499–1507.
[70] M. Lenzi, et al., Liver cytosolic 1 antigen-antibody system in type 2 autoimmune hepatitis and hepatitis C virus infection, Gut 36 (5) (1995) 749–754.
[71] Y. Ma, et al., Antibodies to conformational epitopes of soluble liver antigen define a severe form of autoimmune liver disease, Hepatology 35 (3) (2002) 658–664.
[72] A.J. Czaja, Z. Shums, G.L. Norman, Nonstandard antibodies as prognostic markers in autoimmune hepatitis, Autoimmunity 37 (3) (2004) 195–201.
[73] A.J. Czaja, Z. Shums, G.L. Norman, Frequency and significance of antibodies to soluble liver antigen/liver pancreas in variant autoimmune hepatitis, Autoimmunity 35 (8) (2002) 475–483.
[74] A.J. Czaja, et al., Frequency and significance of antibodies to chromatin in autoimmune hepatitis, Dig. Dis. Sci. 48 (8) (2003) 1658–1664.
[75] A.J. Czaja, G.L. Norman, Autoantibodies in the diagnosis and management of liver disease, J. Clin. Gastroenterol. 37 (4) (2003) 315–329.
[76] P. Milkiewicz, et al., Value of autoantibody analysis in the differential diagnosis of chronic cholestatic liver disease, Clin. Gastroenterol. Hepatol. 7 (12) (2009) 1355–1360.
[77] E.I. Rigopoulou, et al., Asialoglycoprotein receptor (ASGPR) as target autoantigen in liver autoimmunity: lost and found, Autoimmun. Rev. 12 (2) (2012) 260–269.
[78] A.J. Montano-Loza, et al., Prognostic implications of antibodies to Ro/SSA and soluble liver antigen in type 1 autoimmune hepatitis, Liver Int. 32 (1) (2012) 85–92.
[79] A.J. Czaja, et al., Frequency and significance of antibodies to Saccharomyces cerevisiae in autoimmune hepatitis, Dig. Dis. Sci. 49 (4) (2004) 611–618.
[80] A. Montano-Loza, et al., Frequency and significance of antibodies to cyclic citrullinated peptide in type 1 autoimmune hepatitis, Autoimmunity 39 (4) (2006) 341–348.
[81] H. Miyakawa, et al., Detection of antimitochondrial autoantibodies in immunofluorescent AMA-negative patients with primary biliary cirrhosis using recombinant autoantigens, Hepatology 34 (2) (2001) 243–248.
[82] S. Gabeta, et al., Diagnostic relevance and clinical significance of the new enhanced performance M2 (MIT3) ELISA for the detection of IgA and IgG antimitochondrial antibodies in primary biliary cirrhosis, J. Clin. Immunol. 27 (4) (2007) 378–387.
[83] S. Moteki, et al., Use of a designer triple expression hybrid clone for three different lipoyl domain for the detection of antimitochondrial autoantibodies, Hepatology 24 (1) (1996) 97–103.
[84] N. Bizzaro, et al., Overcoming a "probable" diagnosis in antimitochondrial antibody negative primary biliary cirrhosis: study of 100 sera and review of the literature, Clin. Rev. Allergy Immunol. 42 (3) (2012) 288–297.
[85] G. Dahlqvist, et al., Large-scale characterization study of patients with antimitochondrial antibodies but nonestablished primary biliary cholangitis, Hepatology 65 (1) (2017) 152–163.
[86] J. Damoiseaux, et al., Clinical relevance of HEp-2 indirect immunofluorescent patterns: the International Consensus on ANA patterns (ICAP) perspective, Ann. Rheum. Dis. 78 (7) (2019) 879–889.

[87] M. Nakamura, et al., Anti-gp210 and anti-centromere antibodies are different risk factors for the progression of primary biliary cirrhosis, Hepatology 45 (1) (2007) 118–127.
[88] C. Huang, et al., Early prognostic utility of Gp210 antibody-positive rate in primary biliary cholangitis: a meta-analysis, Dis. Markers 2019 (2019) 9121207.
[89] G.L. Norman, et al., The prevalence of anti-hexokinase-1 and anti-kelch-like 12 peptide antibodies in patients with primary biliary cholangitis is similar in Europe and North America: a large international, multi-center study, Front. Immunol. 10 (2019) 662.
[90] A. Reig, et al., Novel anti-hexokinase 1 antibodies are associated with poor prognosis in patients with primary biliary cholangitis, Am. J. Gastroenterol. 115 (10) (2020) 1634–1641.
[91] J.R. Hov, K.M. Boberg, T.H. Karlsen, Autoantibodies in primary sclerosing cholangitis, World J. Gastroenterol. 14 (24) (2008) 3781–3791.
[92] T. Tornai, et al., Loss of tolerance to gut immunity protein, glycoprotein 2 (GP2) is associated with progressive disease course in primary sclerosing cholangitis, Sci. Rep. 8 (1) (2018) 399.
[93] M. Sowa, et al., Mucosal autoimmunity to cell-bound GP2 isoforms is a sensitive marker in PSC and associated with the clinical phenotype, Front. Immunol. 9 (2018) 1959.
[94] E. Wunsch, G.L. Norman, M. Milkiewicz, M. Krawczyk, C. Bentow, Z. Shums, M. Mahler, et al., Anti-glycoprotein 2 (anti-GP2) IgA and anti-neutrophil cytoplasmic antibodies to serine proteinase 3 (PR3-ANCA): antibodies to predict severe disease, poor survival and cholangiocarcinoma in primary sclerosing cholangitis, Aliment. Pharmacol. Ther. (2020), https://doi.org/10.1111/apt.16153.
[95] S.T. Jendrek, et al., Anti-GP2 IgA autoantibodies are associated with poor survival and cholangiocarcinoma in primary sclerosing cholangitis, Gut 66 (1) (2017) 137–144.
[96] T. Tornai, et al., Gut barrier failure biomarkers are associated with poor disease outcome in patients with primary sclerosing cholangitis, World J. Gastroenterol. 23 (29) (2017) 5412–5421.
[97] H. Liu, et al., PBC screen: an IgG/IgA dual isotype ELISA detecting multiple mitochondrial and nuclear autoantibodies specific for primary biliary cirrhosis, J. Autoimmun. 35 (4) (2010) 436–442.
[98] B. Terziroli Beretta-Piccoli, G. Mieli-Vergani, D. Vergani, Serology in autoimmune hepatitis: a clinical-practice approach, Eur. J. Intern. Med. 48 (2018) 35–43.
[99] C. Liaskos, et al., Prevalence of gastric parietal cell antibodies and intrinsic factor antibodies in primary biliary cirrhosis, Clin. Chim. Acta 411 (5–6) (2010) 411–415.
[100] S. Ciesek, et al., Anti-parietal cell autoantibodies (PCA) in primary biliary cirrhosis: a putative marker for recurrence after orthotopic liver transplantation? Ann. Hepatol. 9 (2) (2010) 181–185.
[101] R. Liberal, et al., Diagnostic and clinical significance of anti-centromere antibodies in primary biliary cirrhosis, Clin. Res. Hepatol. Gastroenterol. 37 (6) (2013) 572–585.
[102] G.L. Norman, et al., Is prevalence of PBC underestimated in patients with systemic sclerosis? Dig. Liver Dis. 41 (10) (2009) 762–764.
[103] C. Rigamonti, et al., Clinical features and prognosis of primary biliary cirrhosis associated with systemic sclerosis, Gut 55 (3) (2006) 388–394.
[104] I. Cavazzana, et al., Primary biliary cirrhosis-related autoantibodies in a large cohort of Italian patients with systemic sclerosis, J. Rheumatol. 38 (10) (2011) 2180–2185.
[105] S. Assassi, et al., Primary biliary cirrhosis (PBC), PBC autoantibodies, and hepatic parameter abnormalities in a large population of systemic sclerosis patients, J. Rheumatol. 36 (10) (2009) 2250–2256.
[106] H.H. Nguyen, et al., Evaluation of classical and novel autoantibodies for the diagnosis of primary biliary cholangitis-autoimmune hepatitis overlap syndrome (PBC-AIH OS), PLoS One 13 (3) (2018) e0193960.

[107] S. Joshita, et al., Serum autotaxin is a useful disease progression marker in patients with primary biliary cholangitis, Sci. Rep. 8 (1) (2018) 8159.
[108] A.H. Mulder, et al., Prevalence and characterization of neutrophil cytoplasmic antibodies in autoimmune liver diseases, Hepatology 17 (3) (1993) 411–417.
[109] R. Taubert, N.T. Baerlecken, C. Lalanne, L. Muratori, M.P. Manns, T. Witte, E. Jaeckel, Autoantibodies against Huntingtin-interacting protein 1-related protein are superiortoconventinal autoantibodies in diagnosing autoimmune hepatitis in adults, J. Hepatol. 68 (1) (2018) PS-006.
[110] S. Gabeta, et al., IgA anti-b2GPI antibodies in patients with autoimmune liver diseases, J. Clin. Immunol. 28 (5) (2008) 501–511.
[111] K.R. Reddy, et al., Anti-Saccharomyces cerevisiae antibodies in autoimmune liver disease, Am. J. Gastroenterol. 96 (1) (2001) 252–253.
[112] M.G. Clemente, et al., Enterocyte actin autoantibody detection: a new diagnostic tool in celiac disease diagnosis: results of a multicenter study, Am. J. Gastroenterol. 99 (8) (2004) 1551–1556.
[113] R.T. Stravitz, et al., Autoimmune acute liver failure: proposed clinical and histological criteria, Hepatology 53 (2) (2011) 517–526.
[114] P.S. Leung, et al., Antimitochondrial antibodies in acute liver failure: implications for primary biliary cirrhosis, Hepatology 46 (5) (2007) 1436–1442.
[115] A. Honda, et al., Bezafibrate improves GLOBE and UK-PBC scores and long-term outcomes in patients with primary biliary cholangitis, Hepatology 70 (6) (2019) 2035–2046.
[116] G. Mieli-Vergani, et al., Diagnosis and management of pediatric autoimmune liver disease: ESPGHAN Hepatology Committee position statement, J. Pediatr. Gastroenterol. Nutr. 66 (2) (2018) 345–360.
[117] C.L. Mack, et al., Diagnosis and management of autoimmune hepatitis in adults and children: 2019 practice guidance and guidelines from the American Association for the study of liver diseases, Hepatology 72 (2) (2019) 671–722.
[118] European Association for the Study of the Liver, EASL clinical practice guidelines: autoimmune hepatitis, J. Hepatol. 63 (4) (2015) 971–1004.
[119] R.J. Andrade, M. Robles-Diaz, A. Castiella, Characterizing drug-induced liver injury with autoimmune features, Clin. Gastroenterol. Hepatol. 14 (12) (2016) 1844–1845.
[120] A. Castiella, et al., Drug-induced autoimmune liver disease: a diagnostic dilemma of an increasingly reported disease, World J. Hepatol. 6 (4) (2014) 160–168.
[121] N. Chalasani, et al., Features and outcomes of 899 patients with drug-induced liver injury: the DILIN prospective study, Gastroenterology 148 (7) (2015) 1340–1352.e7.
[122] M. Carbone, et al., Clinical application of the GLOBE and United Kingdom-primary biliary cholangitis risk scores in a trial cohort of patients with primary biliary cholangitis, Hepatol. Commun. 2 (6) (2018) 683–692.
[123] W.J. Lammers, et al., Development and validation of a scoring system to predict outcomes of patients with primary biliary cirrhosis receiving ursodeoxycholic acid therapy, Gastroenterology 149 (7) (2015) 1804–1812. e4.
[124] F. Yang, et al., The risk predictive values of UK-PBC and GLOBE scoring system in Chinese patients with primary biliary cholangitis: the additional effect of anti-gp210, Aliment. Pharmacol. Ther. 45 (5) (2017) 733–743.
[125] K. Boonstra, et al., Population-based epidemiology, malignancy risk, and outcome of primary sclerosing cholangitis, Hepatology 58 (6) (2013) 2045–2055.
[126] P.J. Trivedi, et al., Effects of primary sclerosing cholangitis on risks of cancer and death in people with inflammatory bowel diseases, based on sex, race, and age, Gastroenterology 159 (3) (2020) 915–928.

[127] J. Sabino, et al., Primary sclerosing cholangitis is characterised by intestinal dysbiosis independent from IBD, Gut 65 (10) (2016) 1681–1689.
[128] S. Lemoinne, et al., Fungi participate in the dysbiosis of gut microbiota in patients with primary sclerosing cholangitis, Gut 69 (1) (2020) 92–102.
[129] K.J. Schwenger, N. Clermont-Dejean, J.P. Allard, The role of the gut microbiome in chronic liver disease: the clinical evidence revised, JHEP Rep. 1 (3) (2019) 214–226.
[130] A. Shah, et al., Targeting the gut microbiome as a treatment for primary sclerosing cholangitis: a conceptional framework, Am. J. Gastroenterol. 115 (6) (2020) 814–822.
[131] M.H. Chapman, et al., British Society of Gastroenterology and UK-PSC guidelines for the diagnosis and management of primary sclerosing cholangitis, Gut 68 (8) (2019) 1356–1378.
[132] E.C. Goode, et al., Factors associated with outcomes of patients with primary sclerosing cholangitis and development and validation of a risk scoring system, Hepatology 69 (5) (2019) 2120–2135.
[133] J.E. Eaton, et al., Primary sclerosing cholangitis risk estimate tool (PREsTo) predicts outcomes of the disease: a derivation and validation study using machine learning, Hepatology 71 (1) (2020) 214–224.
[134] R. Chapman, et al., Diagnosis and management of primary sclerosing cholangitis, Hepatology 51 (2) (2010) 660–678.
[135] M. Carbone, et al., The UK-PBC risk scores: derivation and validation of a scoring system for long-term prediction of end-stage liver disease in primary biliary cholangitis, Hepatology 63 (3) (2016) 930–950.
[136] G.L. Norman, E. Wunsch, M. Krawczyk, S. Encabo, J. Milo, C. Bentow, Z. Shums, M. Mahler, P. Milkiewicz, Are the PR3-ANCA autoantibodies present in patients with PSC? Prospective analysis including markers of liver injury and health-related quality of life assessment, Hepatology 66 (2017) 172.
[137] G.L. Norman, E. Wunsch, M. Krawczyk, S. Encabo, J. Milo, C. Bentow, M. Mahler, D. Roggenbuck, P. Milkiewicz, Anti-GP2 IgA autoantibodies are associated with liver cirrhosis and severity of the disease in primary sclerosing cholangitis (PSC), Hepatology 66 (2017) 169.
[138] H. Hagstrom, et al., Ability of noninvasive scoring systems to identify individuals in the population at risk for severe liver disease, Gastroenterology 158 (1) (2020) 200–214.
[139] J.S. Bajaj, A. Khoruts, Microbiota changes and intestinal microbiota transplantation in liver diseases and cirrhosis, J. Hepatol. 72 (5) (2020) 1003–1027.
[140] J.B. Schwimmer, et al., Prevalence of fatty liver in children and adolescents, Pediatrics 118 (4) (2006) 1388–1393.
[141] N.K. Gatselis, et al., Factors associated with progression and outcomes of early stage primary biliary cholangitis, Clin. Gastroenterol. Hepatol. 18 (3) (2019) 684–692.
[142] K. Ohtsubo, J.D. Marth, Glycosylation in cellular mechanisms of health and disease, Cell 126 (5) (2006) 855–867.
[143] X. Verhelst, et al., Protein glycosylation as a diagnostic and prognostic marker of chronic inflammatory gastrointestinal and liver diseases, Gastroenterology 158 (1) (2020) 95–110.
[144] V. Pengo, et al., Antibody profiles for the diagnosis of antiphospholipid syndrome, Thromb. Haemost. 93 (6) (2005) 1147–1152.
[145] Z. Betteridge, et al., Frequency, mutual exclusivity and clinical associations of myositis autoantibodies in a combined European cohort of idiopathic inflammatory myopathy patients, J. Autoimmun. 101 (2019) 48–55.
[146] D.C. Baldo, et al., Evolving liver inflammation in biochemically normal individuals with anti-mitochondria antibodies, Auto Immun. Highlights 10 (1) (2019) 10.

[147] A. Dellavance, et al., Humoral autoimmune response heterogeneity in the spectrum of primary biliary cirrhosis, Hepatol. Int. 7 (2) (2013) 775–784.

[148] K.N. Lazaridis, et al., Increased prevalence of antimitochondrial antibodies in first-degree relatives of patients with primary biliary cirrhosis, Hepatology 46 (3) (2007) 785–792.

[149] W.T. Ma, et al., Development of autoantibodies precedes clinical manifestations of autoimmune diseases: a comprehensive review, J. Autoimmun. 83 (2017) 95–112.

Chapter 7

Precision medicine in autoimmune disease

Kevin D. Deane
Division of Rheumatology, University of Colorado Anschutz Medical Campus, Aurora, CO, United States

Abbreviations

ACPA	autoantibodies to citrullinated protein antigens
G6PD	glucose 6-phosphate dehydrogenase
IA	inflammatory arthritis
ILD	interstitial lung disease
MBDA	multibiomarker disease activity
PM	precision medicine
RA	rheumatoid arthritis
RF	rheumatoid factor
SLE	systemic lupus erythematosus
SSc	systemic sclerosis
T1DM	type 1 diabetes
TPMT	thiopurine *S*-methyltransferase

1. Introduction

Autoimmune diseases are a set of diseases where immune system aberrancies lead to tissue injury, morbidity and decreased quality of life, increased mortality and high financial costs [1]. There are numerous autoimmune diseases including type 1 diabetes (T1DM), hypo- and hyperthyroidism, systemic lupus erythematosus (SLE), rheumatoid arthritis (RA), psoriasis and psoriatic arthritis, autoimmune myositis, vasculitis, inflammatory bowel disease, systemic sclerosis (SSc) and multiple sclerosis, to name several.

Each of these diseases has particular interest to precision medicine (PM) in diagnosis and prognosis of disease, in the identification of appropriate treatments, as well as in identifying response to therapy (Table 1). In particular, because most autoimmune diseases have a wide variety of factors including clinical features, blood (or other biospecimen)-based biomarkers, and imaging findings, approaches that can take large amounts of data and distill that to actionable results are particularly relevant to PM in these diseases. Furthermore,

Precision Medicine and Artificial Intelligence. https://doi.org/10.1016/B978-0-12-820239-5.00005-X

TABLE 1 Key areas where precision medicine is needed in autoimmune diseases.

Diagnosis (including early diagnosis when clear-cut clinical manifestations of disease may not be readily apparent)
Prognosis (and identification of potential co-morbidities, e.g., cancer)
Treatment selection (including initial selection of therapy most likely to be of therapeutic benefit, or minimal toxicity, and selection of additional therapy in case of initial therapeutic failure)
Evaluating response to therapy (including identification of "success" of therapy, perhaps earlier than typical measures may identify improvement; also, to identify when therapy can be reduced and still maintain disease control)
Identification of individuals (and biologic targets) who are candidates for preventive interventions (includes models to predict the likelihood and timing of future clinically-apparent disease in individuals who have findings of autoimmunity yet no clear tissue injury [i.e. pre-autoimmune disease])

there is an emerging understanding that many autoimmune diseases can be identified in an early phase of disease when autoimmunity is present yet there has not been substantial tissue injury; PM approaches may be particular helpful here to identify individuals in this "pre-autoimmune disease stage" who may be targeted for preventive interventions [2]. Each of these aspects of PM in autoimmune disease will be discussed in this chapter.

2. Precision medicine in the diagnosis and prognosis of autoimmune disease

Each autoimmune disease requires a combination of clinical features and often biomarkers in order to identify a specific individual with the condition. For example, in hypothyroidism, an individual may present with symptoms of fatigue, weight gain, and then a "biomarker" test of thyroid function (e.g., abnormal thyroid stimulating hormone) would be used to confirm that this syndrome was due to hypothyroidism. Similarly, for RA, an individual may present with symptoms of joint pain, stiffness and swelling, and on examination be found to have inflammatory arthritis; in addition autoimmune biomarkers that are associated with RA such as rheumatoid factor (RF) or autoantibodies to citrullinated protein antigens (ACPA) may be abnormal [3]; in combination with the symptoms, clinical features and autoantibody tests, the individual may then be diagnosed as having RA, and appropriate treatment begun [4].

For some autoimmune diseases, a combination of clinical symptoms, examination findings and biomarkers is quite effective to establish a clear diagnosis. However, for many autoimmune diseases, due to a broad range of factors, diagnosis can be complicated. In particular, certain diseases, even if they have a similar name, may have great heterogeneity making diagnosis difficult. For example, in SLE, one patient may have skin disease, arthritis and a certain set of biomarkers, while another may have renal disease, seizures and a separate set of biomarkers; these individuals will look clinically very different, yet both will be diagnosed as SLE [5, 6]. As another example, in RA, while the autoantibodies RF and ACPA form a core piece of diagnosis, up to 20–30% of individuals will not have abnormalities of these biomarkers, yet still be classified as RA [4]. As such, if an individual has perhaps early and subtle arthritis yet negative RF and ACPA, it may be difficult to make a diagnosis and implement appropriate therapy.

Furthermore, even within individuals with very similar diagnostic features, the disease course may be very different [7]. For example, some individuals with SLE have concomitant abnormalities of antiphospholipid antibodies which can be associated with venous and arterial thromboses. However, some of these individuals will not ever have a thrombosis, and some will have severe thrombosis. In addition, in autoimmune myositis, certain biomarkers are associated with better, or worse, outcomes. For example, antibodies to Mi-2 typically are seen with more benign disease, while anti-melanoma differentiation-associated protein 5 (anti-MDA5) more severe disease [8].

The ability to readily identify subsets of individuals who will (or will not) have certain disease manifestations—and prior to the clinical appearance of these manifestations—is an area that greatly needs to improve in autoimmune diseases. Indeed, many groups of scientists and clinicians are actively seeking out novel ways to use clinical, genetic and other biomarkers to enhance diagnosis and prognosis of individuals with disease. In particular, scientists and clinicians are seeking ways to develop "molecular classification" of many rheumatic autoimmune diseases that can improve diagnosis, prognosis—as well as improve management [8, 9]. The results of these efforts are of great interest to the field—especially in the area of treatment selection which is discussed further below.

Notably, many of these approaches are now largely using autoantibodies; however, in the future as more is known about the role of certain genetic tests in disease, as well as decreasing costs and turn-around times, genetic testing may gain importance in the diagnosis and "molecular classification" of autoimmune diseases. Indeed, human leukocyte antigen (HLA) testing for the B27 allele is becoming increasing more used to diagnosis and predict outcomes in spondyloarthritis [10]; furthermore, in some rare autoimmune/autoinflammatory diseases (e.g., periodic fever syndromes), genetic testing is becoming a standard of diagnosis [11]. Finally, there is also an emerging role for tissue biopsy in defining disease subtype, prognosis and even potentially in selecting optimal therapies in many autoimmune diseases including, for example, synovial biopsies in RA [12], and renal biopsies in SLE [13].

Importantly, given there are now numerous biomarkers including genetic markers, autoantibodies, cytokines and imaging findings that may be informative in understanding diagnosis in autoimmune disease. However, it is difficult in clinical care to process all of these factors to arrive at an actionable diagnosis. While not yet fully realized and widely implemented into routine clinical care, this is an area that holds great promise for advance approaches such as artificial intelligence to improve diagnoses.

Finally, given the widespread use of electronic medical records, and emerging area is to use these records to identify factors such as symptoms, medication use, and laboratory test to facilitate earlier diagnoses [14]. For example, and individual with certain types of joint symptoms could be identified through electronic algorithms, and then their provider given decisional support to help facilitate a diagnosis of the autoimmune disease RA. This could be particularly important to identify rare diseases that health-care providers are not routinely used to evaluating. This is an area of intense research and may soon become part of main-stream clinical care.

3. Precision medicine and the identification of co-morbidities in autoimmune disease

A unique and emerging area for PM within some autoimmune diseases is the association of certain biomarkers with other conditions such as renal disease, pulmonary hypertension, interstitial lung disease (ILD), or malignancy.

For renal disease, while findings are variable, anti-double stranded DNA antibodies have been associated with nephritis [15], and in systemic sclerosis (SSc), there is a higher risk for renal crisis in individuals who exhibit antibodies to RNA polymerase III [16].

The association of biomarkers with lung disease has been most consistently described in SSc and myositis. For example, in SSc, the presence of anti-topoisomerase I antibodies are associated with PH and ILD [16], and in myositis, the presence of anti-Jo-1 and anti-MDA5 antibodies are associated with ILD [17, 18].

Associations between circulating autoantibodies and cancer is most well-established in two types of autoimmune diseases, SSc [19] and autoimmune myositis [20], where certain autoantibodies are associated with markedly increased risk of cancer. In particular, immunoglobulin G (IgG)2 isotype antibodies to anti-transcription intermediary factor-1γ (anti-TIF1γ) are associated with cancer in patients with dermatomyositis [21].

Importantly, not all biomarkers are associated with increased risk for co-morbidities within autoimmune disease. Indeed, some biomarkers have been associated with decreased risk for certain manifestations. For example, anti-Mi-2 antibodies are associated with decreased risk for ILD in myositis [8]; in addition, non-IgG2 isotypes of anti-TIF1γ have been associated with decreased risk for cancer in dermatomyositis [21].

However, these biomarkers have varying accuracies for a risk, or even protection, against specific conditions within autoimmune diseases. As such, additional studies are needed to determine the exact intended use of biomarkers and other factors to effectively identify and manage specific disease manifestations (e.g., renal disease, lung disease, or cancer), within individual patients.

Importantly, due to the large number of associations of autoantibodies, and other markers, with these conditions (e.g., cancer, ILD), it is becoming increasingly difficult for a single practitioner to interpret findings and make decisions. As such, an area of PM where advanced analytics could be of great benefit would be to take a large number of factors including disease type, autoantibody (or other marker) and create a result where risk for cancer, or other specific disease manifestation would be a clear and actionable readout (e.g., score or visualization).

4. Precision medicine in the treatment of autoimmune diseases

Autoimmune diseases are treated in number of different ways. Importantly, the type of treatment that is used in each disease is largely dependent on three factors: (1) the type and timing of identification of tissue injury, (2) identification and targeting of the important biologic pathways of disease, and (3) regulatory agency-approved [e.g., Food and Drug Administration (FDA)] therapies for the given condition. For example, in T1DM, a "disease state" is identified once autoimmune destruction of the pancreatic islet cells has occurred and blood sugars are elevated; in this situation, the therapy is not directed at the underlying immune response rather at controlling blood glucose through agents such as insulin. In addition, in hypothyroidism, disease is identified once the thyroid has failed, then thyroid hormone is replaced.

However, in other autoimmune diseases, specific pathways of disease may be targeted with immunomodulatory therapy clinically-apparent disease and tissue injury are present yet there is still hope to reduce active autoimmune-mediated tissue injury, preserve function and improve clinical outcomes and well-being. For example, in RA, in order to reduce joint inflammation and potentially prevent future destruction and disability, individuals with disease are treated with immunomodulatory therapy [4]. In particular, agents such as methotrexate and leflunomide are used which have broad effects on dampening down immune responses [22]. In addition, there are several cytokines have been demonstrated as important in the pathophysiology of RA including interleukin (IL)-1, IL-6 and tumor necrosis factor-alpha; as such, each of the cytokines may be targeted [22].

However, while a broad range of therapies exist for autoimmune diseases, the approach to use of these medications is largely based on applying findings obtained in large numbers of subjects from clinical trials as well as regulatory agency-approved approaches, and not specifically tailored to an individual's particular time course, or type, of autoimmune disease. For example, in RA, the standard of care for most all patients with newly diagnosed disease is to start with methotrexate, and escalate to additional therapy based on response to the initial therapy. This is despite a well-established failure rate of methotrexate of up to 70%, especially in individuals with a duration of clinically-apparent RA of more than 1 year [4, 23]. Furthermore, the selection of an agent for RA after methotrexate has failed is not "precision-based" but rather follows clinical practice guidelines that rarely takes into account an individual patient's disease biology. Unfortunately, most autoimmune diseases have a similar approach to treatment where a relatively uniform approach for treatment is taken despite known high failure rates of many agents, and high disease heterogeneity.

Fortunately, there are growing amounts of data that can help identify which treatments may work better for specific patients. For example, in RA, there is data to suggest that the B-cell depleting agent rituximab and the T-cell co-stimulatory blocker abatacept work better in RF or ACPA positive subjects compared to those who are seronegative [24]. Furthermore, in RA there is also growing data that other "personal factors" such as obesity and tobacco use may decrease efficacy of therapy. In addition, there are suggestions that individuals with certain profiles such as high disease activity, low complement and high anti-dsDNA may respond better to the agent belimumab [25, 26].

There is also some understanding in the management of autoimmune diseases of specific factors that may predict that an individual will potentially have toxicity from a particular therapy—or that the therapy may no longer be effective. For example, low levels of the enzyme glucose 6-phosphate dehydrogenase (G6PD) may indicate that an individual will have excessive hemolysis in the setting of the drug dapsone which can be used for SLE-related skin disease [27]. In addition, low levels of the enzyme thiopurine *S*-methyltransferase (TPMT) may indicate an individual is at excessive risk for myelosuppression from the immunosuppressant azathioprine that is commonly used for multiple autoimmune conditions including SLE and inflammatory bowel disease [28]. Drug levels can also be used to guide a specific individual's dosing of a medication. For example, evidence is mounting that measuring blood levels of the drug hydroxychloroquine can help guide dosing, monitor compliance and potentially drug toxicity in patients with SLE [29, 30]. Furthermore, immune responses against drugs such as anti-drug antibodies can be used to determine if an agent may no longer be effective [31]. There are also emerging approaches to use genetic testing to determine efficacy or toxicity of therapies. In particular, certain genes can predict toxicity of non-steroidal anti-inflammatory agents (NSAIDs) which are commonly used across a range of autoimmune diseases [32]. Finally, as mentioned above, tissue biopsy may help guide specific therapies in some diseases such as RA and SLE, in particular as the technology for obtaining and processing tissue is improving; however, more is needed before these approaches are widely useful clinically [12, 13, 33].

However, a broad and successful way to use biomarkers and other factors to help determine initial as well as subsequent therapy, is still an area that needs further development in order to make PM a reality in autoimmune diseases.

5. Precision medicine in the evaluation of outcomes in autoimmune disease

For most autoimmune diseases, the efficacy and safety of treatment is measured by improvement in measure of disease activity, as well as absence of toxicity. For diseases such as SLE, this may be improvement in measures of nephritis such as improving serum creatinine or decreased proteinuria. For psoriasis, it may be decreased skin inflammation identified on physical examination, and for RA it may be decreases in tender and swollen joints as well as decreases in measures of systemic inflammation (e.g., C-reactive protein).

For some diseases there are also now panels of biomarkers that can assess disease activity. For example, in RA, a multi-biomarker disease activity (MBDA) has demonstrated some efficacy in assessing patients, and some studies have shown that it may be more reliable in identifying joint inflammation and damage than other measures of disease activity including physical examination and single laboratory tests such as C-reactive protein, and also avoid confounding from factors such as non-inflammatory pain that may influence other measures of disease activity [34]. In particular, the commercial MBDA assay uses a complex analytic algorithm based on the levels of 12 biomarkers to produce a composite score [35, 36]. This makes results reporting and interpretation relatively straight-forward for patients as well as providers in order to make clinical decisions.

However, many of these biomarker panels require send-out laboratories, are costly and take some time to return, and with the MBDA, it has not uniformly been proven to be superior to other more established measures of disease activity [37]; therefore, such systems have not yet widely and uniformly been adopted in routine clinical practice. However, as the accuracy and efficacy of tests improve, costs improve and these tests are more widely available, their use will likely increase [38, 39]. In addition, biomarker systems that can take complex data and utilize advance analyses to report clear and actionable results will be critical to enhancing the use of such systems in routine care [40, 41]. Importantly, these "biomarker" measures of autoimmune disease activity may become increasing important methods to routinely evaluate individuals with autoimmune diseases as work-force shortage issues may decrease the ability of "experts" in autoimmune disease to routine monitor these patients in-person. Indeed, one could envision a telemedicine visit coupled with high-quality and informative biomarker testing could replace many in-person visits [42, 43].

In addition, in a number of autoimmune diseases include those that involve arthritis and neurologic disease, as well as autoimmune vasculitis, modalities including imaging are becoming increasing used to diagnose as well as follow response to therapy in individual patients [44–47]. Perhaps most dramatic is in the field of rheumatology where the real-time use of ultrasound assessment of the joints in routine clinical practice has become increasingly common, and is particularly useful in patients where physical examination may be difficult [48].

There are still issues of reliability of imaging, and cost, but as technology advances, and skill-sets of practitioners grow, imaging may be a much more commonly used approach for real-time implementation of PM. Importantly, there is growing use of complex analytic techniques (e.g., artificial intelligence) in image interpretation which may enhance the use of imaging in the diagnosis and management of autoimmune diseases—especially in the identification of subtle findings [49–51].

For many diseases, measuring outcomes includes components of patient-reported outcomes (PRO's) that can be used to evaluate response to therapy. For example, in RA, patient-reported pain, fatigue and well-being are key components of multiple outcome measures [52]. In most cases, these PRO's have been collected using paper forms in clinic, or through interviews with health-care providers. However, with the advent of computers, smart phones and electronic "wearables" it is becoming increasing realistic to use electronic devices to capture and record meaningful outcomes [53, 54]. There are still many challenges in this area including how to utilize large volumes of data from electronic capture into succinct, meaningful results; as such, these devices have not yet been widely adopted in health-care. However, their use is growing and the future should see more integration of these into care of patients with autoimmune diseases.

6. Precision medicine and prediction and prevention of autoimmune disease

There is now growing evidence across a number of autoimmune diseases that biomarkers and other factors can be used to identify individuals who are at high-risk of developing future clinically-apparent disease during a period when the individuals are relatively asymptomatic. For example, autoantibodies in T1DM and RA can be used in individuals without current classifiable disease (i.e., elevated blood glucose or inflammatory arthritis, IA, respectively) to predict future disease with fair accuracy, and indeed the predictive power of autoantibodies has led to several clinical prevention trials in these diseases [55, 56]. Furthermore, multiple other autoimmune diseases, including thyroid disease, SLE and vasculitis appear to follow a similar model of development (reviewed in [57]). Notably, however, as RA is one of the most common autoimmune diseases, it serves as an excellent example of how PM may be applicable in the area of autoimmune disease prevention, which is the topic of sections below.

RA is currently diagnosed and treated when an individual has clinical signs and symptoms of IA. However, it is now been identified that most individuals who develop RA have a "Pre-RA" period that can be defined as systemic autoimmunity that can be measured through autoantibodies including RF and ACPA, that appear, on-average, 3–5 years before the onset of clinically-identifiable IA, a key feature of RA [58–60]. Furthermore, there is growing evidence that alterations in autoantibodies including glycosylation patterns, and affinity, are present in Pre-RA [61, 62]. In addition, there are a number of inflammatory markers including C-reactive protein, fatty acid metabolism, and cytokines and chemokines that have been identified as being abnormal in Pre-RA [63–65]. Finally, there are emerging findings that cellular abnormalities may be present in Pre-RA including alterations of B- and T-cell subsets [66].

Overall these findings suggest a model of RA development as pictured in Fig. 1. In this model, genetic and environmental factors combine to trigger initial autoimmunity that has most often been studied thus far as "systemic" autoimmunity, i.e., circulating abnormalities of autoantibodies or other factors. Then, over time this autoimmunity may progress and clinically-apparent IA develops, and a diagnosis of RA is made. Notably, similar models of development are seen in other autoimmune diseases, including SLE, where autoantibodies are also known to develop years prior to future clinically-apparent disease [67].

7. Autoantibodies and other factors are highly predictive of future rheumatoid arthritis (and other autoimmune diseases)

There are now multiple studies that have shown that systemic abnormalities of autoantibodies and other factors such as family history of RA and genetic factors such as the presence of the shared epitope, levels of inflammation, joint symptoms, and factors including tobacco use and obesity all can contribute to prediction of future RA [68].

Of predictive factors, serum elevations of ACPA are the most consistent and strongest predictors of for future IA/RA. In particular, in case-control studies, the specificity of ACPA elevations for future RA have typically been >90% [63, 69, 70]. In addition, ACPA positivity in combination with RF has specificities that are typically >95%.

A caveat is that because of their study design, case-control studies overestimate the specificity of autoantibodies for RA. However, a growing number of prospective studies have shown that 20–70% of individuals with ACPA positivity at baseline progress to RA within the duration of follow-up.

Importantly, this rate of progression to RA can be further refined by incorporating additional factors into prediction. Of these, perhaps the most powerful

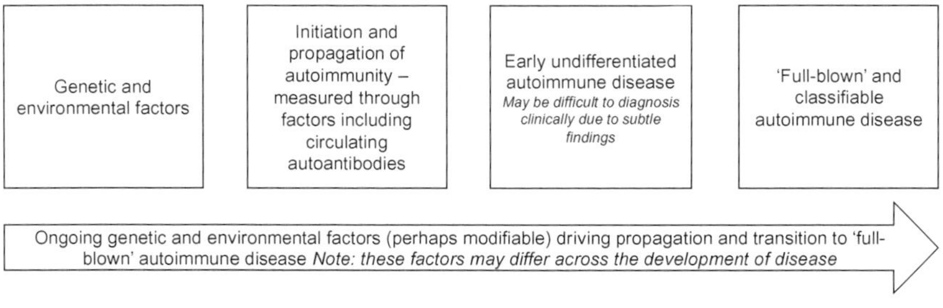

Genetic and environmental factors

Initiation and propagation of autoimmunity – measured through factors including circulating autoantibodies

Early undifferentiated autoimmune disease *May be difficult to diagnosis clinically due to subtle findings*

'Full-blown' and classifiable autoimmune disease

Ongoing genetic and environmental factors (perhaps modifiable) driving propagation and transition to 'full-blown' autoimmune disease *Note: these factors may differ across the development of disease*

FIG. 1 Model of autoimmune disease development.

predictors are high levels of ACPA as well as concomitant ACPA and RF positivity, especially if present in the setting of joint symptoms of pain, stiffness, and/or swelling [71]. Additionally, factors such as ongoing smoking, obesity and the presence of the genetic factor the shared epitope and enhance prediction of RA [68, 72]. In particular, a study by Rakieh and colleagues has demonstrated in prospective follow-up of 100 ACPA positive individuals (50 of whom developed IA/RA), that if >=3 of the following factors were present, 62% of subject progressed to RA within 2 years of follow-up [73]: (1) positive ultrasound doppler signal in a joint (even in absence of IA identified on joint exam), (2) positive RF or ACPA, (3) early morning stiffness for at least 30 min, and (4) tenderness of hand or foot joints.

Notably, there are two aspects of prediction of future RA that are important to define—especially in relationship to PM. The first is the overall likelihood of developing RA, and the second is the timing at which RA will develop. These two aspects are important to individuals who are at-risk for future RA so that they know their personal risk, but also the time frame in which to expect RA to develop since that may influence their decision to take preventive approaches. Furthermore, these two aspects are important to research so that studies can focus on high-risk individuals, as well as be designed in such a way that outcomes of RA are "imminent" and therefore trials can be designed to optimize the identification of outcomes. These latter points are especially important in the design of prevention trials for RA and other autoimmune diseases where it is imperative to have a good estimate of the number of expected events (i.e., incident RA) so that trials can be designed with sufficient events (statistically powered) to demonstrate meaningful interventions.

Fortunately, several of the prospective studies of individuals who are at-risk for future RA have been able to define, to some extent, these two aspects of prediction. As mentioned above, the study by Rakieh and colleagues identified a set of factors that predicted RA onset within 2 year for >60% of subjects [73]. In addition, van de Stadt and colleagues have used a score comprising of 9 variables to identify that individuals with scores of >7 were over 50% likely to develop IA/RA within 2 years [71]. Sokolove and colleagues also combined elevations of cytokines, chemokines and a set of ACPAs to identify a set of factors that if present was ~80% specific for an individual developing RA within 2 years [74]. Finally, Deane and colleagues have also demonstrated that an increasing number of elevated cytokines and chemokines in Pre-RA is associated with a decreased time to diagnosis [63].

However, these models are not perfect, and have only been evaluated in limited numbers of subjects who have progressed from Pre-RA to established, articular disease. As such, a major focus for the field going forward will be refine prediction so that an individual's risk for future RA or other autoimmune diseases, as well as timing of future disease, can be well-quantified. An example of this type of approach is presented in Fig. 2, where a combination of factors, including assessments such as imaging that could enhance an

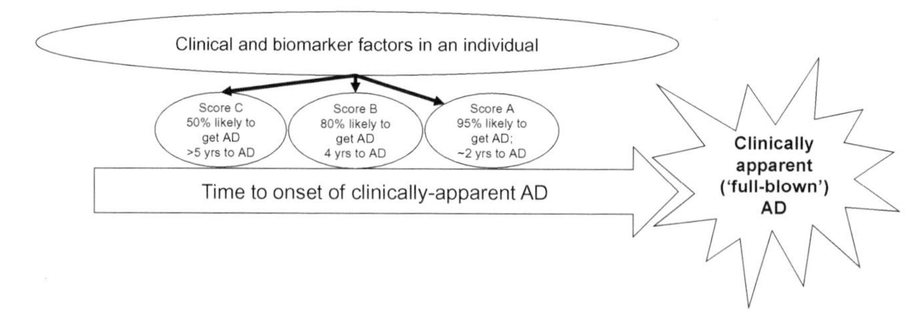

FIG. 2 Model for determining likelihood and timing to diagnosis of future autoimmune disease (AD) (e.g., rheumatoid arthritis). In this approach, clinical and biomarker factors can be used to identify the risk and timing of a future AD. This can be then applied in real-time situations where an individual is found to have profiles most indicative of a certain Pre-AD period. This can guide counseling regarding future risk, as well as identify certain pathways that could be targeted for prevention.

ability to identify early organ injury [73], could be evaluated to identify both the likelihood and timing for developing future RA. Importantly, similar approaches could be taken for other autoimmune diseases. However, it is also important to understand that any measure used to "predict" future clinically-apparent disease needs to be utilized carefully. As an example, in RA, while imaging has improved prediction in some studies, there are also findings suggest "joint inflammation" that may be interpreted as RA are common in populations who do not have or develop future RA, and that imaging findings can be widely variable between clinicians [44, 75–77]. As such, another important PM aspect of Pre-RA, or any Pre-Autoimmune Disease state that needs to be further explored is to better define how to use biomarkers to enhance prediction and prevention, and avoid misclassification of subjects—and perhaps overtreatment.

8. Identification of meaningful biologic targets for prevention

Notably, current understanding of the natural history of RA development, and the predictive power of biomarkers for future RA, while not perfect, have driven the development and implementation of several prevention trials in RA—two of which have been completed. In the first, two doses of high-dose dexamethasone failed to prevent autoantibody positive individuals from progressing to IA/RA [78]. In the second, a single dose of the B-cell depleting agent rituximab did not prevent autoantibody positive individuals from progressing to IA/RA; however, it did delay the onset of IA [79]. In addition, there are several other prevention studies in RA underway, evaluating if agents including hydroxychloroquine, abatacept or methotrexate can prevent or delay future RA in high-risk individuals [80–82].

In aggregate these trials will advance the field, and in particular, while the first two studies did not demonstrate significant benefit in prevention RA, it may be that in the three trials that are underway, a "successful" preventive intervention for RA may be identified.

Nonetheless, it is important to note that all of these trials have used (or are using) agents already established to be effective in established RA. This makes sense as these agents have known benefit, and toxicity, in RA, and are therefore acceptable to clinicians and regulatory agencies for use in the novel area of prevention. However, it may be that these agents are not targeting the appropriate biologic pathways in Pre-RA that may lead to meaningful prevention. In particular, there is an emerging understanding that mucosal inflammation may play a role in the initiation and propagation of autoimmunity in the Pre-RA period (reviewed in Ref. [83]). Furthermore, there is growing understanding in a multitude of other autoimmune diseases that mucosal or microbiologic factors may influence early disease development. As such, it may be that effective targets in Pre-RA and other Pre-Autoimmune Diseases may be related to mucosal inflammation, or other processes that are not well-addressed by current therapies used in established RA. In addition, given the incredible complexities of deciphering the role of the microbiome in human health and disease, as well as other factors that impact the role microbes play in health including metabolomics and inflammation [84–86], PM and advanced analytic techniques such as artificial intelligence are likely of key importance to advance the field.

Furthermore, there is growing evidence that potentially modifiable risk factors such as obesity and tobacco use contribute the evolution of multiple autoimmune diseases. Importantly, if these factors are addressed, they may prove helpful in prevention—and in particular be "PM" by targeting risk factors that are specific to an individual [87].

9. "Personalized medicine"—What should (and will) a person do to treat a current autoimmune disease, or prevent a future autoimmune disease?

Another critically important aspect of PM as it relates to any aspect of care from diagnosis, treatment and prevention, is what is an individual who either has "full-blown" disease or who is at-risk for a future autoimmune diseases is willing to undertake in order to improve their health [88–91]. This is especially important in consideration of prevention because an individual with full-blown autoimmune disease (e.g., RA, SLE) may be willing to take potentially very toxic medications in order to immediately improve their quality of life; however, in contrast, an individual who feels well may be very reluctant to take steps, even if relatively benign, to prevent a future disease that they do not yet have symptoms of.

There is a significant body of literature regarding patient preferences in management of established disease; in addition, there is an emerging understand of these issues regarding prevention in autoimmune diseases. In particular, several themes are that subjects want to know their personal risk, and timing of potential onset of disease, and also want interventions that are tolerable and safe (see also Chapter 10). However, most of these studies are in hypothetical situations and therefore it remains to be seen how the field can effectively communicate with at-risk individuals so that participation in prevention leads to improved public health. This is particularly important when considering one recent prevention trial in RA which was utilizing statin therapy [92]; this trial failed to enroll and had to terminate prematurely because individuals were unwilling to take this medication, even though the safety profile of statins is relatively good and it is commonly used.

Another issue of prevention in both RA as well as other AD's is the ability to find at-risk individuals. While in aggregate, AD's may affect a substantial portion of the population, individually they are quite rare—indeed, RA affects ~0.5–1% of the population. This disease rarity makes identification of individuals who are at-risk for future RA difficult, requiring large amounts of screening to identify few at-risk individuals. For T1DM, decades of work have established extensive networks that can identify at-risk individuals for prevention studies [93]. Building on this, one of the next challenges for PM will be to provide ways to optimally screen populations for at-risk individuals for participation in prevention. Items that could greatly help here are the growing use of at-home collection of samples that can be analyzed for biomarkers (including microbiome) [94, 95], as well as methods to assess early signs and symptoms of diseases through technology. It remains to be seen how these can be leveraged to improve the timing of diagnoses, as well as identify individuals who may be candidates for autoimmune disease prevention.

10. Conclusions

PM has a role across multiple aspects of autoimmune diseases including diagnosis, prognosis, treatment, outcome assessment—and even prevention. However, while clinical features, biomarkers and other factors are currently playing a role in many aspects of care in autoimmune diseases, there are still advances that need to be made in all of these aspects in order to improve PM (Table 2). Fortunately, clinical care is evolving, and studies are underway that should improve our understanding of all of these aspects, including—perhaps most excitingly, how to identify at-risk individuals and prevent future autoimmunity.

Disclosures

Dr. Deane is an investigator on an investigator-initiated grant with Janssen Research and Development. Dr. Deane has served as a consultant to Janssen, Inova Diagnostics, Inc., ThermoFisher, Bristol-Myers Squibb and Microdrop LLC.

TABLE 2 Emerging and future directions in implementation of precision medicine in autoimmune diseases.

Use of at-home biospecimen collection
Telemedicine
Use of electronic medical record and other data for decisional support for early diagnoses of autoimmune diseases
Integration of technology (wearables, other) with diagnosis and management
Imaging including ultrasound, magnetic resonance imaging to identify subclinical tissue injury as well as follow outcomes
Expanding use of biomarker testing for early diagnosis and treatment
Tissue biopsy to guide diagnosis, treatment and evaluate response
Use of advanced analytic techniques (e.g., artificial intelligence) to take large amount of data and present it in clear and actionable fashion

References

[1] L. Wang, F.S. Wang, M.E. Gershwin, Human autoimmune diseases: a comprehensive update, J. Intern. Med. 278 (4) (2015) 369–395.

[2] N.R. Rose, Prediction and prevention of autoimmune disease in the 21st century: a review and preview, Am. J. Epidemiol. 183 (5) (2016) 403–406.

[3] P.F. Whiting, N. Smidt, J.A. Sterne, R. Harbord, A. Burton, M. Burke, et al., Systematic review: accuracy of anti-citrullinated peptide antibodies for diagnosing rheumatoid arthritis, Ann. Intern. Med. 152 (7) (2010) 456–464. W155–66.

[4] J.S. Smolen, D. Aletaha, A. Barton, G.R. Burmester, P. Emery, G.S. Firestein, et al., Rheumatoid arthritis, Nat. Rev. Dis. Primers 4 (2018) 18001.

[5] A. Fava, M. Petri, Systemic lupus erythematosus: diagnosis and clinical management, J. Autoimmun. 96 (2019) 1–13.

[6] T. Dorner, R. Furie, Novel paradigms in systemic lupus erythematosus, Lancet 393 (10188) (2019) 2344–2358.

[7] G.J. Pons-Estel, L. Andreoli, F. Scanzi, R. Cervera, A. Tincani, The antiphospholipid syndrome in patients with systemic lupus erythematosus, J. Autoimmun. 76 (2017) 10–20.

[8] K. Mariampillai, B. Granger, D. Amelin, M. Guiguet, E. Hachulla, F. Maurier, et al., Development of a new classification system for idiopathic inflammatory myopathies based on clinical manifestations and myositis-specific autoantibodies, JAMA Neurol. 75 (12) (2018) 1528–1537.

[9] G. Barturen, L. Beretta, R. Cervera, R. Van Vollenhoven, M.E. Alarcon-Riquelme, Moving towards a molecular taxonomy of autoimmune rheumatic diseases, Nat. Rev. Rheumatol. 14 (3) (2018) 180.

[10] C.S.E. Lim, R. Sengupta, K. Gaffney, The clinical utility of human leucocyte antigen B27 in axial spondyloarthritis, Rheumatology (Oxford) 57 (6) (2018) 959–968.

[11] O. Schnappauf, I. Aksentijevich, Current and future advances in genetic testing in systemic autoinflammatory diseases, Rheumatology (Oxford) 58 (Suppl. 6) (2019) vi44–vi55.

[12] F. Zhang, K. Wei, K. Slowikowski, C.Y. Fonseka, D.A. Rao, S. Kelly, et al., Defining inflammatory cell states in rheumatoid arthritis joint synovial tissues by integrating single-cell transcriptomics and mass cytometry, Nat. Immunol. 20 (7) (2019) 928–942.

[13] I. Ayoub, C. Cassol, S. Almaani, B. Rovin, S.V. Parikh, The kidney biopsy in systemic lupus erythematosus: a view of the past and a vision of the future, Adv. Chronic Kidney Dis. 26 (5) (2019) 360–368.
[14] C.A. Nelson, A.J. Butte, S.E. Baranzini, Integrating biomedical research and electronic health records to create knowledge-based biologically meaningful machine-readable embeddings, Nat. Commun. 10 (1) (2019) 3045.
[15] S. Soliman, C. Mohan, Lupus nephritis biomarkers, Clin. Immunol. 185 (2017) 10–20.
[16] C. Liaskos, E. Marou, T. Simopoulou, M. Barmakoudi, G. Efthymiou, T. Scheper, et al., Disease-related autoantibody profile in patients with systemic sclerosis, Autoimmunity 50 (7) (2017) 414–421.
[17] L. Cavagna, E. Trallero-Araguas, F. Meloni, I. Cavazzana, J. Rojas-Serrano, E. Feist, et al., Influence of antisynthetase antibodies specificities on antisynthetase syndrome clinical spectrum time course, J. Clin. Med. 8 (11) (2019) 2013.
[18] F. Chen, X. Lu, X. Shu, Q. Peng, X. Tian, G. Wang, Predictive value of serum markers for the development of interstitial lung disease in patients with polymyositis and dermatomyositis: a comparative and prospective study, Intern. Med. J. 45 (6) (2015) 641–647.
[19] T. Igusa, L.K. Hummers, K. Visvanathan, C. Richardson, F.M. Wigley, L. Casciola-Rosen, et al., Autoantibodies and scleroderma phenotype define subgroups at high-risk and low-risk for cancer, Ann. Rheum. Dis. 77 (8) (2018) 1179–1186.
[20] X. Lu, Q. Peng, G. Wang, The role of cancer-associated autoantibodies as biomarkers in paraneoplastic myositis syndrome, Curr. Opin. Rheumatol. 31 (6) (2019) 643–649.
[21] A. Aussy, M. Freret, L. Gallay, D. Bessis, T. Vincent, D. Jullien, et al., The IgG2 isotype of anti-transcription intermediary factor 1gamma autoantibodies is a biomarker of cancer and mortality in adult dermatomyositis, Arthritis Rheumatol. 71 (8) (2019) 1360–1370.
[22] D. Aletaha, J.S. Smolen, Diagnosis and management of rheumatoid arthritis: a review, JAMA 320 (13) (2018) 1360–1372.
[23] G.R. Burmester, J.E. Pope, Novel treatment strategies in rheumatoid arthritis, Lancet 389 (10086) (2017) 2338–2348.
[24] P.D. Kiely, Biologic efficacy optimization—a step towards personalized medicine, Rheumatology (Oxford) 55 (5) (2016) 780–788.
[25] L. Iaccarino, L. Andreoli, E.B. Bocci, A. Bortoluzzi, F. Ceccarelli, F. Conti, et al., Clinical predictors of response and discontinuation of belimumab in patients with systemic lupus erythematosus in real life setting. Results of a large, multicentric, nationwide study, J. Autoimmun. 86 (2018) 1–8.
[26] F. Trentin, M. Gatto, M. Zen, M. Larosa, L. Maddalena, L. Nalotto, et al., Effectiveness, tolerability, and safety of belimumab in patients with refractory SLE: a review of observational clinical-practice-based studies, Clin. Rev. Allergy Immunol. 54 (2) (2018) 331–343.
[27] L. Luzzatto, E. Seneca, G6PD deficiency: a classic example of pharmacogenetics with ongoing clinical implications, Br. J. Haematol. 164 (4) (2014) 469–480.
[28] M.J. Coenen, D.J. de Jong, C.J. van Marrewijk, L.J. Derijks, S.H. Vermeulen, D.R. Wong, et al., Identification of patients with variants in TPMT and dose reduction reduces hematologic events during thiopurine treatment of inflammatory bowel disease, Gastroenterology 149 (4) (2015) 907–917.e7.
[29] N. Abdulaziz, A.R. Shah, W.J. McCune, Hydroxychloroquine: balancing the need to maintain therapeutic levels with ocular safety: an update, Curr. Opin. Rheumatol. 30 (3) (2018) 249–255.
[30] C.C. Mok, Therapeutic monitoring of the immuno-modulating drugs in systemic lupus erythematosus, Expert. Rev. Clin. Immunol. 13 (1) (2017) 35–41.

[31] K.P. Pratt, Anti-drug antibodies: emerging approaches to predict, reduce or reverse biotherapeutic immunogenicity, Antibodies (Basel) 7 (2) (2018) 19.
[32] L. Bach-Rojecky, D. Vadunec, K. Zunic, J. Kurija, S. Sipicki, R. Gregg, et al., Continuing war on pain: a personalized approach to the therapy with nonsteroidal anti-inflammatory drugs and opioids, Pers. Med. 16 (2) (2019) 171–184.
[33] G. Lliso-Ribera, F. Humby, M. Lewis, A. Nerviani, D. Mauro, F. Rivellese, et al., Synovial tissue signatures enhance clinical classification and prognostic/treatment response algorithms in early inflammatory arthritis and predict requirement for subsequent biological therapy: results from the pathobiology of early arthritis cohort (PEAC), Ann. Rheum. Dis. 78 (12) (2019) 1642–1652.
[34] J.R. Curtis, C.H. Brahe, M. Ostergaard, M. Lund Hetland, K. Hambardzumyan, S. Saevarsdottir, et al., Predicting risk for radiographic damage in rheumatoid arthritis: comparative analysis of the multi-biomarker disease activity score and conventional measures of disease activity in multiple studies, Curr. Med. Res. Opin. 35 (9) (2019) 1483–1493.
[35] M. Centola, G. Cavet, Y. Shen, S. Ramanujan, N. Knowlton, K.A. Swan, et al., Development of a multi-biomarker disease activity test for rheumatoid arthritis, PLoS One 8 (4) (2013) e60635.
[36] J.R. Curtis, A.H. van der Helm-van Mil, R. Knevel, T.W. Huizinga, D.J. Haney, Y. Shen, et al., Validation of a novel multibiomarker test to assess rheumatoid arthritis disease activity, Arthritis Care Res. 64 (12) (2012) 1794–1803.
[37] T.M. Johnson, K.A. Register, C.M. Schmidt, J.R. O'Dell, T.R. Mikuls, K. Michaud, et al., Correlation of the multi-biomarker disease activity score with rheumatoid arthritis disease activity measures: a systematic review and meta-analysis, Arthritis Care Res. 71 (11) (2019) 1459–1472.
[38] G.M. Oderda, G.D. Lawless, G.C. Wright, S.R. Nussbaum, R. Elder, K. Kim, et al., The potential impact of monitoring disease activity biomarkers on rheumatoid arthritis outcomes and costs, Pers. Med. 15 (4) (2018) 291–301.
[39] J.R. Curtis, D.D. Flake, M.E. Weinblatt, N.A. Shadick, M. Ostergaard, M.L. Hetland, et al., Adjustment of the multi-biomarker disease activity score to account for age, sex and adiposity in patients with rheumatoid arthritis, Rheumatology (Oxford) 58 (5) (2019) 874–883.
[40] W. Ning, S. Chan, A. Beam, M. Yu, A. Geva, K. Liao, et al., Feature extraction for phenotyping from semantic and knowledge resources, J. Biomed. Inform. 91 (2019) 103122.
[41] K.H. Yu, A.L. Beam, I.S. Kohane, Artificial intelligence in healthcare, Nat. Biomed. Eng. 2 (10) (2018) 719–731.
[42] J.A. McDougall, E.D. Ferucci, J. Glover, L. Fraenkel, Telerheumatology: a systematic review, Arthritis Care Res. (Hoboken) 69 (10) (2017) 1546–1557.
[43] L. Lavorgna, F. Brigo, M. Moccia, L. Leocani, R. Lanzillo, M. Clerico, et al., e-Health and multiple sclerosis: an update, Mult. Scler. 24 (13) (2018) 1657–1664.
[44] W.P. Nieuwenhuis, H.W. van Steenbergen, L. Mangnus, E.C. Newsum, J.L. Bloem, T.W.J. Huizinga, et al., Evaluation of the diagnostic accuracy of hand and foot MRI for early rheumatoid arthritis, Rheumatology (Oxford) 56 (8) (2017) 1367–1377.
[45] C. Louapre, B. Bodini, C. Lubetzki, L. Freeman, B. Stankoff, Imaging markers of multiple sclerosis prognosis, Curr. Opin. Neurol. 30 (3) (2017) 231–236.
[46] J.F. Baker, P.G. Conaghan, F. Gandjbakhch, Update on magnetic resonance imaging and ultrasound in rheumatoid arthritis, Clin. Exp. Rheumatol. 36 Suppl. 114 (5) (2018) 16–23.
[47] C. Dejaco, S. Ramiro, C. Duftner, F.L. Besson, T.A. Bley, D. Blockmans, et al., EULAR recommendations for the use of imaging in large vessel vasculitis in clinical practice, Ann. Rheum. Dis. 77 (5) (2018) 636–643.

[48] A.C. Cannella, E.Y. Kissin, K.D. Torralba, J.B. Higgs, G.S. Kaeley, Evolution of musculoskeletal ultrasound in the United States: implementation and practice in rheumatology, Arthritis Care Res. (Hoboken) 66 (1) (2014) 7–13.

[49] S. Gyftopoulos, D. Lin, F. Knoll, A.M. Doshi, T.C. Rodrigues, M.P. Recht, Artificial intelligence in musculoskeletal imaging: current status and future directions, AJR Am. J. Roentgenol. 213 (3) (2019) 506–513.

[50] A. Hosny, C. Parmar, J. Quackenbush, L.H. Schwartz, H. Aerts, Artificial intelligence in radiology, Nat. Rev. Cancer 18 (8) (2018) 500–510.

[51] C. Parmar, J.D. Barry, A. Hosny, J. Quackenbush, H. Aerts, Data analysis strategies in medical imaging, Clin. Cancer Res. 24 (15) (2018) 3492–3499.

[52] B. Fautrel, R. Alten, B. Kirkham, I. de la Torre, F. Durand, J. Barry, et al., Call for action: how to improve use of patient-reported outcomes to guide clinical decision making in rheumatoid arthritis, Rheumatol. Int. 38 (6) (2018) 935–947.

[53] G.R. Burmester, Response to: 'Digital health: a new dimension in rheumatology patient care' by Kataria and Ravindran, Ann. Rheum. Dis. 78 (10) (2019) e104.

[54] S. Kataria, V. Ravindran, Digital health: a new dimension in rheumatology patient care, Rheumatol. Int. 38 (11) (2018) 1949–1957.

[55] K.C. Herold, B.N. Bundy, S.A. Long, J.A. Bluestone, L.A. DiMeglio, M.J. Dufort, et al., An anti-CD3 antibody, teplizumab, in relatives at risk for type 1 diabetes, N. Engl. J. Med. 381 (7) (2019) 603–613.

[56] H.K. Greenblatt, H.A. Kim, L.F. Bettner, K.D. Deane, Preclinical rheumatoid arthritis and rheumatoid arthritis prevention, Curr. Opin. Rheumatol. 32 (3) (2020) 289–296.

[57] K.D. Deane, H. El-Gabalawy, Pathogenesis and prevention of rheumatic disease: focus on preclinical RA and SLE, Nat. Rev. Rheumatol. 10 (4) (2014) 212–228.

[58] K. Raza, V.M. Holers, D. Gerlag, Nomenclature for the phases of the development of rheumatoid arthritis, Clin. Ther. 41 (7) (2019) 1279–1285.

[59] D.M. Gerlag, K. Raza, L.G. van Baarsen, E. Brouwer, C.D. Buckley, G.R. Burmester, et al., EULAR recommendations for terminology and research in individuals at risk of rheumatoid arthritis: report from the study group for risk factors for rheumatoid arthritis, Ann. Rheum. Dis. 71 (5) (2012) 638–641.

[60] K.D. Deane, V.M. Holers, The natural history of rheumatoid arthritis, Clin. Ther. 41 (7) (2019) 1256–1269.

[61] L. Hafkenscheid, E. de Moel, I. Smolik, S. Tanner, X. Meng, B.C. Jansen, et al., N-linked glycans in the variable domain of IgG anti-citrullinated protein antibodies predict the development of rheumatoid arthritis, Arthritis Rheumatol. 71 (10) (2019) 1626–1633.

[62] A. Ercan, J. Cui, D.E. Chatterton, K.D. Deane, M.M. Hazen, W. Brintnell, et al., Aberrant IgG galactosylation precedes disease onset, correlates with disease activity, and is prevalent in autoantibodies in rheumatoid arthritis, Arthritis Rheum. 62 (8) (2010) 2239–2248.

[63] K.D. Deane, C.I. O'Donnell, W. Hueber, D.S. Majka, A.A. Lazar, L.A. Derber, et al., The number of elevated cytokines and chemokines in preclinical seropositive rheumatoid arthritis predicts time to diagnosis in an age-dependent manner, Arthritis Rheum. 62 (11) (2010) 3161–3172.

[64] H. Kokkonen, I. Soderstrom, J. Rocklov, G. Hallmans, K. Lejon, S. Rantapaa Dahlqvist, Up-regulation of cytokines and chemokines predates the onset of rheumatoid arthritis, Arthritis Rheum. 62 (2) (2010) 383–391.

[65] R.W. Gan, E.A. Bemis, M.K. Demoruelle, C.C. Striebich, S. Brake, M.L. Feser, et al., The association between omega-3 fatty acid biomarkers and inflammatory arthritis in an

anti-citrullinated protein antibody positive population, Rheumatology (Oxford) 56 (12) (2017) 2229–2236.
[66] L. Hunt, E.M. Hensor, J. Nam, A.N. Burska, R. Parmar, P. Emery, et al., T cell subsets: an immunological biomarker to predict progression to clinical arthritis in ACPA-positive individuals, Ann. Rheum. Dis. 75 (10) (2016) 1884–1889.
[67] M.R. Arbuckle, M.T. McClain, M.V. Rubertone, R.H. Scofield, G.J. Dennis, J.A. James, et al., Development of autoantibodies before the clinical onset of systemic lupus erythematosus, N. Engl. J. Med. 349 (16) (2003) 1526–1533.
[68] L. van Boheemen, D. van Schaardenburg, Predicting rheumatoid arthritis in at-risk individuals, Clin. Ther. 41 (7) (2019) 1286–1298.
[69] S. Rantapaa-Dahlqvist, B.A. de Jong, E. Berglin, G. Hallmans, G. Wadell, H. Stenlund, et al., Antibodies against cyclic citrullinated peptide and IgA rheumatoid factor predict the development of rheumatoid arthritis, Arthritis Rheum. 48 (10) (2003) 2741–2749.
[70] M.M. Nielen, D. van Schaardenburg, H.W. Reesink, R.J. van de Stadt, I.E. van der Horst-Bruinsma, M.H. de Koning, et al., Specific autoantibodies precede the symptoms of rheumatoid arthritis: a study of serial measurements in blood donors, Arthritis Rheum. 50 (2) (2004) 380–386.
[71] L.A. van de Stadt, B.I. Witte, W.H. Bos, D. van Schaardenburg, A prediction rule for the development of arthritis in seropositive arthralgia patients, Ann. Rheum. Dis. 72 (12) (2013) 1920–1926.
[72] M.J. de Hair, R.B. Landewe, M.G. van de Sande, D. van Schaardenburg, L.G. van Baarsen, D.M. Gerlag, et al., Smoking and overweight determine the likelihood of developing rheumatoid arthritis, Ann. Rheum. Dis. 72 (10) (2013) 1654–1658.
[73] C. Rakieh, J.L. Nam, L. Hunt, E.M. Hensor, S. Das, L.A. Bissell, et al., Predicting the development of clinical arthritis in anti-CCP positive individuals with non-specific musculoskeletal symptoms: a prospective observational cohort study, Ann. Rheum. Dis. 74 (9) (2015) 1659–1666.
[74] J. Sokolove, R. Bromberg, K.D. Deane, L.J. Lahey, L.A. Derber, P.E. Chandra, et al., Autoantibody epitope spreading in the pre-clinical phase predicts progression to rheumatoid arthritis, PLoS One 7 (5) (2012) e35296.
[75] A. Zabotti, S. Finzel, X. Baraliakos, K. Aouad, N. Ziade, A. Iagnocco, Imaging in the preclinical phases of rheumatoid arthritis, Clin. Exp. Rheumatol. 38 (3) (2020) 536–542.
[76] M.H. van Beers-Tas, A.B. Blanken, M.M.J. Nielen, F. Turkstra, C.J. van der Laken, M. Meursinge Reynders, et al., The value of joint ultrasonography in predicting arthritis in seropositive patients with arthralgia: a prospective cohort study, Arthritis Res. Ther. 20 (1) (2018) 279.
[77] A.C. Boer, L.E. Burgers, L. Mangnus, R.M. Ten Brinck, W.P. Nieuwenhuis, H.W. van Steenbergen, et al., Using a reference when defining an abnormal MRI reduces false-positive MRI results—a longitudinal study in two cohorts at risk for rheumatoid arthritis, Rheumatology (Oxford) 56 (10) (2017) 1700–1706.
[78] W.H. Bos, B.A. Dijkmans, M. Boers, R.J. van de Stadt, D. van Schaardenburg, Effect of dexamethasone on autoantibody levels and arthritis development in patients with arthralgia: a randomised trial, Ann. Rheum. Dis. 69 (3) (2010) 571–574.
[79] D.M. Gerlag, M. Safy, K.I. Maijer, M.W. Tang, S.W. Tas, M.J.F. Starmans-Kool, et al., Effects of B-cell directed therapy on the preclinical stage of rheumatoid arthritis: the PRAIRI study, Ann. Rheum. Dis. 78 (2) (2019) 179–185.
[80] Strategy for the Prevention of Onset of Clinically-Apparent Rheumatoid Arthritis (StopRA), ClinicalTrials.gov, identifier NCT02603146 Cited 3-Mar-2018. Available from: https://clinicaltrials.gov/ct2/show/NCT02603146.

[81] Treat Early Arthralgia to Reverse or Limit Impending Exacerbation to Rheumatoid Arthritis (TREAT EARLIER), (Netherlands Trial Register NL4599). Available from: https://www.trialregister.nl/trial/4599.
[82] M. Al-Laith, M. Jasenecova, S. Abraham, A. Bosworth, I.N. Bruce, C.D. Buckley, et al., Arthritis prevention in the pre-clinical phase of RA with abatacept (the APIPPRA study): a multi-centre, randomised, double-blind, parallel-group, placebo-controlled clinical trial protocol, Trials 20 (1) (2019) 429.
[83] V.M. Holers, M.K. Demoruelle, K.A. Kuhn, J.H. Buckner, W.H. Robinson, Y. Okamoto, et al., Rheumatoid arthritis and the mucosal origins hypothesis: protection turns to destruction, Nat. Rev. Rheumatol. 14 (9) (2018) 542–557.
[84] R. Knight, A. Vrbanac, B.C. Taylor, A. Aksenov, C. Callewaert, J. Debelius, et al., Best practices for analysing microbiomes, Nat. Rev. Microbiol. 16 (7) (2018) 410–422.
[85] H.E. Blum, The human microbiome, Adv. Med. Sci. 62 (2) (2017) 414–420.
[86] E. Org, Y. Blum, S. Kasela, M. Mehrabian, J. Kuusisto, A.J. Kangas, et al., Relationships between gut microbiota, plasma metabolites, and metabolic syndrome traits in the METSIM cohort, Genome Biol. 18 (1) (2017) 70.
[87] A. Zaccardelli, H.M. Friedlander, J.A. Ford, J.A. Sparks, Potential of lifestyle changes for reducing the risk of developing rheumatoid arthritis: is an ounce of prevention worth a pound of cure? Clin. Ther. 41 (7) (2019) 1323–1345.
[88] M. Falahee, A. Finckh, K. Raza, M. Harrison, Preferences of patients and at-risk individuals for preventive approaches to rheumatoid arthritis, Clin. Ther. 41 (7) (2019) 1346–1354.
[89] M. Harrison, L. Spooner, N. Bansback, K. Milbers, C. Koehn, K. Shojania, et al., Preventing rheumatoid arthritis: preferences for and predicted uptake of preventive treatments among high risk individuals, PLoS One 14 (4) (2019) e0216075.
[90] A. Finckh, M. Escher, M.H. Liang, N. Bansback, Preventive treatments for rheumatoid arthritis: issues regarding patient preferences, Curr. Rheumatol. Rep. 18 (8) (2016) 51.
[91] A.A. Marshall, A. Zaccardelli, Z. Yu, M.G. Prado, X. Liu, R. Miller Kroouze, et al., Effect of communicating personalized rheumatoid arthritis risk on concern for developing RA: a randomized controlled trial, Patient Educ. Couns. 102 (5) (2019) 976–983.
[92] Statins for the Prevention of Rheumatoid Arthritis, Cited 21 March 2018. Available from: http://www.trialregister.nl/trialreg/admin/rctview.asp?TC=5265.
[93] C.J. Greenbaum, C. Speake, J. Krischer, J. Buckner, P.A. Gottlieb, D.A. Schatz, et al., Strength in numbers: opportunities for enhancing the development of effective treatments for type 1 diabetes—the TrialNet experience, Diabetes 67 (7) (2018) 1216–1225.
[94] J.M. Hall, C.F. Fowler, F. Barrett, R.W. Humphry, M. Van Drimmelen, S.M. MacRury, HbA1c determination from HemaSpot blood collection devices: comparison of home prepared dried blood spots with standard venous blood analysis, Diabet. Med. 37 (9) (2020) 1463–1470, https://doi.org/10.1111/dme.14110.
[95] S.R. Hogue, M.F. Gomez, W.V. da Silva, C.M.A. Pierce, Customized at-home stool collection protocol for use in microbiome studies conducted in cancer patient populations, Microb. Ecol. 78 (4) (2019) 1030–1034.

Chapter 8

Development of multi-omics approach in autoimmune diseases

May Y. Choi[a], Marvin J. Fritzler[a], and Michael Mahler[b]

[a]Cumming School of Medicine, University of Calgary, Calgary, AB, Canada, [b]Inova Diagnostics, Inc., San Diego, CA, United States

Abbreviations

ACPA	anti-citrullinated protein antibodies
ChiP-Seq	chromatin immunoprecipitation-sequencing
LN	lupus nephritis
MHC	major histocompatibility complex
MO	multi-omics
NGS	next generation sequencing
RA	rheumatoid arthritis
SLE	systemic lupus erythematosus

1. Introduction

Multi-omics (MO) refers to the combination of different biological omics data sources and is an analytic approach in which the data sets are derived from multiple "omes," such as the genome, proteome, transcriptome, epigenome, metabolome, and/or microbiome [1, 2]. The MO approach takes advantage of complex biological big data to find novel associations between biological entities, pinpoint relevant biomarkers and build elaborate markers of disease, its pathophysiology and potential interventional therapeutics. In doing so, MO integrates diverse omics data to find coherently matching geno-pheno-enviro-type relationships or associations. In its early applications, MO approaches focused mostly on cancer [3]. However, during the past years, there is a rapid increase in studies and reports on MO in autoimmune diseases [4–7]. This brief chapter aims to provide a high-level overview on the concepts of MO and how it could be applied to autoimmune research and patient management with a few selected examples.

Precision Medicine and Artificial Intelligence. https://doi.org/10.1016/B978-0-12-820239-5.00004-8

2. Translation of multi-omics into clinical application

Autoimmune diseases are a group of heterogeneous disorders that are mostly defined and diagnosed based on clinical signs and symptoms. However, novel approaches are now increasingly used to define such diseases based on molecular/MO classifications [8–10] (Fig. 1). Although it is intriguing to imagine that thousands if not millions of datapoints are assessed and utilized for clinical decision making, the regulatory hurdles (i.e., approval of biomarkers for clinical use) will likely limit the sources and amount of data feeding into decision models.

Statistical analysis and data science are key to the interpretation of MO approaches and are discussed in detail in a different chapter of this book. As a recent and promising example, a MO model was proposed to monitor the drug response in rheumatoid arthritis (RA) based on molecular remission [11], where using high-dimensional phenotyping (transcriptome, serum proteome and molecular profile) identified signatures associated with clinical remission. Although this application has not been translated into clinical practice, it follows successful utilization of approaches in oncology as discussed in Chapter 4.

3. Types of omics technologies

3.1 Epigenomics

Epigenomics refers to DNA or DNA-associated protein modifications such as acetylation, deacetylation, and methylation [12–16] which can be studied using next generation sequencing (NGS). Cell fate and functions can be altered by such modifications in DNA and histones, apart from genetic changes. These changes can be linked to environmental factors that can be passed onto progeny. Epigenetic changes in the genome can also serve as markers for metabolic syndromes, cardiovascular diseases, and physiological disorders. These changes can be cell-and tissue-specific. Thus, it is critical to identify the epigenetic changes during native and diseased states.

3.2 Transcriptomics

Transcriptomics refers to the techniques used to study the sum of all RNA transcripts and their expression levels in an organism or even in single cells [17, 18]. Although only 2% of the DNA is translated into protein via coding mRNA, almost 80% of the genome is transcribed into short RNAs, including microRNA, piwi-interacting RNA, small nuclear and small cytoplasmic RNAs. Apart from acting as an intermediate between DNA and protein, RNA also has structural and regulatory functions during native and altered states. These RNAs have been shown to have a role in myocardial infarction, adipose differentiation, diabetes, endocrine regulation, neuron development, and others [14, 19, 20]. Consequently, it is crucial to understand which transcripts are expressed at a given time. In addition to NGS, probe-based assays, and RNA-seq are also used for transcriptome analyses.

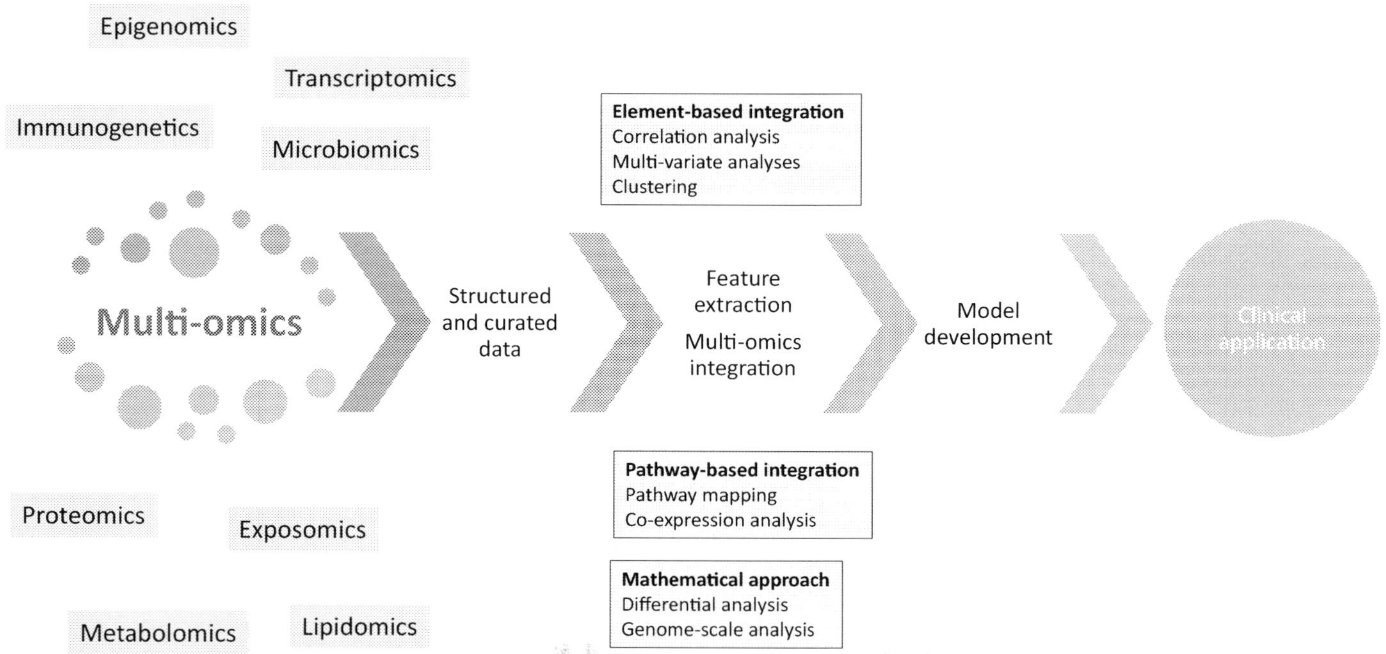

FIG. 1 Potential pathway for the application of multi-omics approach for autoimmune disease management. In this hypothetical pathway, data derived from various omics technologies are combined into a structure and curated database that provides the foundation for feature extraction. Due to the large amount of data, artificial intelligence applications are highly desired. This will eventually allow for feature extraction and data reduction as well as clinical application developments.

3.3 Proteomics

Proteomics involves identification of specific proteins, their functions, levels, modifications, and interactions. Protein-protein interactions can be studied through phage display, classical yeast two hybrid, affinity purification, protein/peptide arrays and chromatin immunoprecipitation-sequencing (ChiP-Seq). The functions, structural roles, and stability of the majority of proteins are regulated through post-translational modifications, such as phosphorylation, acetylation, ubiquitination, nitrosylation, and glycosylation. Mass spectroscopy-based techniques are being used to analyze the global proteomic changes and quantifying the post-translational modifications.

3.4 Metabolomics

Metabolomics comprises the study of all the metabolites present in a cell, tissue (including body fluid), or organism, and utilizes small molecules, carbohydrates, peptides, lipids, nucleosides, and catabolic products [21–27]. It represents the final product of gene transcription and expression and consists of both signaling and structural molecules. Computational methods including machine learning might be key for integration of metabolomic data which is mostly derived from mass spectrometry [28, 29]. In addition, standardization of metabolomic strategies for clinical applications are highly warranted [30].

3.5 Immunogenomics

Immunogenomics focusses on the genetic basis of the immune response. It includes the analyses of integrated immunological pathways and the identification of genetic variations leading to immune defects, which may result in the identification of new drug targets for immune related diseases (such as autoimmune disorders) [31]. Interactions between immune cell receptors and proteins that are linked to disease susceptibility may provide insights on how the immune system is involved in autoimmune diseases.

3.6 Microbiomics

Microbiomics refers to the detection of the microorganisms that reside on or within tissues and biofluids and include bacteria, archaea, fungi, protists and viruses [32–34], along with the corresponding anatomical sites in which they reside. Microbes have been found in virtually every organ system including skin, mucosal surfaces, and especially in the gut. The microbiome present in humans is very complex with over 100 trillion bacteria that are estimated to occupy the gastrointestinal tract. Associations of microbiome with disease development or features have been described in many autoimmune conditions including, but not limited to systemic lupus erythematosus (SLE) [23,35–39], diabetes [40–42], obesity, malignancies, inflammatory bowel disease [43–46], celiac disease [47–49], autoimmune liver disease [50–53], and arthritis [35,36,43,54–63]. Thus, characterization of the microbiome has gained significant attention. The primary approaches to characterization of the microbiome is to sequence the 16S rRNA genes or shotgun sequencing.

3.7 Glycomics

Glycosylation can have major effects on many protein features and characteristics including but not limited to function, stability, activity (for enzymes) or shelf-life. In autoimmunity, glycosylation of antibodies has become a hot topic of research [64–73]. For example, it has been demonstrated that changes in glycan structure in the Fc domain of immunoglobulins can have both, pro- as well as anti-inflammatory properties. In addition, just a couple of years ago, it became evident that the introduction of Fab glycosylation of anti-citrullinated protein antibodies (ACPA) in RA might predict the transition into imminent RA [69,74–76]. Profiling both Fc and Fab glycosylation might provide a novel omics approach to better understand mechanisms in patients with autoimmune diseases.

3.8 Other-omics fields

Besides the omics areas described above, there are other fields that are evolving including exposomics [77], lipidomics [30,78], and others (Table 1). It is likely, that with the expanding utilization of data science, system biology and artificial intelligence, there will be a flood of novel omic definitions and applications. Virtually every data source could be considered a new "omics."

4. Progress and success in autoimmunity

As pointed out in the introduction, progress in applying MO to autoimmune conditions is currently limited, but rapidly growing. A few examples and initiatives providing hope that MO concepts are well suited to improve patient care in autoimmune diseases are summarized in Table 2. Lorenzon et al. introduced the term TRANSIMMUNOM in 2018 which has the goal of cross-phenotyping patients with autoimmune and autoinflammatory diseases [79]. The study includes patients with a wide variety of conditions, including familial Mediterranean fever, ulcerative colitis, Crohn's disease, spondyloarthritis, uveitis, RA, myositis, vasculitis, type 1 diabetes, SLE as well as anti-phospholipid syndrome. Omics applied in this context include deep-immunophenotyping, proteome, transcriptome, T-cell receptor and microbiome which is complemented with other data sources such as biological data (routine biology, serology status, immunochemistry and HLA), disease specific clinical data as well as common data (e.g., demographics, physical examination and many more).

TABLE 1 Overview of omics technologies.

Discipline	Application	Comment	Methods/approach
Genomics	Genome-wide mutational analysis Exosome mutational analysis Gene exon analysis	Whole exosome sequencing or targeted gene sequencing	NGS GWAS
Epigenomics	Genome-wide mapping of methylation patterns or epigenetic markers	Methylomics ChIP-sequencing	Array based, bisulfite-seq
Transcriptomics	Genome-wide differential gene expression analysis Differential gene expression analysis	RNA-sequencing Microarray	RNA-sequencing and quantitation NGS Hybridization microRNA analysis
Proteomics	Differential protein abundance or expression analysis	RPPA Deep-proteomics	MS, RPPA chips
Metabolomics	Metabolite expression analysis	Deep-metabolomics	GC/MS, nuclear magnetic resonance, LC, MS/MS
Peptidomics	Similar to proteomics, but on peptide level	Epitope mapping	MS
Glycomics	Glycan profiling of proteins (e.g., antibodies)	Description of the glycan structures on proteins	HPLC, MS, lectin arrays
Interactomics	Mapping of DNA-protein interaction sites Histone modification maps	ChIP-sequencing	ChIP-sequencing
Lipidomics	Interaction analysis of lipids with other lipids, proteins and metabolites	Extracted and separated subjected to MS analysis using chromatography techniques, such as thin layer chromatography (TLC), gas chromatography (GC) or HPLC	Several HPLC, MS
Exposomics	Studies internal and external exposure		Several HPLC, MS

Abbreviations: *GC/MS*, gas chromatography-mass spectrometry; *HPLC*, high pressure liquid chromatography; *LC*, liquid chromatography; *MS*, mass spectrometry; *MS/MS*, tandem mass spectrometry; *NGS*, next generation sequencing; *RPPA*, reverse phase protein arrays.

TABLE 2 Overview of applied omics technologies in autoimmune diseases with selected examples and areas of focus.

Disease area	Comment	References
Rheumatoid arthritis	Molecular signatures for treatment response and selection Disease activity Prediction of imminent disease (e.g., glycan profiles)	[8,11,15,43,64,74,75, 79–82]
Systemic lupus erythematosus	Disease activity, Lupus nephritis	[4,8,15,16,23,24,79, 83–86]
Autoimmune liver disease	Molecular diagnostics, risk stratification, prognosis	[5,50,87,88]
Celiac disease	Exposome, microbiome, antibiotic exposure	[47–49]
Inflammatory bowel disease (IBD)	Screen to diagnose IBD	[43,44,54,89,90]
Sjögren's syndrome	Transcriptome analysis, inflammation clusters Microbiome	[10,16,35,36,45,91,92]
Autoimmune diabetes	Exposome, microbiome Disease prevention	[40,93–95]

Another MO approach in autoimmunity focusses on drug-response monitoring. For example, a study by Tasaki et al., demonstrated that treatment in RA altered the transcriptome, serum proteome and immunophenotype level closer to healthy individuals [11]. In addition, the study also revealed molecular profiles that are resistant to drug treatment.

Similarly, MO has also been studied in SLE especially to comprehend the intrinsic molecular mechanism and underlying pathogenesis [6,8,16,83]. Through genomics, multiple susceptibility genes have been identified such as major histocompatibility complex (MHC) [96], ITGAM, PXK, KIAA1542, and other loci [97]. A lot of work has been conducted in immunogenetics in SLE especially around the type I interferon response including mono- and polygenetic factors as well as epigenetic influences [98,99]. DNA hypomethylation, abnormal microRNA and mRNA expression are thought to important in the pathogenesis of SLE and could be potential markers of disease [100,101]. Expanding our knowledge on the functional mechanisms of causal genetic variants of SLE will allow us to incorporate genetic associations into personalized medicine approaches. More specifically, proteomic studies have helped identify abnormal expression of proteins in SLE that can distinguish between patients with inactive or active lupus nephritis (LN) [102]. Urine metabolomics studies have also suggested that different taurine and citric acid levels in the urine can discriminate between classes of LN [103]. Sera metabolomic studies have demonstrated that SLE patients have dysregulated lipid metabolism and amino acids compared to healthy individuals [21]. Although many molecules have been identified through different omic methods, a single molecule or omics cannot fully explain the pathogenesis of a complex and heterogeneous disease such as SLE. A novel approach to studying MO in SLE is called "integrative MO analysis," an approach that tries to identify networks and relationships among different types of omics data.

In conclusion, MO approaches hold promise to improve management of autoimmune diseases. However, the current benefits are mainly found in clinical and translational research. It will require cross-functional efforts to translate system biology findings into clinical practice of autoimmunity.

Competing interests

M.Y. Choi is supported by the Lupus Foundation of America Gary S. Gilkeson Career Development Award.

M.J. Fritzler is a consultant to and has received honoraria from Inova Diagnostics, Janssen Pharmaceutics, Grifols, and Alexion Canada.

M. Mahler is employee of Inova Diagnostics.

References

[1] I. Subramanian, et al., Multi-omics data integration, interpretation, and its application, Bioinf. Biol. Insights 14 (2020). 1177932219899051.

[2] C. Wu, et al., A selective review of multi-level omics data integration using variable selection, High Throughput 8 (1) (2019) 4.

[3] G. de Anda-Jáuregui, E. Hernández-Lemus, Computational oncology in the multi-omics era: state of the art, Front. Oncol. 10 (2020) 423.

[4] M.J. Fritzler, M. Mahler, Redefining systemic lupus erythematosus—SMAARTT proteomics, Nat. Rev. Rheumatol. 14 (8) (2018) 451–452.

[5] V. Ronca, et al., Precision medicine in primary biliary cholangitis, J. Dig. Dis. 20 (7) (2019) 338–345.

[6] W. Song, et al., Advances in applying of multi-omics approaches in the research of systemic lupus erythematosus, Int. Rev. Immunol. 39 (4) (2020) 163–173.

[7] T.Y. Wang, et al., Identification of regulatory modules that stratify lupus disease mechanism through integrating multi-omics data, Mol. Ther. Nucleic Acids 19 (2020) 318–329.

[8] G. Barturen, et al., Moving towards a molecular taxonomy of autoimmune rheumatic diseases, Nat. Rev. Rheumatol. 14 (2) (2018) 75–93.

[9] M.J. Lewis, et al., Autoantibodies targeting TLR and SMAD pathways define new subgroups in systemic lupus erythematosus, J. Autoimmun. 91 (2018) 1–12.

[10] J.A. James, et al., Unique Sjogren's syndrome patient subsets defined by molecular features, Rheumatology (Oxford) 59 (4) (2020) 860–868.

[11] S. Tasaki, et al., Multi-omics monitoring of drug response in rheumatoid arthritis in pursuit of molecular remission, Nat. Commun. 9 (1) (2018) 2755.

[12] T.H. Akbaba, et al., Epigenetics for clinicians from the perspective of pediatric rheumatic diseases, Curr. Rheumatol. Rep. 22 (8) (2020) 46.

[13] M. Dolcino, et al., Editorial: role of epigenetics in autoimmune diseases, Front. Immunol. 11 (2020) 1284.

[14] L. Zhang, et al., Clinical significance of miRNAs in autoimmunity, J. Autoimmun. 109 (2020) 102438.

[15] E. Ballestar, A.H. Sawalha, Q. Lu, Clinical value of DNA methylation markers in autoimmune rheumatic diseases, Nat. Rev. Rheumatol. 16 (9) (2020) 514–524.

[16] E. Carnero-Montoro, et al., Epigenome-wide comparative study reveals key differences between mixed connective tissue disease and related systemic autoimmune diseases, Front. Immunol. 10 (2019) 1880.

[17] C. Buccitelli, M. Selbach, mRNAs, proteins and the emerging principles of gene expression control, Nat. Rev. Genet. 21 (10) (2020) 630–644.

[18] M. Efremova, et al., Immunology in the era of single-cell technologies, Annu. Rev. Immunol. 38 (2020) 727–757.

[19] Q. Luo, et al., Identification of circular RNAs hsa_circ_0044235 and hsa_circ_0068367 as novel biomarkers for systemic lupus erythematosus, Int. J. Mol. Med. 44 (4) (2019) 1462–1472.

[20] Q. Luo, et al., Circular RNAs hsa_circ_0000479 in peripheral blood mononuclear cells as novel biomarkers for systemic lupus erythematosus, Autoimmunity 53 (3) (2020) 167–176.

[21] X. Ouyang, et al., ^{1}H NMR-based metabolomic study of metabolic profiling for systemic lupus erythematosus, Lupus 20 (13) (2011) 1411–1420.

[22] R.A. Kellogg, J. Dunn, M.P. Snyder, Personal omics for precision health, Circ. Res. 122 (9) (2018) 1169–1171.

[23] Q. Zhang, et al., Fecal metabolomics and potential biomarkers for systemic lupus erythematosus, Front. Immunol. 10 (2019) 976.

[24] W. Song, et al., Advances in applying of multi-omics approaches in the research of systemic lupus erythematosus, Int. Rev. Immunol. (2020) 1–11.

[25] X. Zhang, et al., Gut microbiome and metabolome were altered and strongly associated with platelet count in adult patients with primary immune thrombocytopenia, Front. Microbiol. 11 (2020) 1550.

[26] E.E. Balashova, et al., Metabolomic diagnostics and human digital image, Pers. Med. 16 (2) (2019) 133–144.

[27] G. Carneiro, et al., Novel strategies for clinical investigation and biomarker discovery: a guide to applied metabolomics, Horm. Mol. Biol. Clin. Invest. 38 (3) (2019), https://doi.org/10.1515/hmbci-2018-0045.

[28] T. Eicher, et al., Metabolomics and multi-omics integration: a survey of computational methods and resources, Metabolites 10 (5) (2020) 202.

[29] U.W. Liebal, et al., Machine learning applications for mass spectrometry-based metabolomics, Metabolites 10 (6) (2020) 243.

[30] N.P. Long, et al., Toward a standardized strategy of clinical metabolomics for the advancement of precision medicine, Metabolites 10 (2) (2020) 51.

[31] N. Hagberg, C. Lundtoft, L. Rönnblom, Immunogenetics in systemic lupus erythematosus: transitioning from genetic associations to cellular effects, Scand J Immunol 92 (4) (2020), https://doi.org/10.1111/sji.12894.PMID: 32428248, e12894.

[32] E. Dekaboruah, et al., Human microbiome: an academic update on human body site specific surveillance and its possible role, Arch. Microbiol. 202 (8) (2020) 2147–2167.

[33] J.A. Gilbert, et al., Current understanding of the human microbiome, Nat. Med. 24 (4) (2018) 392–400.

[34] X. Su, et al., Method development for cross-study microbiome data mining: challenges and opportunities, Comput. Struct. Biotechnol. J. 18 (2020) 2075–2080.

[35] M.F. Konig, The microbiome in autoimmune rheumatic disease, Best Pract. Res. Clin. Rheumatol. 34 (1) (2020) 101473.

[36] F. De Luca, Y. Shoenfeld, The microbiome in autoimmune diseases, Clin. Exp. Immunol. 195 (1) (2019) 74–85.

[37] C. Dehner, R. Fine, M.A. Kriegel, The microbiome in systemic autoimmune disease: mechanistic insights from recent studies, Curr. Opin. Rheumatol. 31 (2) (2019) 201–207.

[38] N. Katz-Agranov, G. Zandman-Goddard, The microbiome and systemic lupus erythematosus, Immunol. Res. 65 (2) (2017) 432–437.

[39] Y. Shoenfeld, M. Ehrenfeld, O. Perry, The kaleidoscope of autoimmunity—from genes to microbiome, Clin. Immunol. 199 (2019) 1–4.

[40] P. Zheng, Z. Li, Z. Zhou, Gut microbiome in type 1 diabetes: a comprehensive review, Diabetes Metab. Res. Rev. 34 (7) (2018) e3043.

[41] A. Giongo, et al., Toward defining the autoimmune microbiome for type 1 diabetes, ISME J. 5 (1) (2011) 82–91.

[42] A. Paun, C. Yau, J.S. Danska, Immune recognition and response to the intestinal microbiome in type 1 diabetes, J. Autoimmun. 71 (2016) 10–18.

[43] J.C. Clemente, J. Manasson, J.U. Scher, The role of the gut microbiome in systemic inflammatory disease, BMJ 360 (2018) j5145.

[44] S.A. Tabatabaeizadeh, et al., Vitamin D, the gut microbiome and inflammatory bowel disease, J. Res. Med. Sci. 23 (2018) 75.

[45] D.P. Bogdanos, L.I. Sakkas, From microbiome to infectome in autoimmunity, Curr. Opin. Rheumatol. 29 (4) (2017) 369–373.

[46] R. Vangoitsenhoven, G.A.M. Cresci, Role of microbiome and antibiotics in autoimmune diseases, Nutr. Clin. Pract. 35 (3) (2020) 406–416.

[47] M.M. Leonard, et al., Celiac disease genomic, environmental, microbiome, and metabolomic (CDGEMM) study design: approach to the future of personalized prevention of celiac disease, Nutrients 7 (11) (2015) 9325–9336.

[48] J.T. Russell, et al., Genetic risk for autoimmunity is associated with distinct changes in the human gut microbiome, Nat. Commun. 10 (1) (2019) 3621.

[49] S. Krishnareddy, The microbiome in celiac disease, Gastroenterol. Clin. N. Am. 48 (1) (2019) 115–126.

[50] Y. Wei, et al., Alterations of gut microbiome in autoimmune hepatitis, Gut 69 (3) (2020) 569–577.

[51] W. Cai, et al., Intestinal microbiome and permeability in patients with autoimmune hepatitis, Best Pract. Res. Clin. Gastroenterol. 31 (6) (2017) 669–673.

[52] K. Glassner, et al., Autoimmune liver disease and the enteric microbiome, AIMS Microbiol. 4 (2) (2018) 334–346.

[53] B. Li, et al., The microbiome and autoimmunity: a paradigm from the gut-liver axis, Cell. Mol. Immunol. 15 (6) (2018) 595–609.

[54] E.A. Yamamoto, T.N. Jorgensen, Relationships between vitamin D, gut microbiome, and systemic autoimmunity, Front. Immunol. 10 (2019) 3141.

[55] A.D. Proal, P.J. Albert, T.G. Marshall, The human microbiome and autoimmunity, Curr. Opin. Rheumatol. 25 (2) (2013) 234–240.

[56] M. Cojocaru, B. Chicos, The human microbiome in autoimmune diseases, Rom. J. Intern. Med. 52 (4) (2014) 285–288.

[57] K.D. Deane, et al., Genetic and environmental risk factors for rheumatoid arthritis, Best Pract. Res. Clin. Rheumatol. 31 (1) (2017) 3–18.

[58] S. Dudics, et al., Natural products for the treatment of autoimmune arthritis: their mechanisms of action, targeted delivery, and interplay with the host microbiome, Int. J. Mol. Sci. 19 (9) (2018) 2508.

[59] A.S. Bergot, R. Giri, R. Thomas, The microbiome and rheumatoid arthritis, Best Pract. Res. Clin. Rheumatol. 33 (6) (2019) 101497.

[60] R. Bodkhe, B. Balakrishnan, V. Taneja, The role of microbiome in rheumatoid arthritis treatment, Ther. Adv. Musculoskelet. Dis. 11 (2019). 1759720X19844632.

[61] P.M. Wells, et al., RA and the microbiome: do host genetic factors provide the link? J. Autoimmun. 99 (2019) 104–115.

[62] M. du Teil Espina, et al., Talk to your gut: the oral-gut microbiome axis and its immunomodulatory role in the etiology of rheumatoid arthritis, FEMS Microbiol. Rev. 43 (1) (2019) 1–18.

[63] J. Manasson, R.B. Blank, J.U. Scher, The microbiome in rheumatology: where are we and where should we go? Ann. Rheum. Dis. 79 (6) (2020) 727–733.

[64] E.M. Vletter, et al., A comparison of immunoglobulin variable region N-linked glycosylation in healthy donors, autoimmune disease and lymphoma, Front. Immunol. 11 (2020) 241.

[65] R. Goulabchand, et al., Impact of autoantibody glycosylation in autoimmune diseases, Autoimmun. Rev. 13 (7) (2014) 742–750.

[66] M.H. Biermann, et al., Sweet but dangerous—the role of immunoglobulin G glycosylation in autoimmunity and inflammation, Lupus 25 (8) (2016) 934–942.

[67] M. Collin, Antibody glycosylation predicts relapse in autoimmune vasculitis, EBioMedicine 17 (2017) 15.

[68] A. Visser, et al., Acquiring new N-glycosylation sites in variable regions of immunoglobulin genes by somatic hypermutation is a common feature of autoimmune diseases, Ann. Rheum. Dis. 77 (10) (2018) e69.

[69] R.D. Vergroesen, et al., N-glycosylation site analysis of citrullinated antigen-specific B-cell receptors indicates alternative selection pathways during autoreactive B-cell development, Front. Immunol. 10 (2019) 2092.

[70] Y. Reinke, et al., Sugars make the difference—glycosylation of cardiodepressant antibodies regulates their activity in dilated cardiomyopathy, Int. J. Cardiol. 292 (2019) 156–159.

[71] Y. Kronimus, et al., IgG Fc N-glycosylation: alterations in neurologic diseases and potential therapeutic target? J. Autoimmun. 96 (2019) 14–23.

[72] X. Zhou, et al., Glycomic analysis of antibody indicates distinctive glycosylation profile in patients with autoimmune cholangitis, J. Autoimmun. 113 (2020) 102503.

[73] S. Boune, et al., Principles of N-linked glycosylation variations of IgG-based therapeutics: pharmacokinetic and functional considerations, Antibodies (Basel) 9 (2) (2020) 22.

[74] T. Kissel, et al., On the presence of HLA-SE alleles and ACPA-IgG variable domain glycosylation in the phase preceding the development of rheumatoid arthritis, Ann. Rheum. Dis. 78 (12) (2019) 1616–1620.

[75] L. Hafkenscheid, et al., N-linked glycans in the variable domain of IgG anti-citrullinated protein antibodies predict the development of rheumatoid arthritis, Arthritis Rheumatol. 71 (10) (2019) 1626–1633.

[76] L.A. Trouw, R.E. Toes, Rheumatoid arthritis: autoantibody testing to predict response to therapy in RA, Nat. Rev. Rheumatol. 12 (10) (2016) 566–568.

[77] P. Vineis, et al., The exposome in practice: design of the EXPOsOMICS project, Int. J. Hyg. Environ. Health 220 (2 Pt. A) (2017) 142–151.

[78] Y. Qi, et al., Bile acid signaling in lipid metabolism: metabolomic and lipidomic analysis of lipid and bile acid markers linked to anti-obesity and anti-diabetes in mice, Biochim. Biophys. Acta 1851 (1) (2015) 19–29.
[79] R. Lorenzon, et al., Clinical and multi-omics cross-phenotyping of patients with autoimmune and autoinflammatory diseases: the observational TRANSIMMUNOM protocol, BMJ Open 8 (8) (2018) e021037.
[80] H.H. Chang, et al., A molecular signature of preclinical rheumatoid arthritis triggered by dysregulated PTPN22, JCI Insight 1 (17) (2016) e90045.
[81] Q. Wang, R. Xu, Data-driven multiple-level analysis of gut-microbiome-immune-joint interactions in rheumatoid arthritis, BMC Genomics 20 (1) (2019) 124.
[82] J.W. Whitaker, et al., Integrative omics analysis of rheumatoid arthritis identifies non-obvious therapeutic targets, PLoS One 10 (4) (2015) e0124254.
[83] G. Barturen, M.E. Alarcon-Riquelme, SLE redefined on the basis of molecular pathways, Best Pract. Res. Clin. Rheumatol. 31 (3) (2017) 291–305.
[84] B. Chen, L. Sun, X. Zhang, Integration of microbiome and epigenome to decipher the pathogenesis of autoimmune diseases, J. Autoimmun. 83 (2017) 31–42.
[85] M.R.W. Barber, et al., Economic evaluation of lupus nephritis in the systemic lupus international collaborating clinics inception cohort using a multistate model approach, Arthritis Care Res. 70 (9) (2018) 1294–1302.
[86] H. Wu, C. Chang, Q. Lu, The epigenetics of lupus erythematosus, Adv. Exp. Med. Biol. 1253 (2020) 185–207.
[87] W.H. Robinson, L. Steinman, P.J. Utz, Protein and peptide array analysis of autoimmune disease, Biotechniques Suppl (2002) 66–69.
[88] L. Cristoferi, et al., Prognostic models in primary biliary cholangitis, J. Autoimmun. 95 (2018) 171–178.
[89] J.A. Berinstein, et al., The IBD SGI diagnostic test is frequently used by non-gastroenterologists to screen for inflammatory bowel disease, Inflamm. Bowel Dis. 24 (5) (2018) e18.
[90] M. Kumar, M. Garand, S. Al Khodor, Integrating omics for a better understanding of inflammatory bowel disease: a step towards personalized medicine, J. Transl. Med. 17 (1) (2019) 419.
[91] N.G. Nikitakis, et al., The autoimmunity-oral microbiome connection, Oral Dis. 23 (7) (2017) 828–839.
[92] T.A. van der Meulen, et al., Microbiome in Sjogren's syndrome: here we are, Ann. Rheum. Dis. (2020). Jul 22, Online ahead of print.
[93] C.J. Rosen, J.R. Ingelfinger, Traveling down the long road to type 1 diabetes mellitus prevention, N. Engl. J. Med. 381 (7) (2019) 666–667.
[94] C.B. Lambring, et al., Impact of the microbiome on the immune system, Crit. Rev. Immunol. 39 (5) (2019) 313–328.
[95] E. Marietta, et al., Role of the intestinal microbiome in autoimmune diseases and its use in treatments, Cell. Immunol. 339 (2019) 50–58.
[96] Z. Fronek, et al., Major histocompatibility complex genes and susceptibility to systemic lupus erythematosus, Arthritis Rheum. 33 (10) (1990) 1542–1553.
[97] J.B. Harley, et al., Genome-wide association scan in women with systemic lupus erythematosus identifies susceptibility variants in ITGAM, PXK, KIAA1542 and other loci, Nat. Genet. 40 (2) (2008) 204–210.
[98] T.L. Wampler Muskardin, et al., Lessons from precision medicine in rheumatology, Mult. Scler. 26 (5) (2020) 533–539.

[99] Y. Ghodke-Puranik, T.B. Niewold, Immunogenetics of systemic lupus erythematosus: a comprehensive review, J. Autoimmun. 64 (2015) 125–136.

[100] B. Richardson, DNA methylation and autoimmune disease, Clin. Immunol. 109 (1) (2003) 72–79.

[101] S. Ding, et al., Decreased microRNA-142-3p/5p expression causes CD4+ T cell activation and B cell hyperstimulation in systemic lupus erythematosus, Arthritis Rheum. 64 (9) (2012) 2953–2963.

[102] K. Mosley, et al., Urinary proteomic profiles distinguish between active and inactive lupus nephritis, Rheumatology (Oxford) 45 (12) (2006) 1497–1504.

[103] L.E. Romick-Rosendale, et al., Identification of urinary metabolites that distinguish membranous lupus nephritis from proliferative lupus nephritis and focal segmental glomerulosclerosis, Arthritis Res. Ther. 13 (6) (2011) R199.

Chapter 9

N-of-1 trials: Implications for clinical practice and personalized clinical trials

Joanne Bradbury[a] and Michael Mahler[b]
[a]*Southern Cross University, Gold Coast, QLD, Australia,* [b]*Inova Diagnostics, Inc., San Diego, CA, United States*

Abbreviations

AI	artificial intelligence
HMGCR	3-hydroxy-3-methyl-glutaryl-coenzyme A reductase
ICMJE	International Committee of Medical Journal editors
IMNM	immune mediated necrotizing myopathies
irAE	immune related adverse events
IVIG	intravenous immunoglobulin
RCT	randomized clinical trial
SCED	single case experimental design
SPSS	Statistical Package for the Social Sciences
SRP	signal recognition particle

1. Introduction

Traditionally, most studies in autoimmune research are based on randomized control trials (RCT) or cohort studies. More recently, several studies attempted to gather evidence from a single individual to study how biomarkers and other measures change over time or how baseline results might impact outcomes [1–6]. These types of single case observational or experimental designs allow for multiple baseline monitoring of individual effects and then prospectively tracking changes over time on outcomes of interest. The single case experimental design (SCED), also known as an N-of-1 trial, could have implications for future clinical trial designs [6–14]. More specifically, conventional observational studies and clinical trials are based on large number of patients but limited datapoints per individual. Most of the features designed to minimize bias in clinical trial design can be applied to N-of-1 trials.

Precision Medicine and Artificial Intelligence. https://doi.org/10.1016/B978-0-12-820239-5.00008-5

2. Lessons learned from case reports and case series

In case of rare diseases (like many autoimmune diseases), clinical case reports have been used to document and educate clinicians [10, 15–17]. Many of those case reports have been instrumental in understanding the disease and helping to manage rare conditions. This can be very helpful in rare autoimmune diseases such as immune mediated necrotizing myopathies (IMNM) which is associated with statin use [18, 19]. Since the identification of statin associated IMNM, several case reports have been published describing not only the clinical presentations, but also the pathological findings as well as the treatment responses and outcomes [16, 20–22]. Based on experience of leading experts in the field of IMNM as well as compilations of case series publications it was concluded that immune modulation using intravenous immunoglobulin (IVIG) admiration provides good efficacy in this rare condition. However, more standardized data recording as well as capturing more biomarker data has the potential to provide more insights in IMNM and might help to understand the underlying mechanism leading to better treatments. Just recently, the first specific trial for IMNM has been initiated which utilizes anti-HMGCR and anti-SRP antibodies as inclusion criteria (NCT04025632, https://clinicaltrials.gov/ct2/show/NCT04025632). Such case reports have also been helpful in understanding immune related adverse events (irAE) as described elsewhere in this book [9, 23, 24]. However, extrapolation of outcomes of such case findings can be misleading and need to be considered with care. Combining data in case series might prevent over-interpretation of results or bias [23]. In addition, it is important to mention that those reports are mostly based on retrospective data collection and not on a proper data collection protocol.

In clinical practice, N-of-1 clinical trials can be implemented to include off-label use of diagnostic tests or drugs. Although not encouraged by regulatory agencies, those trials can provide significant insights into the treatment options of individuals, especially those with rare conditions [24]. Perhaps the biggest benefits can be expected in the evaluation of individual responses to treatments [25]; specifically whether an individual patient is a responder or non-responder to a drug or treatment that has been proven to be safe and efficacious through early phases of clinical trials. There may also be a role for N-of-1 trials in piloting early clinical drug development as well in repurposing drugs, although this aspect has been controversial [25].

3. Complementary medicine approach

In autoimmune diseases, complementary medicine might be under-utilized and could provide significant potential for individual patients. N-of-1 trials that seek to find the optimal intervention for an individual patient can accommodate routine care as the baseline or control condition [26]. In these cases, the hypothesis tested is whether the treatment is a safe and effective adjunct treatment for the individual patient, given their unique set of circumstances. By placing the person at the center of the research question, the research promotes patient-centered outcomes. Further, complementary medicines are frequently used by patients with serious, chronic diseases who are reluctant to disclose this to their conventional medical practitioner, posing concerns about the safety of untested treatment combinations [27, 28]. Research designs that can incorporate patient preferences have the potential to improve the safety and quality of care.

Many complementary treatments have demonstrated safe and effective adjuncts in autoimmune diseases. For instance, a meta-analysis of RCTs on the effects of a Traditional Chinese Medicine (L. *Paeonia lactiflora*; common name: White Paeony) for pain and inflammation in rheumatoid arthritis found that the addition of paeony with usual treatment (methotrexate) was more effective and was associated with lower adverse effects than methotrexate alone [29]. In another example, a randomized, multicentre, double-blind, placebo-controlled trial in 89 patients with inflammatory bowel disease [IBD, namely ulcerative colitis (UC)] demonstrated that 2 g/d curcumin alongside the standard medications (sulfasalazine or mesalamin) significantly improved clinical and endoscopic outcomes compared with placebo ($P < 0.05$) after 6 months of adjunctive treatment [30]. These studies indicate that traditional and complementary medicines may represent safe and effective adjunct therapies in patients with autoimmune diseases.

N-of-1 trials methodologies are increasingly used to assess the individual effects of complementary treatments as adjunct therapies. In a study [31] of pediatric IBD, a series of 50 N-of-1 trials were undertaken to assess the comparative effects of two different diets. Patient reported outcomes of pain and symptom severity were the primary outcomes. Objective measures such as fecal calprotectin for changes in intestinal inflammation were also included. All patients received personalized results about the most effective diet for their own healthcare. The results across all trials were also aggregated to estimate population effects.

The challenges of using N-of-1 trials to assess complementary medicines as adjunct therapies include the longer duration of such designs, due to the multiple cross-over periods [32] and the delay in treatment effect for most complementary therapy interventions [26]. These factors increase the length of the trial, increasing the burden on the patient. Nevertheless, given that the patient has a central role in the research design on an N-of-1 trial, they are able to terminate the trial at any time, especially if it becomes clear which intervention is more effective. Of course, the advantage of N-of-1 trials for patients is that they have the potential to spare patients from exposure to unnecessary medications [33] or ineffective, expensive adjunct therapies.

4. Study designs

An N-of-1 trial is a cross-over clinical trial for one participant at a time (see Fig. 1). The order of exposure to the comparative treatments is randomized and the participant and researchers/clinicians should be blinded as to the sequence of exposures. Thus, the N-of-1 trial design incorporates the elements of rigor

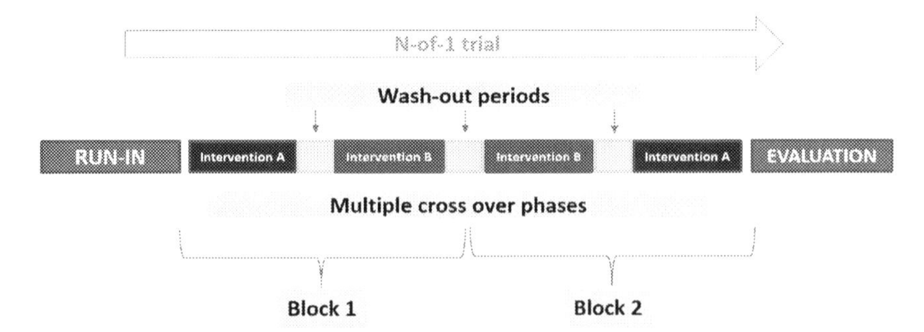

FIG. 1 Schematic drawing of N-of-1 clinical trial design. In this AB BA design, there is a run-in period followed by a wash out period. During the next phase, different interventions are studied using appropriate washout periods as well as multiple cross over approaches. Finally, conclusion will be generated during the evaluation/follow-up phase.

that characterizes the RCT methodology. When more than one participant is included in the same trial design, it is called an N-of-1 series or, more correctly, a series of N-of-1 trials. An N-of-1 series is essentially a cross-over RCT, but with more cross-overs and less participants [34].

Indications for an N-of-1 trial include a treatment with rapid action and withdrawal properties in a chronic condition that is symptomatic but stable over time [34]. In interventions where there may be a slow effect for treatment and a prolonged carry over effect, there are SCED variants that are more appropriate [35]. Principles of conventional trials such as blinding patients to the treatment as well as "wash-out" periods in between the different drugs can be applied to allow the treatment effects to wear off. Comparison of N-of-1 and conventional trials is presented in Table 1.

TABLE 1 Comparison of clinical trial designs.

	Randomized clinical trial (RCT)	N-of-1 trials
Inferences	To determine the treatment effect size in the population	To determine the optimal treatment intervention for an individual patient at a given point in time
Efficacy vs effectiveness	Measures efficacy of treatment in a tightly controlled clinical environment, i.e., strict inclusions/exclusions criteria to increase internal validity, often at the expense of low external validity	Measures effectiveness of treatment through a pragmatic patient-centered approach. For instance, usual care can be incorporated into the design as a baseline or placebo condition

TABLE 1 Comparison of clinical trial designs—cont'd

	Randomized clinical trial (RCT)	N-of-1 trials
Randomization	Patients are randomly allocated to treatment groups	The order of treatment exposures per patient is randomized
Blinding	The process and sequence of group allocation is concealed from researchers and patients; double-blinding implies that group allocation is masked from both patient and researchers throughout the trial, normally by including a placebo	The process and sequence of the randomization process is concealed from the researchers and patients; double blinding implies that the sequence allocation is masked from both patient and researchers throughout the trial, normally using a placebo
Analysis	Inferential statistics are used to enable conclusions based on hypothesis testing or confidence intervals	Data analysis involves visual observation of time series plots for one individual. When analyzing a series of N-of-1 trials data, hierarchical or Bayesian modeling approaches are used to estimate population effects
Power	Based on number of participants	Based on number of observations
Aggregating data	Can be included in meta-analyses of RCTs	Can be included in meta-analyses of N-of-1 trials
Resources	Expensive costs and time consuming to run, out of reach for individual practitioners	Within reach of individual health practitioners, but can be time consuming for practitioner and patient

5. Statistical considerations

The approach to best analyze data collected in N-of-1 trial designs represents one of the most challenging aspects of this type of clinical trials [36]. Most authors reporting N-of-1 trials use the approach of visually inspecting a graphical display of the data for individual effects and variance (see Fig. 2). Indeed, visualizing the data is the first step of all good data analysis. Clinicians have reported taking a consensus approach with the patient through discussion and interpretation of the plots [34], while statisticians have pointed out that visual interpretation alone is too prone to bias in interpretation [37].

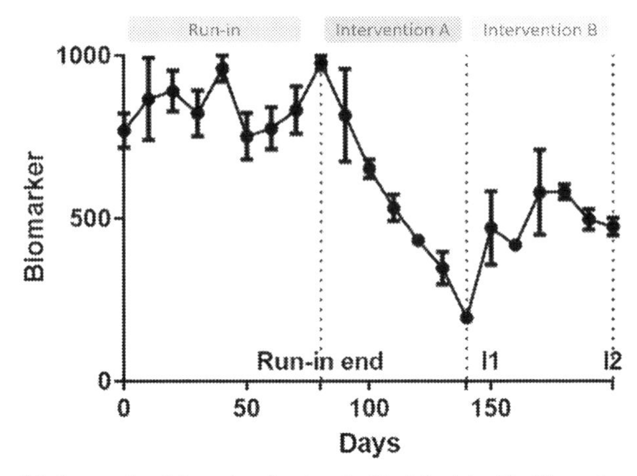

FIG. 2 Simplified example of data visualization for N-of-1 trials. The biomarker data is plotted (y-axis) against the time of the study (days here). Often, dose levels are indicated on secondary y-axis. In addition, clinical disease activity scores can be used instead of the surrogate biomarkers. For simplicity, wash-out periods are not indicated.

The early papers pioneering N-of-1 trials with medical practitioners describe using a sign test to assess whether the patient correctly identified the active treatment and a paired t-test to compare the difference in the average scores between active and placebo conditions [38]. However, due to the low power and high probability of false negatives applying parametric t-tests to N-of-1 trials data, Senn [39] cautions against relying on P-values in this context. When more than one patient is included in an N-of-1 trial, most agree that the sources of variation need to be identified and modeled [40]. Chen and Chen [41] proposed four analytic approaches to N-of-1 series data (see Fig. 3), including (1) paired t-tests and (2) mixed effects modeling in data with potential carry over effects, (3) mixed effects modeling with no carry over effects, (4) meta-analysis. Further, a summary measures approach has been proposed that reduces to a paired t-test for each patient, rather than for each condition [42].

Krone et al. [1] extends this work by looking at the question of Bayesian versus frequentist approaches to the mixed modeling. They compared three modeling approaches: mixed effects models using (1) frequentist approach, (2) Baysean approach, and (3) meta-analysis. Bayesian and frequentist mixed models were found to be superior to meta-analysis and both provided similar estimates if no informative prior knowledge is included. However, the Bayesian mixed model allows the inclusion of prior knowledge and gives potential for population-based and personalized inference.

Bayesian approaches to modeling N-of-1 trials data has found support among even the strongest defenders of the frequentist approach [1]. This field is in its infancy and many other creative modeling approaches are being debated in the literature [43]. A recent paper provides a guided 10-step SPSS (Statistical Package

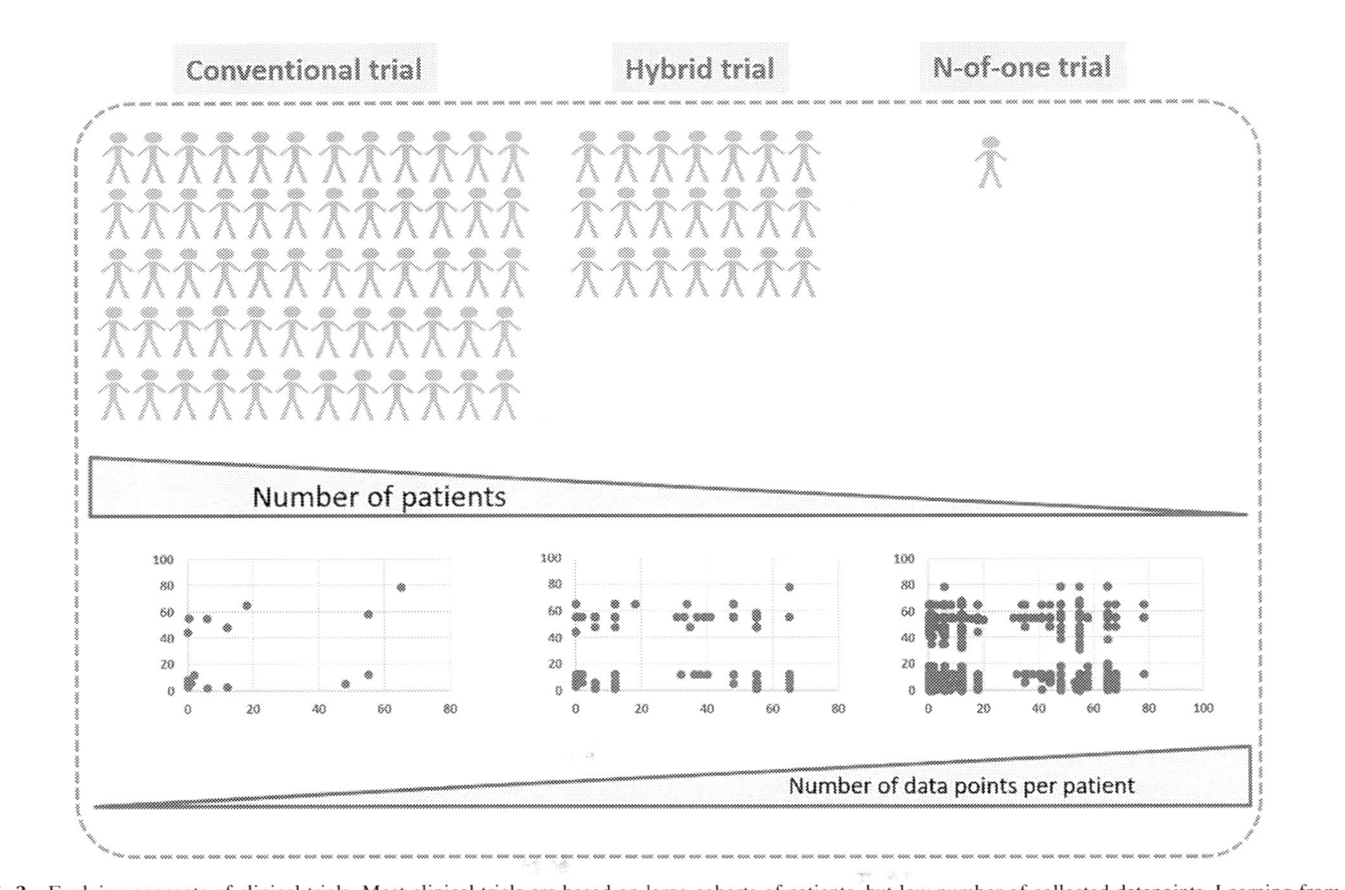

FIG. 3 Evolving concepts of clinical trials. Most clinical trials are based on large cohorts of patients, but low number of collected datapoints. Learning from the N-of-1 concept, future clinical trials might benefit from generation of large datasets that might allow for the reduction of clinical trials size and potentially costs.

for the Social Sciences) tutorial on basic modeling of N-of-1 data for beginners [30]. These relatively simple techniques are often more interpretable for general readers, clinicians and journal reviewers. Advanced modeling approaches can be inaccessible for many general readers, but are critically important for the field in order to further establish the validity of N-of-1 trials data to estimate population parameters. Artificial intelligence (AI) approaches such as machine learning (ML) hold promise to improve clinical trial design and outcome [44,45]. However, a lot is still hype and validation is highly required. Recently, an AI based system was successfully used to define dosing of bromodomain inhibitor in a patient with prostate cancer [46]. Theoretically, data derived from N-of-1 trials, if recorded and managed correctly, could be merged into larger datasets and analyzed using AI/ML approaches. This could either compose a synthesized clinical trial or provide the foundation for the next clinical trials.

6. Ethics and insurance

N-of-1 trials can be used in both research and clinical practice. When used in clinical practice as an extension of usual care, there is normally no requirement for prior institutional ethical approval. However, the onus is on the practitioner to be aware of local institutional and legal requirements for various treatment options and variations in requirements by location and institution. If in doubt, it is recommended to seek advice from a local ethics committee. If the intention is to use the data as evidence to publish or disseminate or in a secondary analysis, prior approval of an institutional human research ethics committee is required to use the data for research purposes. Ethical governance processes that form the legal framework for all research with human participants are covered under national and international legislations, as first set out under the Nuremberg Code and the Declaration of Helsinki [47]. Practitioners intending to conduct research in practice should seek prior advice and ethical approval from a hospital or university Human Research Ethics Committee. Practitioners wishing to undertake N-of-1 trial research should ensure they maintain professional indemnity insurance that includes clinical trial research. If there is a commercial sponsor, the sponsor normally provides clinical trial insurance in addition to product liability insurance. However, it is critical that an agreement regarding insurance and indemnity is reached as part of the approvals process in setting up a clinical trial [48]. In scenarios without a commercial sponsor, the practitioner should seek advice from a local ethics committee and confirm that the trial would be included under the corresponding hospital or university clinical trial insurance scheme.

7. Regulatory aspects

All clinical trials, including N-of-1 trials should be pre-registered on a clinical trial register before the first participant is recruited [49]. This is a prerequisite for publication in many journals, as set out by the International Committee of

Medical Journal editors (ICMJE) [50]. In addition, N-of-1 clinical trials must follow good clinical trials governance processes, as expected from all clinical trials. The specific governance framework will be determined by the affiliating or collaborating organization. However, there are many processes in common to all governance frameworks [51]. These include that structures and processes for on-going monitoring of the trial, including immediate reporting of serious adverse events and regular reporting to the ethics committee, are established.

An independent trial monitoring committee is also recommended for all clinical trials [52], especially in the case of a series of N-of-1 trials. The main objective of this committee is to provide an independent monitor of the data that is collected during the trial and to provide oversight for patient safety, data quality and treatment effects (or lack thereof). In an N-of-1 trial, this committee may comprise 2–3 members, such as statisticians, clinicians or researchers, who can bring relevant experience or expertise. This committee can make independent decisions about the whether to terminate the trial early.

8. Conclusions

In conclusion, SCEDs, such as N-of-1 trials, might provide valuable insights on individual effects that can be used to optimize treatments for individual patients. In addition, results from a series of N-of-1 trials can be combined and analyzed to estimate population effects. While N-of-1 clinical trials might not represent a feasible option for a broad range of diseases, in cases of rare diseases N-of-1 trials provide a clinical assessment methodology to produce high quality clinical evidence of safety and effectiveness at an individual level. Further, evidence from such designs could be aggregated to estimate population effects. Individual-level data may also provide useful insights for future trial design, of course considering the regulatory aspects involved with producing research evidence in practice. Lastly, for complementary medicine N-of-1 trails may help to determine the safety and effectiveness of a proposed therapy as an adjunct therapy. The strength of N-of-1 trials is that they represent a promising methodology to enable and facilitate personalized, precision medicine.

References

[1] T. Krone, R. Boessen, S. Bijlsma, R. van Stokkum, N.D. Clabbers, W.J. Pasman, The possibilities of the use of N-of-1 and do-it-yourself trials in nutritional research, PLoS One 15 (5) (2020), e0232680.

[2] G.S. Cooper, M.L. Bynum, E.C. Somers, Recent insights in the epidemiology of autoimmune diseases: improved prevalence estimates and understanding of clustering of diseases, J. Autoimmun. 33 (3–4) (2009) 197–207.

[3] G.S. Cooper, B.C. Stroehla, The epidemiology of autoimmune diseases, Autoimmun. Rev. 2 (3) (2003) 119–125.

[4] S.J. Walsh, L.M. Rau, Autoimmune diseases: a leading cause of death among young and middle-aged women in the United States, Am. J. Public Health 90 (9) (2000) 1463–1466.

[5] Y. Shoenfeld, et al., The mosaic of autoimmunity: hormonal and environmental factors involved in autoimmune diseases—2008, Isr. Med. Assoc. J. 10 (1) (2008) 8–12.
[6] C.M. Brickman, Y. Shoenfeld, The mosaic of autoimmunity, Scand. J. Clin. Lab. Invest. 61 (7) (2001) 3–15.
[7] A.D. Proal, P.J. Albert, T. Marshall, Autoimmune disease in the era of the metagenome, Autoimmun. Rev. 8 (8) (2009) 677–681.
[8] T.G. Marshall, R.E. Lee, F.E. Marshall, Common angiotensin receptor blockers may directly modulate the immune system via VDR, PPAR and CCR2b, Theor. Biol. Med. Model. 3 (2006) 1.
[9] A. Kuek, B.L. Hazleman, A.J.K. Ostor, Immune-mediated inflammatory diseases (IMIDs) and biologic therapy: a medical revolution, Postgrad. Med. J. 83 (978) (2007) 251–260.
[10] A. Davidson, B. Diamond, Autoimmune diseases, N. Engl. J. Med. 345 (5) (2001) 340–350.
[11] G. Lettre, J.D. Rioux, Autoimmune diseases: insights from genome-wide association studies, Hum. Mol. Genet. 17 (R2) (2008) R116–R121.
[12] S.V. Ramagopalan, et al., Expression of the multiple sclerosis-associated MHC class II allele HLA-DRB1*1501 is regulated by vitamin D, PLoS Genet. 5 (2) (2009) e1000369.
[13] A. Torkamani, et al., High-definition medicine, Cell 170 (5) (2017) 828–843.
[14] J.A. McKay, J.C. Mathers, Diet induced epigenetic changes and their implications for health, Acta Physiol. 202 (2) (2011) 103–118.
[15] I.R. Mackay, Science, medicine, and the future: tolerance and autoimmunity, BMJ 321 (7253) (2000) 93–96.
[16] N. Hill, N. Sarvetnick, Cytokines: promoters and dampeners of autoimmunity, Curr. Opin. Immunol. 14 (6) (2002) 791–797.
[17] L.G. Delogu, et al., Infectious diseases and autoimmunity, J. Infect. Dev. Ctries. 5 (10) (2011) 679–687.
[18] M. Sospedra, R. Martin, Immunology of multiple sclerosis, Annu. Rev. Immunol. 23 (1) (2005) 683–747.
[19] C.M. Porth, Pathophysiology: Concepts of Altered Health States, fourth ed., J.B. Lippincott Company, Philadelphia, 1994.
[20] J. Kountouras, et al., Challenge in the pathogenesis of autoimmune pancreatitis: potential role of helicobacter pylori infection via molecular mimicry, Gastroenterology 133 (1) (2007) 368–369.
[21] M. Falcone, N. Sarvetnick, Cytokines that regulate autoimmune responses, Curr. Opin. Immunol. 11 (6) (1999) 670–676.
[22] R.L. Coffman, Origins of the T(H)1-T(H)2 model: a personal perspective, Nat. Immunol. 7 (6) (2006) 539–541.
[23] I.J. Elenkov, et al., Cytokine dysregulation, inflammation and well-being, NeruroImmunoModulation 12 (2005) 255–269.
[24] C.T. Weaver, et al., IL-17 family cytokines and the expanding diversity of effector T cell lineages, Annu. Rev. Immunol. 25 (1) (2007) 821–852.
[25] K.H.G. Mills, Induction, function and regulation of IL-17-producing T cells, Eur. J. Immunol. 38 (10) (2008) 2636–2649.
[26] J. Bradbury, S. Grace, C. Avila, N-of-1 trials in CAM: the new frontier for generating practice-based evidence? Monthly Newsletter (2019).
[27] C. Avila, S. Grace, J. Bradbury, How do patients integrate complementary medicine with mainstream healthcare? A survey of patients' perspectives, Complement. Ther. Med. 49 (2020) 102317, https://doi.org/10.1016/j.ctim.2020.102317.
[28] S. Grace, et al., 'The healthcare system is not designed around my needs': how healthcare consumers self-integrate conventional and complementary healthcare services, Complement. Ther. Clin. Pract. 32 (2018) 151–156.

[29] N. Kong, et al., Antigen-specific transforming growth factor-induced Treg cells, but not natural Treg cells, ameliorate autoimmune arthritis in mice by shifting the Th17/Treg cell balance from Th17 predominance to Treg cell predominance, Arthritis Rheum. 64 (8) (2012) 2548–2558.
[30] H. Hanai, et al., Curcumin maintenance therapy for ulcerative colitis: randomized, multicenter, double-blind, placebo-controlled trial, Clin. Gastroenterol. Hepatol. 4 (12) (2006) 1502–1506.
[31] H.C. Kaplan, et al., Evaluating the comparative effectiveness of two diets in pediatric inflammatory bowel disease: a study protocol for a series of N-of-1 trials, in: Healthcare (Basel), Multidisciplinary Digital Publishing Institute, 2019. PMID: 31683925.
[32] L. Sung, B.M. Feldman, N-of-1 trials: innovative methods to evaluate complementary and alternative medicines in pediatric cancer, J. Pediatr. Hematol. Oncol. 28 (4) (2006) 263–266.
[33] J. Mahon, et al., Randomised study of n of 1 trials versus standard practice, BMJ 312 (7038) (1996) 1069–1074.
[34] M.A. Dooley, S.L. Hogan, Environmental epidemiology and risk factors for autoimmune disease, Curr. Opin. Rheumatol. 15 (2) (2003) 99–103.
[35] A. Reckner Olsson, T. Skogh, G. Wingren, Comorbidity and lifestyle, reproductive factors, and environmental exposures associated with rheumatoid arthritis (Erratum appears in Ann. Rheum. Dis. 2001 Dec;60(12):1161), Ann. Rheum. Dis. 60 (10) (2001) 934–939.
[36] G.S. Cooper, et al., Smoking and use of hair treatments in relation to risk of developing systemic lupus erythematosus, J. Rheumatol. 28 (12) (2001) 2653–2656.
[37] G.S. Cooper, F.W. Miller, D.R. Germolec, Occupational exposures and autoimmune diseases, Int. Immunopharmacol. 2 (2–3) (2002) 303–313.
[38] A.M. Landtblom, et al., Organic solvents and multiple sclerosis: a synthesis of the current evidence, Epidemiology 7 (4) (1996) 429–433.
[39] S. Senn, Suspended judgment n-of-1 trials, Control. Clin. Trials 14 (1) (1993) 1–5.
[40] G.E. Miller, E. Chen, K.J. Parker, Psychological stress in childhood and susceptibility to the chronic diseases of aging: moving toward a model of behavioral and biological mechanisms, Psychol. Bull. 137 (6) (2011) 959–997.
[41] X. Chen, P. Chen, A comparison of four methods for the analysis of N-of-1 trials, PLoS One 9 (2) (2014) e87752.
[42] S. Srivastava, J.L. Boyer, Psychological stress is associated with relapse in type 1 autoimmune hepatitis, Liver Int. 30 (10) (2010) 1439–1447.
[43] P.M. Brooks, R.O. Day, COX-2 inhibitors, Med. J. Aust. 173 (8) (2000) 433–436.
[44] S. Harrer, et al., Artificial intelligence for clinical trial design, Trends Pharmacol. Sci. 40 (8) (2019) 577–591.
[45] K.K. Mak, M.R. Pichika, Artificial intelligence in drug development: present status and future prospects, Drug Discov. Today 24 (3) (2019) 773–780.
[46] A. Pantuck, et al., Modulating BET bromodomain inhibitor ZEN-3694 and enzalutamide combination dosing in a metastatic prostate cancer patient using CURATE.AI, an artificial intelligence platform, Adv. Ther. 1 (2018) 1800104.
[47] B.A.T. Fischer, A summary of important documents in the field of research ethics, Schizophr. Bull. **32** (1) (2006) 69–80.
[48] Research governance, 2019. 15 Feb. [cited 2020 23 Jul]; Available from: https://www.australianclinicaltrials.gov.au/researchers/research-governance.
[49] International Clinical Trials Registry Platform (ICTRP), 2020. 15 Feb [cited 2020 23 Jul]; Available from: https://www.who.int/ictrp/en/.

[50] C.D. DeAngelis, et al., Is this clinical trial fully registered?: a statement from the International Committee of Medical Journal Editors, JAMA 293 (23) (2005) 2927–2929.
[51] P. Lowthian, Research governance, in: Textbook of Medical Administration and Leadership, Springer, Singapore, 2019, pp. 249–254.
[52] S. Piantadosi, Treatment effects monitoring, in: Clinical Trials; A Methodological Perspective, John Wiley & Sons, New Jersey, 2005.

Chapter 10

Health risk assessment and family history: Toward disease prevention

Lily W. Martin[a,*], Lauren C. Prisco[a,*], Laura Martinez-Prat[b], Michael Mahler[b], and Jeffrey A. Sparks[a,b,c]
[a]*Division of Rheumatology, Inflammation, and Immunity, Brigham and Women's Hospital, Boston, MA, United States,* [b]*Inova Diagnostics, Inc., San Diego, CA, United States,* [c]*Harvard Medical School, Boston, MA, United States*

Abbreviations

ACPA	anti-citrullinated protein antibody
APIPPRA	arthritis prevention in the preclinical phase of RA with abatacept
BMI	body mass index
DMARD	disease-modifying antirheumatic drug
FDR	first-degree relative
HCQ	hydroxychloroquine
MTX	Methotrexate
NSAID	non-steroidal anti-inflammatory drug
PAD	protein-larginine deiminase
PD	periodontitis
PRE-RA	personalized risk estimator for rheumatoid arthritis
RA	rheumatoid arthritis
RF	rheumatoid factor
SE	Shared epitope
SLE	systemic lupus erythematosus
UA	Undifferentiated arthritis

1 Background

1.1 Genetic and environmental risk factors for autoimmune diseases

The interaction of genetic and environmental risk factors underlies the model for pathogenesis of many autoimmune diseases. In this paradigm, individuals

[*] Contributed equally.

Precision Medicine and Artificial Intelligence. https://doi.org/10.1016/B978-0-12-820239-5.00013-9

genetically predisposed to an autoimmune disease are exposed to environmental risk factors throughout the life course, which may eventually manifest as clinical disease. Since many autoimmune diseases are more likely to occur within the same family, this suggests both shared genetic and environmental components for autoimmune disease susceptibility. Twin studies have shown that most autoimmune diseases have moderate to strong hereditability [1].

Most autoimmune diseases are polygenic, meaning they are linked to many genes, each of which usually has only a modest association with a specific condition. In this chapter, we will often use rheumatoid arthritis (RA) as the example that may apply to many other autoimmune diseases. Over 100 independent genetic loci are associated with risk of RA, although the risk of any one of these single nucleotide polymorphisms is modest [2]. Unlike monogenic diseases, the genetic components of complex diseases are not usually deterministic. Rather, complex chronic diseases such as autoimmune diseases alter the probability of disease development only slightly. For example, the strongest genetic risk factor for RA is the "shared epitope (SE)" at *HLA-DRB1* and it is linked to a threefold increased RA risk compared to not having any shared epitope allele [3]. However, the shared epitope is relatively common even in the general population, so the absolute risk of RA is relatively low even among individuals who do have this genetic factor. Therefore, there is currently relatively low clinical utility of testing for a common and strongly linked genetic factor (such as the shared epitope status) for autoimmune disease risk (such as RA) at an individual level.

Family history of RA is a strong risk factor for personal development of RA. Having a first-degree relative (FDR) with RA increases personal risk of RA by about threefold compared to not having a relative with RA [4]. Environmental factors, including cigarette smoking, excess body weight, periodontitis, and low fish intake, may also contribute to RA risk [4]. Other environmental factors such as infections and microbiota changes, drugs, and stress are hypothesized to be associated with other autoimmune diseases, but the evidence is lacking for many of these relatively uncommon and heterogeneous diseases that are therefore difficult to study.

1.2 Pathogenesis of autoimmune diseases

Understanding the natural history of each particular autoimmune disease is important for the development of prevention strategies for each condition. Autoimmune diseases may be a result of genetic and environmental triggers that progress through a series of preclinical phases, involving dysregulation of the immune system and subsequent tissue inflammation prior to clinical signs and symptoms. The first preclinical phase of the development of many autoimmune diseases involves the presence of genetic and environmental risk factors in the absence of detectable biomarkers, either in local tissue or in systemic circulation [5]. The second preclinical phase may be a period of asymptomatic, or subclinical disease during which pathogenic biologic processes are initiated at some local tissues without resulting in overt clinical manifestations [6]. One of the best studied paradigms for autoimmune disease development relates to RA pathogenesis whereby cigarette smoking induces citrullination and aberrant antigen presentation in individuals with the shared epitope. This may result in local pulmonary mucosal inflammation and production of autoantibodies against citrullinated proteins years prior to articular onset. At-risk individuals with elevated levels of anti-citrullinated protein antibodies (ACPA), highly specific to RA, have about 50% increased risk of developing RA in the next 5 years [7]. Since many autoimmune diseases are characterized by elevation of autoantibodies prior to disease onset, this paradigm may apply to other diseases as well.

The final phase in autoimmune disease development is the onset of clinical symptoms. For example, the initial presenting symptoms of RA may include fatigue, stiffness, swelling, and pain in the joints. However, some individuals may have subtle, non-specific, or atypical symptoms such that clinical presentations may be insidious and challenging to identify. This makes autoimmune diseases particularly difficult to diagnose and manage since some patients may present with years of subclinical or atypical symptoms. Even among those with established disease, identifying flares or monitoring disease activity can be challenging. For example, RA is a disease that typically involves pain, swelling and fatigue. However, these symptoms are common in the general population and often may be due to etiologies other than RA, even in patients known to have the disease. Overall, the pathogenesis of autoimmune diseases likely involves the interaction of genetic and environmental risk factors to produce an inflammatory response resulting in tissue damage.

2 Family history as a construct for prevention

2.1 Assessing family history

A family history of RA is known to be a risk factor for development of this disease, likely due to shared genetic and environmental factors [8]. However, RA and other autoimmune diseases can often have ambiguous clinical presentations, often making the diagnosis challenging. Most autoimmune diseases are also relatively uncommon. Family history of RA in the context of FDR, is often referred to an individual's biologic parents, siblings, and children. FDRs may have varying involvement with their affected relative's medical care—ranging from very involved to not involved at all. Therefore, a self-report of RA or another autoimmune disease, may be prone to error if the family member's true diagnosis is not able to be verified.

2.2 Scenarios for prevention based on family history

FDRs are a specific group of at-risk individuals that may be particularly amenable to prevention strategies since they are familiar with the index disease and may be interested in its prevention. These strategies include lifestyle modification and pharmacologic interventions. Some examples of lifestyle modifications to incorporate healthy habits in order to prevent RA or other autoimmune diseases can include regular exercise, proper oral hygiene, regular visits to the physician and dentist, healthy diet, and abstaining from smoking.

However, the unaffected relative's views toward prevention are affected by many factors such as convenience, risk, benefit, healthcare recommendations, and personal preference [9]. In addition, without adequate knowledge of the autoimmune disease that is being attempted to be prevented, this can result in a discordance between one's perceived and actual risk of developing these diseases.

Making lifestyle changes is largely dependent on convenience and personal preference. For example, if someone does not have the financial resources, and time to spend regularly exercising and eating healthfully, it is unlikely they could make these positive health changes. Even with adequate resources and time, it may be difficult to motivate change beyond the normal routine. Some FDRs may be willing to optimize their diet and lose weight after learning this is a risk factor for a disease that they care about preventing. Another preventative target includes regular visits to a primary care provider and a dental health professional. This can be challenging for some individuals due to health care availability, inconvenience, stress, finances, and insurance status. Regular visits may not be realistic for some individuals, while others may have extreme fear of regular check-ups, wanting to avoid this all together, perhaps fearful of being diagnosed with the very disease they may be interested in preventing. Therefore, some FDRs may be highly motivated to attempt to alter their risk, while others may prefer a watch-and-wait approach. Some FDRs may also be anxious about receiving test results related to autoimmune disease risk, while others may feel empowered to receive this information as long as they understand the implications [10]. Many FDRs are hesitant to take medications for primary disease prevention [10]. Unless FDRs are at very elevated risk, many prefer to modify their lifestyle rather than medication [11]. Given these complex psychosocial factors, it is unlikely that a single preventative strategy for an autoimmune disease will be applicable to FDRs. Therefore, precision medicine offers the opportunity to tailor prevention strategies to FDRs based on their personal preferences related to screening and lifestyle/pharmacologic interventions for prevention or early detection of autoimmune diseases.

2.3 Increased awareness and earlier diagnosis

Individuals with family history may be more aware of early signs and symptoms of the autoimmune disease, resulting in less delay in diagnosis and earlier treatment, which has been proven to be associated with better disease outcomes. Therefore, this could sometimes result in a relatively more benign disease trajectory. However, individuals with family history may have a stronger genetic component to their disease which may result in an earlier onset, more severe manifestations, and relatively longer time living with the disease. On the other hand, approaching a discussion about an illness is a complicated area, and researchers found that the willingness to have these conversations depends on the individual's perception of the degree of negative emotion to arise [11]. Having conversations with family members with autoimmune diseases can help with prevention, diagnosis and management of the disease itself.

Family history may have both positive and negative impacts on the diagnosis and course of a disease. Related to RA, family history could lead to early diagnosis and appropriate treatment that may prevent chronic pain, discomfort, and disability as well as prevent joint and organ damage that may have otherwise accrued from chronic systemic inflammation. Those with familial RA who develop RA themselves are diagnosed about 4.5 years earlier than those with sporadic (non-familial) RA [12]. However, family history could also lead to a misdiagnosis of RA, unnecessary treatment, and chronic anxiety that the diagnosis could occur. Relatives with equivocal signs and symptoms or abnormal laboratory values may be diagnosed as having early RA, but may have never actually progressed to *bona fide* clinical RA if the family history were not present. Therefore, family history brings with it complex biologic and psychologic factors that could impact the entire disease course, including prevention, diagnosis, management, and prognosis.

2.4 Relative and absolute risk

Understanding the concepts of relative and absolute risk are important when considering prevention of relatively rare conditions such as autoimmune diseases. For example, the prevalence of RA is just under 1% of adults, but the annual incidence rate is only about 40 per 100,000 persons in the United States. Therefore, the absolute risk is low, even when considering the entire life. Since the absolute risk is low, even relatively large relative risks still confer only modest absolute risks. For example, a threefold relative risk from positive family history of RA may only increase the absolute residual lifetime risk of RA from 1% to 3%. While the threefold relative risk may seem alarming, the 3% absolute risk may seem trivial to some even though this describes an identical scenario. Therefore, understanding risk perception and how individuals value and interpret risk estimates are paramount when weighing pros and cons of pursuing preventive strategies on individual and societal levels. Since RA is relatively uncommon, the general population pool without family history of RA is much larger than the pool of individuals who have a family history of RA. Therefore, on a population level, RA is much more likely to occur sporadically rather than in families. Hence, focusing prevention efforts only among those with family history for complex diseases such as RA or other autoimmune diseases will be unlikely to eradicate these diseases, even if successful among those with a positive family history.

Other groups besides FDRs at elevated risk for RA are based on genetic predisposition (such as with the shared epitope), environmental exposures (such as cigarette smoking), development of systemic autoimmunity (such as elevated ACPA), and pre-disease clinical states (such as undifferentiated inflammatory arthritis) [13]. While most studies are focused on family history of the index disease of interest (e.g., family history of RA for personal risk of developing RA), family history of other diseases may also impact the index disease as well. For example, family history of pulmonary fibrosis, inflammatory bowel disease, hyper/hypothyroidism and obstructive sleep apnea were all recently shown to be associated with increased risk for development of RA [14]. This highlights how genetic factors may have pleiotropic effects such that family history of one autoimmune disease may impact risk for developing other types of autoimmune diseases. For example, family history of systemic lupus erythematosus (SLE) not only increases personal risk of developing SLE, but also increases risk for other autoimmune diseases including RA [15].

There may also be differences in how family history may affect personal risk of an autoimmune disease, related to the age and sex of the affected relative and the proband. For example, having a sister affected with SLE may greatly impact the personal risk of SLE for another sister, but may not have much impact on their father since SLE typically affects younger women [16]. Truly comprehensive, personalized assessments of autoimmune disease risk using family history would consider these complex contributions related to age, sex, relationship, number of relatives, and conditions other than index disease. Therefore, thorough and accurate family history could offer deep insight into personalized disease risk through precision medicine concepts.

3 Prevention strategies

3.1 Primordial and primary prevention

There are four levels of prevention in the field of public health: primordial, primary, secondary, and tertiary. We will continue focusing on RA as an example for autoimmune disease development (Fig. 1). For the purpose of this chapter, we focus mostly on primordial and primary prevention, rather than secondary and tertiary prevention (since the latter two stages are often related to disease-specific management). Primordial prevention is the prevention of the risk factors themselves that are later associated with the disease. This step begins in the social and behavioral environment that an individual resides. For example, cigarette smoking is a known RA risk factor, so primordial prevention may focus on implementing methods to stop adolescents and young adults from ever starting smoking. Many of these interventions may be focused on societal levels, such as education at schools, advertisements, taxes, or laws against smoking at a certain age. Primary prevention is the prevention of the disease itself. This can be achieved by addressing the risk factors that already affect individuals that place them at increased risk for RA. For example, primary prevention of RA might focus on smoking cessation among those who are current smokers. Other RA risk factors include excess weight, low quality diet, and dental health (Table 1). In the context of RA development, there are distinct phases in which RA progresses and risk factors are thought to impact progressions throughout

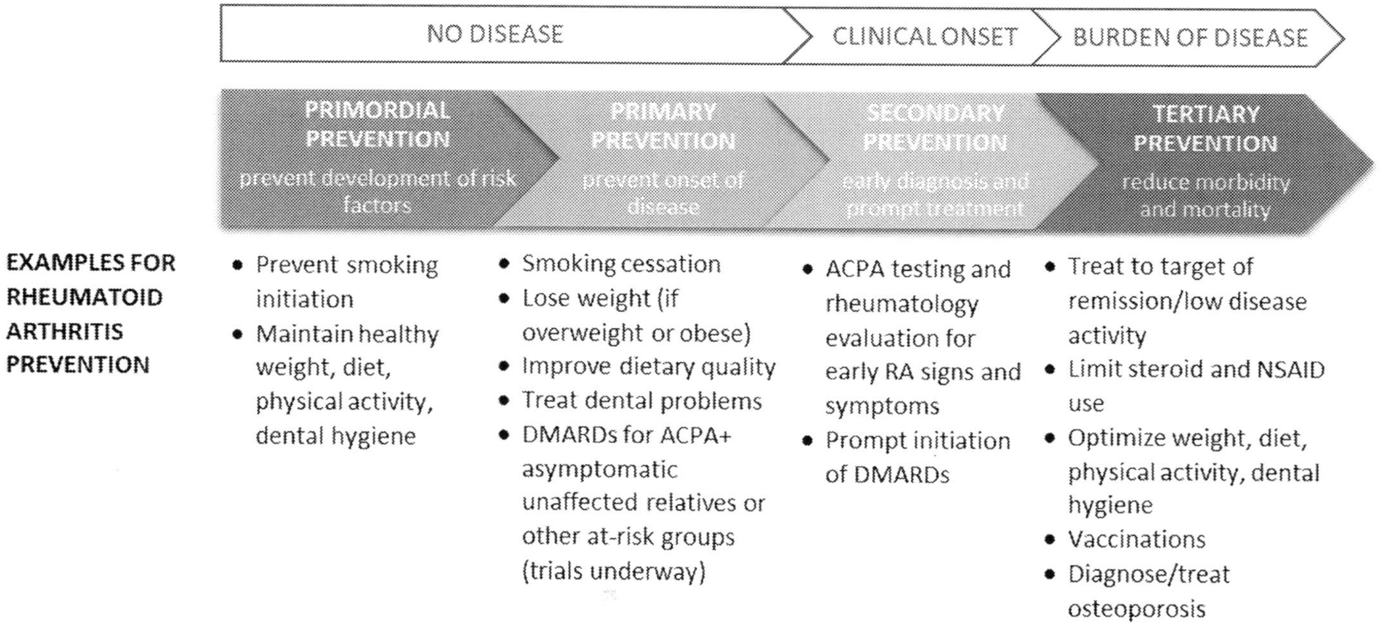

NO DISEASE | CLINICAL ONSET | BURDEN OF DISEASE

	PRIMORDIAL PREVENTION prevent development of risk factors	PRIMARY PREVENTION prevent onset of disease	SECONDARY PREVENTION early diagnosis and prompt treatment	TERTIARY PREVENTION reduce morbidity and mortality
EXAMPLES FOR RHEUMATOID ARTHRITIS PREVENTION	• Prevent smoking initiation • Maintain healthy weight, diet, physical activity, dental hygiene	• Smoking cessation • Lose weight (if overweight or obese) • Improve dietary quality • Treat dental problems • DMARDs for ACPA+ asymptomatic unaffected relatives or other at-risk groups (trials underway)	• ACPA testing and rheumatology evaluation for early RA signs and symptoms • Prompt initiation of DMARDs	• Treat to target of remission/low disease activity • Limit steroid and NSAID use • Optimize weight, diet, physical activity, dental hygiene • Vaccinations • Diagnose/treat osteoporosis

FIG. 1 Stages of prevention for rheumatoid arthritis incorporating family history. *ACPA*, anti-citrullinated protein antibody; *DMARD*, disease-modifying anti-rheumatic drug; *NSAID*, non-steroidal anti-inflammatory drug; *RA*, rheumatoid arthritis.

TABLE 1 Selected examples of lifestyle risk factors for primordial and primary prevention of rheumatoid arthritis.

Lifestyle risk factor	Direction of risk for RA (comparison)	Primordial prevention	Primary prevention
Smoking	Increased (compared to not smoking)	• Smoking health risk awareness education programs • Barriers to access smoking products (e.g., age restrictions and taxation)	• Smoking cessation (pharmacologic and non-pharmacologic) • Support groups for smokers • Nicotine replacement
Low quality diet	Increased (compared to high quality diet)	• Public health education programs • Accessibility of healthy and affordable food options • Healthy meal/voucher programs for schools	• Encourage moderate alcohol intake and high fish intake • Decrease sugar sweetened soda and red meat consumption • Omega-3 fatty acid rich diet and supplements
Poor dental health	Increased (compared to good dental health)	• Provide education on optimal dental hygiene • Encourage routine dental visits	• Treat periodontitis • Increase brushing/flossing for those with poor dental health
Excess weight	Increased (compared to normal weight)	• Educate on healthy eating and exercise options • Maintain healthy weight	• Lose weight if overweight or obese • Exercise classes • Support groups for overweight • Pharmacologic therapy for weight loss if indicated

RA, rheumatoid arthritis.

these preclinical phases. Therefore, it is possible that the preclinical phases of RA development could be halted or delayed by specific behavioral changes of these risk factors [17].

3.2 Cigarette smoking

Though it is not yet proven through clinical trials, researchers suspect that changing behaviors and modifying lifestyles could delay or even prevent RA onset. Many observational studies strongly suggest that cigarette smoking is a risk factor for RA and may be one of the most important out of the modifiable lifestyle factors [18]. Smoking is thought to possibly impact RA risk throughout every phase of preclinical RA development [17]. The biologic mechanism of smoking influencing risk for RA may be due to the chronic inflammation in the lungs of smokers that eventually leads to production of ACPA [19]. Smoking seems to be specifically associated with increased risk of seropositive RA, characterized by elevated ACPA and rheumatoid factor (RF), rather than seronegative RA characterized by the absence of ACPA and RF. After smoking cessation, RA risk may be reduced, suggesting that behavior change may delay or even prevent RA compared to continuing smoking [20]. Since smokers tend to have family members who also smoke, even environmental factors such as smoking have hereditary components that may be difficult to disentangle from true genetic effects. Primordial and primary prevention of RA may manifest as societal efforts to persuade non-smokers to never start smoking and smokers to quit. However, despite steady decreases in the rate of smoking in the United States, the incidence and prevalence of RA has remained stable or even increased. This argues either that smoking may not actually be a (strong) causal RA risk factor or that changes in other RA risk factors (such as lower quality dietary intake and obesity) have contributed in the opposite direction, resulting in an overall stable incidence and prevalence of the disease. Rigorously conducted clinical trials would be necessary to prove a causal link between smoking and RA risk. However, this would require large sample size and lengthy follow-up and would have ethical and feasibility concerns given the known negative health impact of smoking on other diseases.

3.3 Low quality diet

Diet as a risk factor for RA can be investigated by individual food and beverage items or by the overall pattern of intake of foods/beverages. Several dietary patterns have been examined in the field of autoimmune disease prevention. The Mediterranean diet has been of interest due to its anti-inflammatory effects, overall health benefits, and lower prevalence of RA in areas of the world that adhere to this diet, although studies have not found a strong relationship between this diet and RA [21]. The Alternative Healthy Eating Index is a measure of long-term healthy eating patterns and it has been associated with reduced

RA risk. The Empirical Dietary Inflammatory Pattern which classifies anti- and pro-inflammatory food/beverage groups in relation to serum inflammatory markers found an inverse relationship between following a healthy diet and systemic inflammation [17]. The type of healthy diet can be classified as a prudent diet (fruits, vegetables, legumes, fish/poultry) [17], while an unhealthy diet can be classified as a Western diet that includes high caloric intake, sugar, total and saturated fats, and other pro-inflammatory foods [22]. Regarding individual food/beverage items, omega-3 fatty acid/fish consumption have been thought to reduce RA risk, though this is controversial [17, 23]. Moderate alcohol consumption has also been suggested to protect against RA development compared to no intake [17]. Increased intake of sugar-sweetened soda has been associated with increased RA risk [17]. However, investigating dietary factors and disease outcomes is highly susceptible to confounding and measurement error, so dietary factors have relatively low-quality evidence for RA risk. Further research and clinical trials are needed to firmly establish the relationship of dietary factors with RA. Because eating is a social event, particularly among family members, diet assessment and interventions may be a promising future direction for unaffected FDRs of patients with RA or other autoimmune diseases that are known to be related to dietary factors.

3.4 Excess weight

Obesity is considered an inflammatory condition that also may increase estrogen levels, which both play important roles in RA pathogenesis [17]. Depending on the age and sex, excess weight has been suggested to contribute to RA risk [17]. Excess RA risk could be due to complex interactions between excess weight, low quality dietary intake, and sedentary physical activity [17]. Since these are all intrinsically related, it is unclear which behavior may contribute to RA risk and the relative weight of associations particularly when considering interactions [24]. Similar to smoking, primordial prevention of RA would be to implement societal strategies to maintain normal weight, while primary prevention of RA would be to implement weight loss strategies for those with obesity or overweight. Of course, weight and body mass index (BMI) are relatively crude measures of adiposity, but these categories have been linked to increased RA risk on a population level. Recent studies suggest that BMI outperforms waist circumference for prediction of RA risk. However, individuals with particular muscle and bone compositions may be in healthy ranges despite abnormal BMI. Most of the current research has focused on static BMI categories for RA risk. A recent study showed that bariatric surgery had no impact on future risk of RA [25]. Therefore, the relationship between excess weight and RA is still not clear. While weight loss may benefit those with established RA, there is less robust evidence examining the impact of weight gain or non-surgical weight loss on the risk of RA [17]. Similar to other factors, excess weight tends to congregate in families. Therefore, strategies focusing on weight loss, optimizing dietary intake, and increasing physical activity among unaffected family members could be a promising strategy to modify RA risk as well as risk for other autoimmune diseases related to excess weight. Having close family members as support while introducing a new goal of weight loss and increasing awareness of what foods are being consumed is an important resource to have throughout a lifestyle modification.

3.5 Dental health management

Poor dental health, specifically periodontitis (PD), has been examined as a possible risk factor for RA. PD is a major oral health issue that causes a chronic inflammatory disease of the teeth and gums. PD and RA share similar characteristics regarding risk factors, immunity, and tissue destruction pathways [26]. The biologic link between RA and PD may be due to the oral bacteria *Porphyromonas gingivalis* [27], the most common pathogen in PD. This bacterium is the only prokaryote described to date that expresses a functional bacterial protein-arginine deiminase (PAD), the enzyme catalyzing citrullination of proteins, so it has been suggested to contribute to ACPA development at inflamed oral mucosa. The presence of antibodies against *P. gingivalis* has been shown to be increased years before RA diagnosis [28]. While the literature linking *P. gingivalis* and periodontitis to RA pathogenesis is growing, the association between dental hygiene behaviors (such as brushing, flossing, and regular teeth cleanings by a dental hygienist or dentist) and RA are less clear. The dental hygiene behaviors may be considered primordial prevention of RA, while treatment of PD would be considered primary prevention of RA. Oral health is a complex result of genetics, familial environment, and learned behaviors such as daily dental hygiene care formed through familial influence. Therefore, this is an emerging area to target, especially in families that are at higher risk for RA. Cigarette smoking, dietary intake, excess weight, and physical activity are already associated with many negative health outcomes so family members at risk for RA may already be attempting to optimize these health behaviors. However, family members may be less aware of the link between poor oral health and RA and education on this novel risk factor could motivate improvement in this health practice. Therefore, precision medicine approaches to educate at-risk FDRs on risk factors and providing their own risk estimates may be a strategy to motivate health positive behavior maintenance and improvements for the primordial and primary prevention of RA.

4 PRE-RA family study

The Personalized Risk Estimator for RA (PRE-RA) was a randomized controlled trial among FDRs without RA that investigated the effectiveness of disclosing RA risk through personalized genetics, biomarkers, and lifestyle factors on health behavior intentions. This intervention study investigated personalized

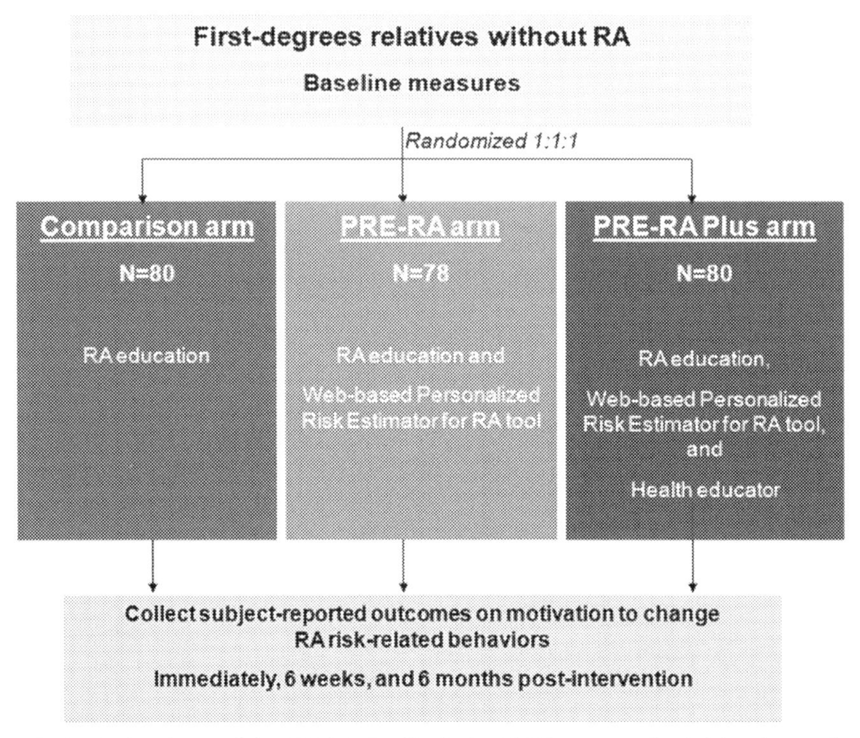

FIG. 2 Study schema of the PRE-RA Family Study. *PRE-RA*, personalized risk estimator for rheumatoid arthritis; *RA*, rheumatoid arthritis.

medicine for improving health behaviors among FDRs. Subjects were randomly assigned to either the PRE-RA group that received personalized risk estimation for RA or to standard (non-personalized) RA risk education (Fig. 2) [29]. The PRE-RA Plus arm had the addition of a health educator to provide risk interpretation and navigate the psychosocial factors that may be limiting improvement in health behaviors. The personalized RA risk estimate was based on demographics (age, sex, relationship to affected RA relative and family history of other autoimmune conditions), four RA risk-associated behaviors, genetics (shared epitope), and biomarkers (ACPA and RF results). The four RA risk-associated behaviors included cigarette smoking, low fish intake, excess weight, and poor oral health [4]. Those randomly assigned to the PRE-RA group were provided both relative and absolute risk estimates that emphasized the behavior changes that could potentially impact their future risk of RA (Fig. 3). Therefore, the intervention was finely tuned to the subject's personal characteristics to test whether a precision medicine approach could encourage positive changes in RA risk-related behaviors.

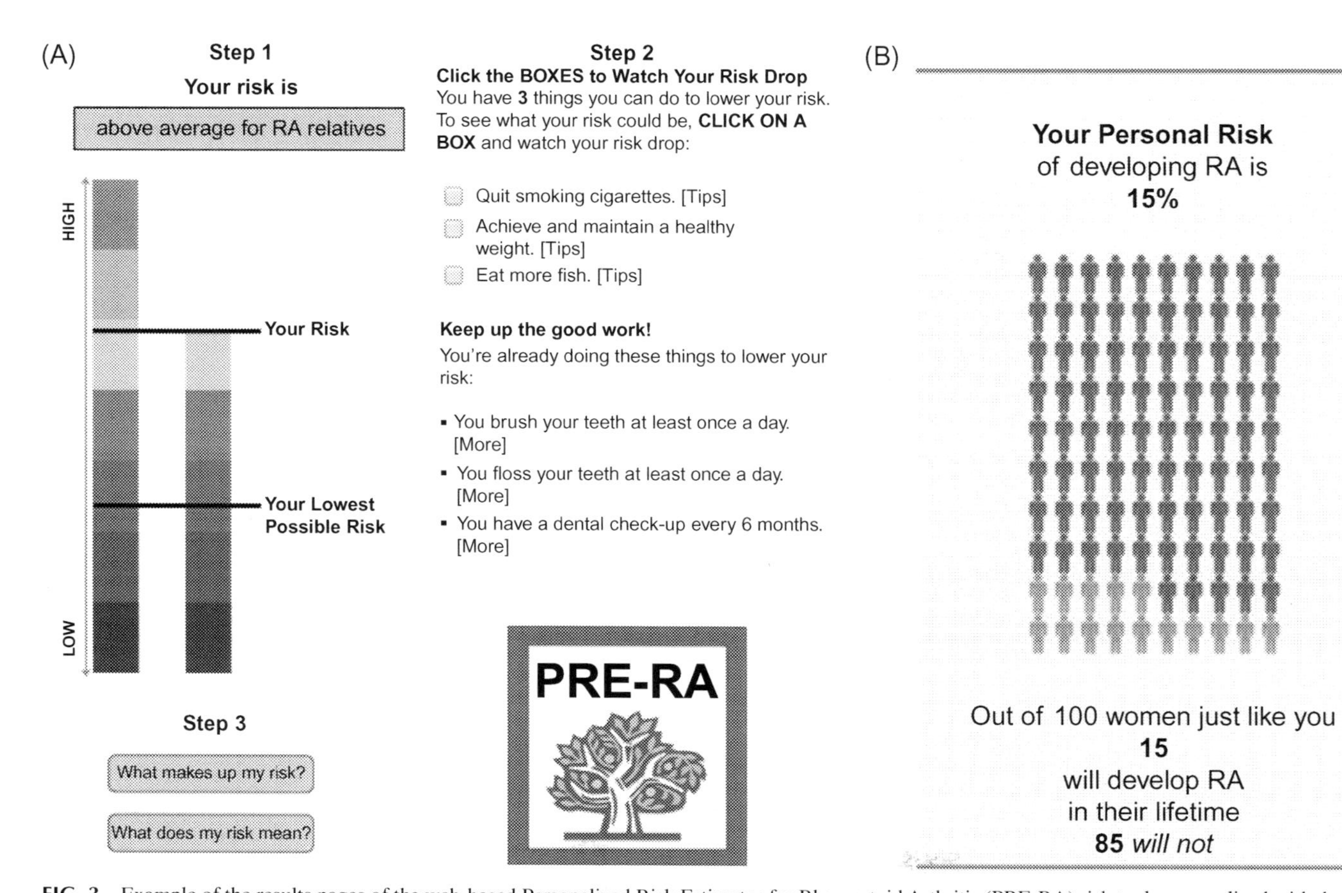

FIG. 3 Example of the results pages of the web-based Personalized Risk Estimator for Rheumatoid Arthritis (PRE-RA) risk tool, personalized with demographics, genetics, RA-related autoantibodies, and behaviors using A, an interactive relative RA risk display, and B, a pictogram displaying absolute lifetime RA risk.

All subjects were verified that their relative was truly affected with RA since they were recruited through RA patients that were being seen clinically at an academic rheumatology center. All subjects had a physical examination by a rheumatologist verifying that they did not have RA or another similar autoimmune disease. Subjects were assessed at baseline as well as immediately, 6 weeks, 6 months, and 12 months post-intervention. At baseline, less than half of the participants correctly identified smoking, low fish intake, obesity, or poor oral health to be risk factors for RA despite having relatives affected with RA and enrolling in this prevention trial. After the educational intervention, the personalized RA education PRE-RA tool increased the participants' understanding of RA risk compared to standard education [30]. At the 6-month follow-up, individuals assigned to the PRE-RA group were also more likely to quit smoking, increase fish intake, and increase brushing/flossing than the standard education group [29]. A high proportion of subjects increased physical activity and attempted to lose weight, but there was no difference between the groups.

The results of the PRE-RA Family Study suggest that FDRs may be more interested in improving health behaviors that they did not know were related to chronic disease development. Many were already attempting to lose weight and increase physical activity (known to be associated with many common chronic diseases). The study also disclosed genetic and biomarker results and provided estimates of both relative and absolute risk of RA. Interestingly, those who were told that they were at high risk through genetics and biomarkers were most likely to be motivated to optimize health behaviors. Those who were told to be at low risk had similar changes to health behaviors as the standard group. Therefore, the study suggests that the highest impact of health behavior improvement for primordial/primary prevention would be those who are deemed to be at highest risk. Even so, the intervention tended to reassure subjects regardless of whether they were deemed to be at high or low risk, rather than induce anxiety [31]. However, it is possible that risk perception and actions may differ based on the underlying disease. For example, unaffected FDRs may have different reactions to a similar intervention that was focused on relatively more severe diseases such as Alzheimer's disease or cancer. The PRE-RA Family Study showed that personalized medicine could motivate at-risk individuals by family history to improve lifestyle behaviors that could impact the risk for RA [29]. These findings demonstrate proof-of-concept of how a personalized medicine approach can motivate those at increased chronic disease risk due to family history to make health behavior improvements.

5 Pharmacologic prevention

Several autoimmune diseases have had clinical trials of drugs for primary prevention. For instance, a recent randomized controlled trial recently found that treatment with teplizumab, an anti-CD3 antibody, delayed progression to type 1 diabetes in high-risk participants due to family history and serum elevation of diabetes-related autoantibodies [32]. This section will focus on prevention trials and pharmacologic treatment options for RA and how family members may be a population to target.

5.1 Drugs for primary prevention of RA

Several studies have investigated the potential use of disease-modifying antirheumatic drugs (DMARDs) for primary prevention of RA (Table 2). DMARDs are drugs that have been shown to change the disease course of RA and improve radiographic outcomes. The currently available DMARDs vary by mechanism of action, method of administration, efficacy, time to effect, and side effect profile. Traditional DMARDs, such as methotrexate and hydroxychloroquine (HCQ), can be administered orally and bluntly suppress the entire immune system in a non-targeted manner. Methotrexate (MTX) is the typical first-line DMARD for patients with moderate or severe RA. MTX has been suggested to delay progression to RA of individuals with undifferentiated arthritis, however only in ACPA-positive individuals and as a secondary analysis in a prior clinical trial [33]. HCQ is a drug originally used as an antimalarial and now repurposed as first-line treatment in patients with mild RA or as add-on therapy in patients with RA who have inadequate response to other DMARDs. HCQ is also often used in many other connective tissue diseases, such as SLE, making this an attractive option for primary prevention across many different autoimmune diseases [34]. The ongoing placebo-controlled StopRA trial is investigating the efficacy and safety of HCQ for primary RA prevention in individuals at high risk of developing RA due to elevated ACPA.

Biologic DMARDs target specific molecules important in inflammatory pathways and are also being evaluated for primary prevention of RA. These drugs were developed more recently than conventional DMARDs and are administered through either injection or infusion. They are typically only considered after failure of one or more conventional DMARDs in patients with active RA. One of these biologic DMARDs, abatacept, disrupts T-cell co-stimulation through CTLA4. Therefore, abatacept works as a targeted anti-inflammatory agent that may result in attenuated immunologic responses at sites of immune dysregulation that also modulate B-cell function, resulting in decreased autoantibody titers. Studies have shown that patients with undifferentiated arthritis or very early RA who were treated with abatacept had slower radiographic and magnetic resonance imaging progression to RA and lower ACPA levels than those on placebo, suggesting an effect on the underlying autoimmune process [35]. The ongoing Arthritis prevention in the pre-clinical phase of RA with abatacept (APIPPRA) study is testing the efficacy and safety of this drug for primary RA prevention in individuals at high risk of developing RA due to elevated ACPA and clinically suspicious arthralgias [36]. On the other hand, rituximab is a CD20 antagonist that works as a B-cell inhibitor, attractive as possible prevention in many autoimmune diseases that are characterized by elevated autoantibodies secreted by these immune cells, including RA. Rituximab

TABLE 2 Disease-modifying antirheumatic drugs (DMARDs) being considered for primary rheumatoid arthritis (RA) prevention among individuals with anti-citrullinated protein antibodies (ACPA) positivity and other high-risk groups.

DMARD	Mechanism of action	Route(s) of administration	Benefits	Risks	Trials for RA prevention
Hydroxy-chloroquine	Multiple	Oral	• Well established side effect profile and tolerability • Inexpensive • Already used off-label for undifferentiated CTD • Already used as first line for mild RA • No infectious risk	• Relatively common side effects: nausea, diarrhea, headache photosensitivity, rash • Serious but rare risks usually with long-term use: retinopathy, myopathy, heart failure	StopRA
Methotrexate	Antimetabolite of folate; multiple	Oral subcutaneous	• Already used as initial therapy for moderate/severe RA • Inexpensive • Well-established side effect profile	• Common tolerability side effects: fatigue, nausea, mucositis • Potential serious organ toxicity (liver, kidney, lung, bone marrow) • Infectious risk	PROMPT TREAT EARLIER
Abatacept	CTLA4 antagonist (blocks T cell co-stimulation)	Subcutaneous intravenous	• Strong biologic plausibility • Subcutaneous administration can occur at home • Infusion form complete in <1 h	• Expensive • Only given by parenteral administration • Risk of infection • Not approved as initial RA therapy	APIPPRA ADJUST ARIAA

Rituximab	Anti-CD20 (B cell inhibitor)	Intravenous	• Strong biologic plausibility • Long half-life, infrequent dosing	• Expensive • Only available by infusion over many hours • Pre-infusion medications required • High rate of infusion reactions • Risk of infection and PML • Risk of hypogammaglobulinemia • Not approved as initial RA therapy	PRAIRI

ACPA, anti-citrullinated protein antibody; *ADJUST*, abatacept study to determine the effectiveness in preventing the development of rheumatoid arthritis in patients with undifferentiated inflammatory arthritis and to evaluate safety and tolerability; *APIPPRA*, arthritis prevention in the pre-clinical phase of RA with abatacept; *ARIAA*, abatacept reversing subclinical inflammation as measured by MRI in ACPA-positive arthralgia; *CTD*, connective tissue disease; *CTLA4*, cytotoxic T-lymphocyte-associated protein 4; *DMARD*, disease-modifying antirheumatic drug; *PML*, progressive multifocal leukoencephalopathy; *PRAIRI*, prevention of clinically manifest rheumatoid arthritis by B cell directed therapy in the earliest phase of the disease; *PROMPT*, probable rheumatoid arthritis methotrexate versus placebo treatment; *RA*, rheumatoid arthritis; *TREAT EARLIER*, treat early arthralgia to reverse or limit the exacerbation of RA.

has been suggested to delay the progression to RA compared to placebo in individuals at risk due to elevated RA-related autoantibodies [37]. While RA prevention could be possible using targeted drugs such as abatacept and rituximab, these treatments are expensive and not currently used as first-line even for patients with diagnosed RA. Therefore, it may be difficult for them to meet the rigorous thresholds needed in cost-benefit analyses to be widely implemented in routine clinical care for primary prevention of RA. Generic drugs with well-known safety profiles, such as HCQ and methotrexate, may be more likely to have clinical uptake, but the efficacy for primary prevention for RA has yet to be firmly established. Although tremendous progress has been made to understand RA pathogenesis and perform clinical trials for RA prevention, further research is needed to determine pharmacologic methods for primary prevention of RA and other autoimmune diseases.

5.2 Family history of RA and perception of pharmacologic interventions

Unaffected FDRs of patients with RA are an attractive target population for primary prevention strategies since they are both at increased risk and familiar with RA. However, many individuals with a family history of RA tend to prefer lifestyle modifications, rather than pharmacologic strategies [11]. Symptomatic relatives may be more likely to consider pharmacologic prevention methods. They may also be further along the phases of preclinical RA development [38]. Some research studies [39] have advocated screening FDRs for ACPA, but the clinical utility is not yet firmly established. Even with the elevated relative risk of RA, only about 3%–6% of FDRs have elevated ACPA, so even broad screening may be difficult to identify many individuals at risk [40]. Even those who are identified by elevated ACPA may be reluctant to participate in clinical trials for primary prevention of RA unless symptoms have already started. In addition to treatment efficacy and safety, FDRs consider multiple factors when deciding whether to take preventative medication, including method of administration, opinion of health care professionals, and the reversibility of side effects [13]. Thus, it is important to educate FDRs on the risks, benefits, and administration route of preventative medications even after efficacy is established. Further research on the connection between family history and attitudes of at-risk individuals toward primary prevention is required to develop personalized prevention plans.

5.3 Drugs for secondary and tertiary prevention of RA

Although there is no cure for RA, patients with clinically-diagnosed RA have many treatment options for drugs that have known efficacy in secondary and tertiary prevention. Non-steroidal anti-inflammatory drugs (NSAIDs) reduce pain and improve function through anti-inflammatory activity. Corticosteroids have anti-inflammatory and immunoregulatory effects. However, neither NSAIDs nor corticosteroids alter the disease course of RA. Due to their short onset of action, they are typically used as bridging therapy while awaiting onset of DMARD or as adjunctive therapy for flares. DMARDs are the only drugs that have been shown to change the disease course of RA and improve radiographic outcomes.

While individuals with a family history of RA may learn about prevention options from their relatives, family history has not been shown to predict response to treatment with methotrexate or tumor necrosis factor inhibitors [41]. The optimal treatment for RA involves a combination of pharmacologic and lifestyle interventions with the goals of achieving the lowest level of arthritis disease activity, preventing further joint damage, and improving physical functionality and quality of life. Patients should work with their doctors to create treatment plans tailored to their lifestyles and needs.

6 Conclusions

Familial aggregation of autoimmune diseases suggests a genetic predisposition and/or shared environmental factors in relatives of patients with autoimmune diseases. Thus, family history may be considered a risk-factor for autoimmune diseases. However, care should be made to verify the family history status. Those with family history may have varying interest in pursuing prevention strategies. Many may only be interested in lifestyle modifications. A subset of at-risk individuals may be particularly suited for primary prevention using medications. In addition to biomarkers such as autoantibodies, family history may be useful in identifying candidates for pharmacologic prevention treatments. Future research should focus on whether family history is a predictor of treatment response and adverse reactions.

Financial support and sponsorship

Dr. Sparks is supported by the National Institute of Arthritis and Musculoskeletal and Skin Diseases (grant numbers K23 AR069688, R03 AR075886, L30 AR066953, P30 AR070253, and P30 AR072577), the Rheumatology Research Foundation K Supplement Award, the Brigham Research Institute, and the R. Bruce and Joan M. Mickey Research Scholar Fund. Dr. Sparks has received research support from Amgen and Bristol-Myers Squibb and performed consultancy for Bristol-Myers Squibb, Gilead, Inova, Janssen, and Optum unrelated to this work. The funders had no role in the decision to publish or preparation of this manuscript. The content is solely the responsibility of the authors and does not necessarily represent the official views of Harvard University, its affiliated academic health care centers, or the National Institutes of Health.

Conflict of interest

Dr. Martinez-Prat and Dr. Mahler are employees of Inova Diagnostics. All other authors declare no conflict of interest.

References

[1] D.P. Bogdanos, D.S. Smyk, E.I. Rigopoulou, M.G. Mytilinaiou, M.A. Heneghan, C. Selmi, et al., Twin studies in autoimmune disease: genetics, gender and environment, J. Autoimmun. 38 (2–3) (2012) J156–J169.

[2] Y. Okada, S. Eyre, A. Suzuki, Y. Kochi, K. Yamamoto, Genetics of rheumatoid arthritis: 2018 status, Ann. Rheum. Dis. 78 (4) (2019) 446–453.

[3] L. Klareskog, P. Stolt, K. Lundberg, H. Kallberg, C. Bengtsson, J. Grunewald, et al., A new model for an etiology of rheumatoid arthritis: smoking may trigger HLA-DR (shared epitope)-restricted immune reactions to autoantigens modified by citrullination, Arthritis Rheum. 54 (1) (2006) 38–46.

[4] J.A. Sparks, M.D. Iversen, R. Miller Kroouze, T.G. Mahmoud, N.A. Triedman, S.S. Kalia, et al., Personalized risk estimator for rheumatoid arthritis (PRE-RA) family study: rationale and design for a randomized controlled trial evaluating rheumatoid arthritis risk education to first-degree relatives, Contemp. Clin. Trials 39 (1) (2014) 145–157.

[5] K.D. Deane, V.M. Holers, The natural history of rheumatoid arthritis, Clin. Ther. 41 (7) (2019) 1256–1269.

[6] A. Finckh, K.D. Deane, Prevention of rheumatic diseases: strategies, caveats, and future directions, Rheum. Dis. Clin. N. Am. 40 (4) (2014) 771–785.

[7] J.A. Ford, X. Liu, A.A. Marshall, A. Zaccardelli, M.G. Prado, C. Wiyarand, et al., Impact of cyclic citrullinated peptide antibody level on progression to rheumatoid arthritis in clinically tested cyclic citrullinated peptide antibody-positive patients without rheumatoid arthritis, Arthritis Care Res. 71 (12) (2019) 1583–1592.

[8] T. Frisell, S. Saevarsdottir, J. Askling, Family history of rheumatoid arthritis: an old concept with new developments, Nat. Rev. Rheumatol. 12 (6) (2016) 335–343.

[9] M. Harrison, L. Spooner, N. Bansback, K. Milbers, C. Koehn, K. Shojania, et al., Preventing rheumatoid arthritis: preferences for and predicted uptake of preventive treatments among high risk individuals, PLoS One 14 (4) (2019), e0216075.

[10] G. Simons, R.J. Stack, M. Stoffer-Marx, M. Englbrecht, E. Mosor, C.D. Buckley, et al., Perceptions of first-degree relatives of patients with rheumatoid arthritis about lifestyle modifications and pharmacological interventions to reduce the risk of rheumatoid arthritis development: a qualitative interview study, BMC Rheumatol. 2 (2018) 31.

[11] M. Falahee, G. Simons, C.D. Buckley, M. Hansson, R.J. Stack, K. Raza, Patients' perceptions of their relatives' risk of developing rheumatoid arthritis and of the potential for risk communication, prediction, and modulation, Arthritis Care Res. 69 (10) (2017) 1558–1565.

[12] X.Y. Zhang, J.Y. Jin, J. He, Y.Z. Gan, J.L. Chen, X.Z. Zhao, et al., Family history of rheumatic diseases in patients with rheumatoid arthritis: a large scale cross-sectional study, Beijing Da Xue Xue Bao 51 (3) (2019) 439–444.

[13] M. Falahee, A. Finckh, K. Raza, M. Harrison, Preferences of patients and at-risk individuals for preventive approaches to rheumatoid arthritis, Clin. Ther. 41 (7) (2019) 1346–1354.

[14] V.L. Kronzer, C.S. Crowson, J.A. Sparks, E. Myasoedova, J. Davis 3rd., Family history of rheumatologic, autoimmune, and non-autoimmune diseases and risk of rheumatoid arthritis, Arthritis Care Res. (2019), https://doi.org/10.1002/acr.24115. Online ahead of print.

[15] A. Ganapati, G. Arunachal, S. Arya, D. Shanmugasundaram, L. Jeyaseelan, S. Kumar, et al., Study of familial aggregation of autoimmune rheumatic diseases in Asian Indian patients with systemic lupus erythematosus, Rheumatol. Int. 39 (12) (2019) 2053–2060.

[16] Y. Koumantaki, E. Giziaki, A. Linos, A. Kontomerkos, P. Kaklamanis, G. Vaiopoulos, et al., Family history as a risk factor for rheumatoid arthritis: a case-control study, J. Rheumatol. 24 (8) (1997) 1522–1526.

[17] A. Zaccardelli, H.M. Friedlander, J.A. Ford, J.A. Sparks, Potential of lifestyle changes for reducing the risk of developing rheumatoid arthritis: is an ounce of prevention worth a pound of cure? Clin. Ther. 41 (7) (2019) 1323–1345.

[18] K. Chang, S.M. Yang, S.H. Kim, K.H. Han, S.J. Park, J.I. Shin, Smoking and rheumatoid arthritis, Int. J. Mol. Sci. 15 (12) (2014) 22279–22295.

[19] R. Anderson, P.W. Meyer, M.M. Ally, M. Tikly, Smoking and air pollution as pro-inflammatory triggers for the development of rheumatoid arthritis, Nicotine Tob. Res. 18 (7) (2016) 1556–1565.

[20] A.K. Hedstrom, L. Stawiarz, L. Klareskog, L. Alfredsson, Smoking and susceptibility to rheumatoid arthritis in a Swedish population-based case-control study, Eur. J. Epidemiol. 33 (4) (2018) 415–423.

[21] A. Linos, V.G. Kaklamani, E. Kaklamani, Y. Koumantaki, E. Giziaki, S. Papazoglou, et al., Dietary factors in relation to rheumatoid arthritis: a role for olive oil and cooked vegetables? Am. J. Clin. Nutr. 70 (6) (1999) 1077–1082.

[22] E. Philippou, E. Nikiphorou, Are we really what we eat? Nutrition and its role in the onset of rheumatoid arthritis, Autoimmun. Rev. 17 (11) (2018) 1074–1077.

[23] J.A. Sparks, E.J. O'Reilly, M. Barbhaiya, S.K. Tedeschi, S. Malspeis, B. Lu, et al., Association of fish intake and smoking with risk of rheumatoid arthritis and age of onset: a prospective cohort study, BMC Musculoskelet. Disord. 20 (1) (2019) 2.

[24] L. Ljung, S. Rantapaa-Dahlqvist, Abdominal obesity, gender and the risk of rheumatoid arthritis—a nested case-control study, Arthritis Res. Ther. 18 (1) (2016) 277.

[25] C. Maglio, Y. Zhang, M. Peltonen, J. Andersson-Assarsson, P.A. Svensson, C. Herder, et al., Bariatric surgery and the incidence of rheumatoid arthritis—a Swedish obese subjects study, Rheumatology 59 (2) (2020) 303–309.

[26] Y.H. Lee, P.H. Lew, C.W. Cheah, M.T. Rahman, N.A. Baharuddin, R.D. Vaithilingam, Potential mechanisms linking periodontitis to rheumatoid arthritis, J. Int. Acad. Periodontol. 21 (3) (2019) 99–110.

[27] R.S. de Molon, C. Rossa Jr., R.M. Thurlings, J.A. Cirelli, M.I. Koenders, Linkage of periodontitis and rheumatoid arthritis: current evidence and potential biological interactions, Int. J. Mol. Sci. 20 (18) (2019) 4541.

[28] Z. Cheng, J. Meade, K. Mankia, P. Emery, D.A. Devine, Periodontal disease and periodontal bacteria as triggers for rheumatoid arthritis, Best Pract. Res. Clin. Rheumatol. 31 (1) (2017) 19–30.

[29] J.A. Sparks, M.D. Iversen, Z. Yu, N.A. Triedman, M.G. Prado, R. Miller Kroouze, et al., Disclosure of personalized rheumatoid arthritis risk using genetics, biomarkers, and lifestyle factors to motivate health behavior improvements: a randomized controlled trial, Arthritis Care Res. 70 (6) (2018) 823–833.

[30] M.G. Prado, M.D. Iversen, Z. Yu, R. Miller Kroouze, N.A. Triedman, S.S. Kalia, et al., Effectiveness of a web-based personalized rheumatoid arthritis risk tool with or without a health educator for knowledge of rheumatoid arthritis risk factors, Arthritis Care Res. 70 (10) (2018) 1421–1430.

[31] A.A. Marshall, A. Zaccardelli, Z. Yu, M.G. Prado, X. Liu, R. Miller Kroouze, et al., Effect of communicating personalized rheumatoid arthritis risk on concern for developing RA: a randomized controlled trial, Patient Educ. Couns. 102 (5) (2019) 976–983.

[32] K.C. Herold, B.N. Bundy, S.A. Long, J.A. Bluestone, L.A. DiMeglio, M.J. Dufort, et al., An anti-CD3 antibody, teplizumab, in relatives at risk for type 1 diabetes, N. Engl. J. Med. 381 (7) (2019) 603–613.

[33] J. van Aken, L. Heimans, H. Gillet-van Dongen, K. Visser, H.K. Ronday, I. Speyer, et al., Five-year outcomes of probable rheumatoid arthritis treated with methotrexate or placebo during the first year (the PROMPT study), Ann. Rheum. Dis. 73 (2) (2014) 396–400.

[34] L. Gonzalez-Lopez, J.I. Gamez-Nava, G. Jhangri, A.S. Russell, M.E. Suarez-Almazor, Decreased progression to rheumatoid arthritis or other connective tissue diseases in patients with palindromic rheumatism treated with antimalarials, J. Rheumatol. 27 (1) (2000) 41–46.
[35] P. Emery, P. Durez, M. Dougados, C.W. Legerton, J.C. Becker, G. Vratsanos, et al., Impact of T-cell costimulation modulation in patients with undifferentiated inflammatory arthritis or very early rheumatoid arthritis: a clinical and imaging study of abatacept (the ADJUST trial), Ann. Rheum. Dis. 69 (3) (2010) 510–516.
[36] M. Al-Laith, M. Jasenecova, S. Abraham, A. Bosworth, I.N. Bruce, C.D. Buckley, et al., Arthritis prevention in the pre-clinical phase of RA with abatacept (the APIPPRA study): a multi-centre, randomised, double-blind, parallel-group, placebo-controlled clinical trial protocol, Trials 20 (1) (2019) 429.
[37] D.M. Gerlag, M. Safy, K.I. Maijer, M.W. Tang, S.W. Tas, M.J.F. Starmans-Kool, et al., Effects of B-cell directed therapy on the preclinical stage of rheumatoid arthritis: the PRAIRI study, Ann. Rheum. Dis. 78 (2) (2019) 179–185.
[38] E. Mosor, M. Stoffer-Marx, G. Steiner, K. Raza, R.J. Stack, G. Simons, et al., I would never take preventive medication! perspectives and information needs of people who underwent predictive tests for rheumatoid arthritis, Arthritis Care Res. 72 (2020) 360–368.
[39] M. Mahler, Population-based screening for ACPAs: a step in the pathway to the prevention of rheumatoid arthritis? Ann. Rheum. Dis. 76 (11) (2017) e42.
[40] D. Alpizar-Rodriguez, L. Brulhart, R.B. Mueller, B. Möller, J. Dudler, A. Ciurea, U.A. Walker, I. Von Mühlenen, D. Kyburz, P. Zufferey, M. Mahler, S. Bas, D. Gascon, C. Lamacchia, P. Roux-Lombard, K. Lauper, M.J. Nissen, D.S. Courvoisier, C. Gabay, A. Finckh, The prevalence of anticitrullinated protein antibodies increases with age in healthy individuals at risk for rheumatoid arthritis, Clin. Rheumatol. 36 (3) (2017) 677–682. https://doi.org/10.1007/s10067-017-3547-3. PMID: 28110385.
[41] T. Frisell, S. Saevarsdottir, J. Askling, Does a family history of RA influence the clinical presentation and treatment response in RA? Ann. Rheum. Dis. 75 (6) (2016) 1120–1125.

Chapter 11

Regulatory aspects of artificial intelligence and machine learning-enabled software as medical devices (SaMD)

Michael Mahler[a], Carolina Auza[a], Roger Albesa[a], Carlos Melus[a], and Jungen Andrew Wu[b]

[a]Inova Diagnostics, Inc., San Diego, CA, United States, [b]Rook Quality Systems Taiwan Branch, New Taipei, Taiwan

Abbreviations

ACP	algorithm change protocol
AI	Artificial Intelligence
CAD	computer-aided design
CDRH	center for devices and radiological health
CDSS	clinical decision software support
CLS	continuous learning systems
CQOE	culture of quality and organizational excellence
CT	computer tomography
DICOM	digital imaging and communications in medicine
DL	deep learning
DMF	device master file
FDA	food and drug administration
GDPR	general data protection regulation
GMLP	good machine learning practice
HIPAA	health insurance portability and accountability act
IMDRF	international medical device regulators forum
IVD	in vitro diagnostic
IVDR	in vitro diagnostic regulation
MDR	medical device regulation
ML	machine learning
MR	magnetic resonance
NB	notified bodies
NIST	national institute for standard and technology
NMPA	national medical products administration
PHI	protected health information
PII	personally identifiable information

Precision Medicine and Artificial Intelligence. https://doi.org/10.1016/B978-0-12-820239-5.00010-3

PMA	premarket approval
RASE	reasonable assurance of safety and effectiveness
SaMD	software as medical device
SPP	software pre-cert program
SPS	SaMD pre-specifications
TPLC	total product life cycle

1 Introduction

Artificial intelligence (AI) and machine learning (ML) are believed to revolutionize healthcare in many areas [1–20] including drug development [5, 7, 16], diagnosis of patients [2, 4, 6, 8, 9, 11] and triaging treatment decisions as software technologies evolve rapidly. AI/ML-enabled software as medical devices (SaMD) has the potential to allow earlier disease detection (e.g., triage and workflow optimization), more accurate diagnosis (e.g., quantification work such as segmentation), new insights into human physiology (e.g., deep learning magnetic resonance (MR) reconstruction or super-resolution computer tomography (CT)/MR), and personalized diagnostics and therapeutics [5, 7, 16, 21]. In the booming digital health device world, regulations are evolving rapidly to ensure that integration of AI/ML SaMD to current standard of care would not introduce safety and effectiveness concerns [5, 7, 16]. The regulatory requirements for AI/ML in SaMD are dependent on many aspects, including, but not limited to the criticality of the device, the geographic location of the market, user's qualification or ability to independently review diagnosis or patient management recommendations. The Food and Drug Administration (FDA) has been working with many other regulatory agencies and voluntary groups such as International Medical Device Regulators Forum (IMDRF) to build regulatory frameworks to accelerate international medical device regulatory harmonization and convergence for AI/ML in SaMD [5, 7, 16]. In this book chapter, we discuss some of the most relevant aspects around regulatory pathways with strong focus on the main medical device and in-vitro diagnostic markets (United States, FDA; Europe, CE and China, NMPA) for AI/ML in SaMD and the challenges ahead for both regulators and manufacturers. In addition, we summarize some examples of SaMD that achieved regulatory clearance (Table 1).

2 The Food and Drug Administration (FDA)'s perspective

2.1 Regulatory aspects on artificial intelligence (AI)-enabled clinical decision support software

The 21st Century Cures Act was released on December of 2016. The policy was an inflection point of FDA's commitment to regulating digital health devices. New digital health guidances were issued and pre-existing documents were revised to capture the fast-changing digital health landscape. The FDA's Clinical Decision Support Software (CDSS) guidance drew particular attention to the public as many critical aspects pertaining to the regulations of AI/ML SaMD (e.g., qualification and classification of software devices) are explained by the FDA [22] (see also Table 2).

TABLE 1 Examples of regulatory clearances of medical devices based on artificial intelligence (status January 2021; extracted from https://medicalfuturist.com/fda-approved-ai-based-algorithms/).

Name of device or algorithm	Name of parent company	Short description	FDA approval number	Type of FDA approval	Date	Medical specialty
Arterys Cardio DL	Arterys Inc	software analyzing cardiovascular images from MR	K163253	510k	2016 11	Radiology
EnsoSleep	EnsoData, Inc	diagnosis of sleep disorders	K162627	510k	2017 03	Neurology
Arterys Oncology DL	Arterys Inc	medical diagnostic application	K173542	510k	2017 11	Radiology
Idx	IDx LLC	detection of diabetic retinopathy	DEN180001	de-novo	2018 01	Ophthalmology
Koios DS for Breast	Koios Medical, Inc	diagnostic software for lesions suspicious for cancer	K190442	510k	2019 06	
ContaCT	Viz.AI	stroke detection on CT	DEN170073	de-novo	2018 02	Radiology
OsteoDetect	Imagen Technologies, Inc.	X-ray wrist fracture diagnosis	DEN180005	de-novo	2018 02	Radiology
Guardian Connect System	Medtronic	predicting blood glucose changes	P160007	PMA	2018 03	Endocrinology
EchoMD Automated Ejection Fraction Software	Bay Labs, Inc.	echocardiogram analysis	K173780	510k	2018 05	Radiology
DreaMed	DreaMed Diabetes, Ltd	managing Type 1 diabetes.	DEN170043	de-novo	2018 06	Endocrinology

Continued

TABLE 1 Examples of regulatory clearances of medical devices based on artificial intelligence (status January 2021; extracted from https://medicalfuturist.com/fda-approved-ai-based-algorithms/)—cont'd

Name of device or algorithm	Name of parent company	Short description	FDA approval number	Type of FDA approval	Date	Medical specialty
LungQ	Thirona Corporation	Quantitative analysis of chest CT scans	K173821	PMA	2018 06	Radiology
BriefCase	Aidoc Medical, Ltd.	triage and diagnosis of time sensitive patients	K180647	510k	2018 07	Radiology
ProFound™ AI Software V2.1	iCAD, Inc	breast density via mammogprahy	K191994	510k	2018 07	Radiology
SubtlePET	Subtle Medical, Inc	radiology image processing software	K182336	510k	2018 09	Radiology
Arterys MICA	Arterys Inc	Liver and lung cancer diagnosis on CT and MRI	K182034	510k	2018 08	Radiology
AI-ECG Platform	Shenzhen Carewell Electronics., Ltd	ECG analysis support	K180432	510k	2018 09	Cardiology
Accipiolx	MaxQ-AI Ltd	acute intracranial hemorrhage triage algorithm	K182177	510k	2018 10	Radiology
icobrain	icometrix NV	MRI brain interpretation	K181939	510k	2018 10	Radiology
FerriSmart Analysis System	Resonance Health Analysis Service Pty Ltd	measure liver iron concentration	K182218	510k	2018 11	Internal Medicine

cmTriage	CureMetrix, Inc.	Mammogram workflow	K183285	510k	2019 03	Radiology
Deep Learning Image Reconstruction	GE Medical Systems, LLC.	CT image reconstruction	K183202	510k	2019 04	Radiology
HealthPNX	Zebra Medical Vision Ltd.	Chest X-Ray assessment pneumothorax	K190362	510k	2019 05	Radiology
Advanced Intelligent Clear-IQ Engine (AiCE)	Canon Medical Systems Corporation	Noise reduction algorithm	K183046	510k	2019 06	Radiology
SubtleMR	Subtle Medical, Inc	radiology image processing software	K191688	510k	2019 07	Radiology
AI-Rad Companion (Pulmonary)	Siemens Medical Solutions USA, Inc.	CT image reconstruction - pulmonary	K183271	510k	2019 07	Radiology
Critical Care Suite	GE Medical Systems, LLC.	Chest X-Ray assessment pneumothorax	K183182	510k	2019 08	Radiology
AI-Rad Companion (Cardiovascular)	Siemens Medical Solutions USA, Inc.	CT image reconstruction - cardiovascular	K183268	510k	2019 09	Radiology
EchoGo Core	Ultromics Ltd.	quantification and reporting of results of cardiovascular function	K191171	510k	2019 11	Cardiology
TransparaTM	Screenpoint Medical B.V.	Mammogram workflow	K192287	510k	2019 12	Radiology
QuantX	Quantitative Insights, Inc	Radiological software for lesions suspicious for cancer	DEN170022	de-novo	2020 01	Radiology
Eko Analysis Software	Eko Devices Inc.	Cardiac Monitor	K192004	510k	2020 01	Cardiology

Continued

TABLE 1 Examples of regulatory clearances of medical devices based on artificial intelligence (status January 2021; extracted from https://medicalfuturist.com/fda-approved-ai-based-algorithms/)—cont'd

Name of device or algorithm	Name of parent company	Short description	FDA approval number	Type of FDA approval	Date	Medical specialty
BodyGuardian Remote Monitoring System	Preventice	Remote monitoring device for patients with cardiac arrhythmias	K121197	510k	2012 08	Cardiology
Temporal Comparison	Riverain Technologies	Chest X-Ray Scanning Software	K123526	510k	2012 12	Radiology
Ahead 100	BrainScope	Device to interprete the structural condition of the patient's brain after head injury.	DEN140025	de-novo	2014 08	Neurology
AliveCor	AliveCor, Inc	detection of atrial fibrillation	K140933	510k	2014 08	Cardiology
QbCheck	QbTech AB	diagnosis and treatment of ADHD	K143468	510k	2016 03	Psychiatry
Steth IO	Stratoscientific, Inc.	acoustic device to collect heart and lung sounds	K160016	510k	2016 07	General medicine
Cantab Mobile	CAMBRIDGE COGNITION LTD.	memory assessment for the elderly	K161328	510k	2017 01	Neurology
AmCAD-US	AmCad BioMed Corporation	analysis of thyroid nodules	K162574	510k	2017 05	Radiology

Rooti Rx System ECG Event Recorder, Rooti Link APP Software	Rooti Labs Ltd	wearable continuous ECG monitor	K163694	510k	2017 11	Cardiology
BioFlux	Biotricity Inc.	detecting arrhythmias	K172311	510k	2017 12	Cardiology
WAVE Clinical Platform	Excel Medical Electronics, LLC	Monitoring vital signs	K171056	510k	2018 01	Hospital monitoring
DM-Density	Densitas, Inc.	Breast density via mammogprahy	K170540	510k	2018 02	Radiology
NeuralBot	Neural Analytics, Inc.	transcranial Doppler probe positioning	K180455	510k	2018 04	Radiology
HealthCCS	Zebra Medical Vision Ltd.	coronary artery calcification algorithm	K172983	510k	2018 05	Radiology
MindMotion GO	MindMaze SA	software with rehabilitation exercises for the elderly	K173931	510k	2018 05	Orthopedics
ECG App	Apple, Inc.	detection of atrial fibrillation	DEN180044	de-novo	2018 08	Cardiology
FibriCheck	Qompium NV	Cardiac Monitor	K173872	510k	2018 09	Cardiology
Irregular Rhythm Notification Feature	Apple, Inc.	detection of atrial fibrillation	DEN180042	de-novo	2018 09	Cardiology
RightEye Vision System	RightEye, LLC	identifying visual tracking impairment	K181771	510k	2018 09	Ophthalmology
FluoroShield™	Omega Medical Imaging, LLC	Radiation dosage reduction	K191713	510k	2018 10	Radiology
Embrace	Empatica Srl	wearable for seizure monitoring	K181861	510k	2018 11	Neurology

Continued

TABLE 1 Examples of regulatory clearances of medical devices based on artificial intelligence (status January 2021; extracted from https://medicalfuturist.com/fda-approved-ai-based-algorithms/)—cont'd

Name of device or algorithm	Name of parent company	Short description	FDA approval number	Type of FDA approval	Date	Medical specialty
iSchemaView RAPID	iSchemaView, Inc.	stroke detection on CT and MRI	K182130	510k	2018 12	Radiology
Quantib ND	Quantib BV	Neurodegenerative disease MRI brain reading	K182564	510k	2018 12	Radiology
Study Watch	Verily Life Sciences LLC	ECG feature of the Study Watch	K182456	510k	2019 01	Cardiology
KardiaAI	AliveCor, Inc	six-lead smartphone ECG	K181823	510k	2019 03	Cardiology
Loop System	Spry Health, Inc.	Monitoring vital signs	K181352	510k	2019 03	Hospital monitoring
RhythmAnalytics	Biofourmis Singapore Pte. Ltd	Monitoring cardiac arrhythmias	K182344	510k	2019 03	Cardiology
eMurmer ID	CSD Labs GmbH	Heart murmur detection	K181988	510k	2019 04	Cardiology
ReSET-O	Pear Therapeutics, Inc.	adjuvant treatment of substance abuse disorder	K173681	510k	2019 5	Psychiatry
ACR | LAB Urine Analysis Test System	Healthy.io Ltd	Urinary tract infection diagnosis	K182384	510k	2019 07	Urology / General practise
Current Wearable Health Monitoring System	Current Health Ltd.	Monitoring vital signs	K191272	510k	2019 07	Hospital monitoring

physIQ Heart Rhythm and Respiratory Module	physIQ, Inc	detection of atrial fibrillation	K183322	510k	2019 07	Cardiology
RayCare 2.3	RaySearch Laboratories AB	Medical charged-particle radiation therapy system	K191384	510k	2019 07	Radiology
Biovitals Analytics Engine	Biofourmis Singapore Pte. Ltd	Cardiac Monitor	K183282	510k	2019 08	Cardiology
Caption Guidance	Caption Health, Inc.	Software to assist medical professionals in the acquisition of cardiac ultrasound images.	DEN190040	de-novo	2019 08	Radiology
BrainScope TBI	BrainScope Company, Inc	EEG-based Concussion Index	K190815	510k	2019 09	Neurology
AIMI-Triage CXR PTX	RADLogics, Inc.	Chest X-ray prioritization service	K193300	510k	2020 03	Radiology
EyeArt	Eyenuk, Inc	Automated detection of diabetic retinopathy	K200667	510k	2020 06	Ophthalmology
qER	Qure.ai Technologies	Computer aided triage and notification software for CT images	K200921	510k	2020 06	Radiology
AVA (Augmented Vascular Analysis)	See-Mode Technologies Pte. Ltd.	Analysis and reporting of vascular ultrasound scans	K201369	510k	2020 08	Radiology

TABLE 2 Overview of regulatory documents for relevant of software as medical devices (SaMD) using artificial intelligence.

ID	Document/source title
ISO 14971	Medical devices—Application of risk management to medical devices
ISO 13485	Medical devices—Quality management systems—Requirements for regulatory purposes
MEDDEV 2.7.1 Revision 4	Clinical evaluation: A guide for manufacturers and notified bodies under directives 93/42/EEC and 90/385/EEC
Cybersecurity guidance for pre market	Content of premarket submissions for management of cybersecurity in medical devices (FDA-2013-D-0616)
Cybersecurity guidance for post-market	Postmarket management of cybersecurity in medical devices FDA-2015-D-5105
IEC 80002-1:2009(en)	Medical device software—Part 1: Guidance on the application of ISO 14971 to medical device software
ISO/TR 80002-2:2017(en)	Medical device software—Part 2: Validation of software for medical device quality systems
IEC80002-3:2014(en)	Medical device software—Part 3: Process reference model of medical device software life cycle processes (IEC 62304)
IEC/DIS 80001-1(en)	Safety, effectiveness and security in the implementation and clinical use of connected medical devices or connected health software—Part 1: Application of risk management
ISO/TR 24971:2020(en)	Medical devices—Guidance on the application of ISO 14971
ISO/TR 17791:2013(en)	Health informatics—Guidance on standards for enabling safety in health software
ISO/TR27809:2007(en)	Health informatics—Measures for ensuring patient safety of health software
IEC/DIS 80001-1(en)	Safety, effectiveness and security in the implementation and clinical use of connected medical devices or connected health software—Part 1: Application of risk management
IEC60601-4-1:2017(en)	Medical electrical equipment—Part 4-1: Guidance and interpretation—Medical electrical equipment and medical electrical systems employing a degree of autonomy

From the FDA perspective, the requirements for AI/ML in SaMD are dependent mostly on the criticality of the device and the impact on clinical decisions. In addition, according to the FDA, it is imperative to ensure that the training data is sufficient and unbiased because the input data can be highly influencing recommendations or clinical information presented to the health care professional to make decisions. Consequently, the training data has to reflect the anticipated target population for clinical use which includes different age, gender and ethnicity groups (etc.) to allow generalizability among patient subgroups. However, FDA has yet to issue a guidance that defines data management practices (e.g., data set selection, annotation, training, configuration management, maintenance, etc.) and test dataset requirements (e.g., dataset type, size, etc.). Currently, manufacturers can only leverage the pre-submission program, which takes in average 75-90 days, to receive feedback from the agency on the proposed subjects.

One significant challenge is that an AI/ML-based system can only be as good as the gold standard that was used for training. Consequently, the definition and selection of a reference or 'gold standard' is critical. Therefore, this "gold standard" is most important for defining the performance characteristics of the AI/ML based systems and the selection of the "gold standard" often relies on experienced clinicians and stakeholders. The situation becomes even more complex in case of continuous learning software (CLS) [20]. Individual clinicians using AI/ML-enabled software should not be expected or relied upon to comprehend the capabilities and capacity of AI/ML systems in general, or its ability to aid with a patient's diagnosis in particular. The certainty of the recommendation and any limitations of the model that may pertain to their patient, in a way he or she can quickly understand during a patient visit/assessment, will likely be preferred over otherwise equivalent product.

2.2 Inadequacy of traditional *510(k)* process for clearing AI/ML-enabled SaMD

Traditionally, the FDA reviews medical device submissions through the most appropriate premarket route based on device risk profile (i.e., *510(k)* premarket notifications, *de-novo* classification, or premarket approval (PMA)). Modifications to medical device software are treated following the FDA guidance for risk-based approach for *510(k)* software modifications which is based on the risk posed to patients due to the change [23]. Under the FDA's current approach to software modifications, it is expected that many of these AI/ML-driven software changes to a device may need a premarket review and consequently cause burden to manufacturers.

FDA chaired the International Medical Device Regulators Forum (IMDRF) SaMD committee and co-developed the SaMD framework with the goal to allow independent implementation by regulatory jurisdictions over the world using converged IMDRF principles [24]. The risk framework from SaMD N12 [25] is adopted by the FDA in the Pre-Certification (Pre-Cert) Program and the risk categories of software is highly dependent on the clinical scenario (e.g., state of

TABLE 3 Classification of software as medical device per IMDRF framework.

	Significance of information provided by SaMD to health care decision		
Care situation or condition	**Treat or diagnose**	**Drive clinical management**	**Inform clinical management**
Critical	IV	III	II
Serious	III	II	I
Non-serious	II	I	I

healthcare situation or condition, and significance of information provided by SaMD to healthcare decision) as shown in Table 3. In addition, the FDA published a discussion paper [26] in April 2019 that describes the FDA's foundation for a potential approach to premarket review for AI/ML-driven software modifications. The ideas described in this discussion paper are derived from the current premarket programs and rely on IMDRF's risk categorization principles, the FDA's benefit-risk framework, risk management principles described in the software modifications guidance [23], and the organization-based total product life cycle (TPLC) approach, which is also envisioned in the Digital Health Software Pre-Cert Program. Here, the FDA introduces the organizational excellence metrics and product/process level evaluation approach for manufacturers to follow as the paradigm shifts to reflect the fast-evolving SaMD landscape, particularly with AL/ML. As the FDA has mentioned during several workshops [21], the framework was built from regulating computer-aided detection (CAD) software in the past decades and it's worth reflecting the evolution of CAD regulatory pathway to better inform FDA's thinking on regulating AI/ML in SaMD. CAD devices can be classified for detection (CADe), diagnosis (CADx), detection and diagnosis (CADe/x), triage and notification (CADt), and radiological acquisition and optimization guidance (CADa/o). The early PMAs for CADe were in mammography, special dental, ultrasound indications, and chest radiograph. As the applications for CAD expands rapidly to other uses, the FDA's objective is to tailor regulatory framework with reasonable assurance of safety and effectiveness (RASE). For instance, special controls required by the FDA in *de-novo(s)* or some *510(k)s* define the pathway which leverages past experience and knowledge to establish device-specific risk mitigations. Special controls provide consistency and transparency on the expectations for the validation of safety, effectiveness, and risk mitigation in future *510(k)s* [21]. In essence, SaMD manufacturers are encouraged to leverage the software technology's capability to capture real world performance to understand user interactions with the SaMD, and to conduct on-going monitoring of analytical and technical performance to support future intended users. This is particularly critical for continuous adaptive algorithm [20], in contrast to "locked" algorithm with discrete updates.

2.3 *De-novo* pathway for AI/ML SaMD

FDA released guidance intended to explain the framework for the Pre-Cert Program under the agency's current regulatory authorities [27]. Specifically, the program will be implemented under the *de-novo* pathway, traditionally used for approval of low to moderate risk medical devices without appropriate device designation identified. Under this pathway, new types of low to moderate risk devices can obtain market authorization as a class I or class II device rather than being designated as a class III device, which requires a PMA. Pilot participants with a SaMD product eligible for the *de-novo* classification may have the opportunity to participate in an excellence appraisal wherein FDA intends to evaluate the excellence principals of the organization that correspond to *de novo* request content. The results of this assessment will be collated in a device master file (DMF) to support the *de-novo* request and to support future premarket submissions. The excellence-appraised manufacturer could then participate in a streamlined Pre-Cert *de-novo* request in which the manufacturer submits any additional information not already captured in the DMF to support that the device is safe and effective. After substantive review of the DMF, the FDA would classify the device by written order and for a class II device establish special controls such as post-market data collection, changes in appraisal data, or post-market real-world performance data necessary to assure safety and effectiveness of the device type. The program is in early stages, but its goal is to "determine the contours of a possible regulatory model that provides efficient regulatory oversight of certain SaMD from manufacturers who have demonstrated a robust culture of quality and organizational excellence (CQOE) and are committed to monitoring real-world performance while assuring device safety and efficacy."

2.4 The Pre-Certification (Pre-Cert) Program

In 2017, the FDA established the Pre-Cert Program to streamline the approval process for SaMD, including AI and ML-based medical technologies. This novel program is designed for manufacturers who have demonstrated a robust CQOE and are committed to monitoring real-world performance while maintaining efficient oversight of the SaMD. The program launched in 2017 as part of the Digital Health Innovation Action Plan, and was limited to FDA-regulated SaMD, defined as software intended to be used for one or more medical purposes without being part of a hardware medical device. Instead of the traditional FDA approach for the regulation of hardware-based medical devices, the program focuses on evaluating the manufacturer's capability to response to real-world performance and continue to ensure that consumers have access to safe and effective SaMDs. The FDA used selective metrics such as product quality, patient safety, clinical responsibility, clinical responsibility, cybersecurity responsibility, and proactive culture to identify companies that exhibited organizational excellence. A total of nine out of more than 100 companies were selected as trusted SaMD manufacturers to participate in the program including Apple, Fitbit, Johnson & Johnson, Pear Therapeutics, Phosphorus, Roche, Samsung, Tidepool, and Verily (Fig. 1).

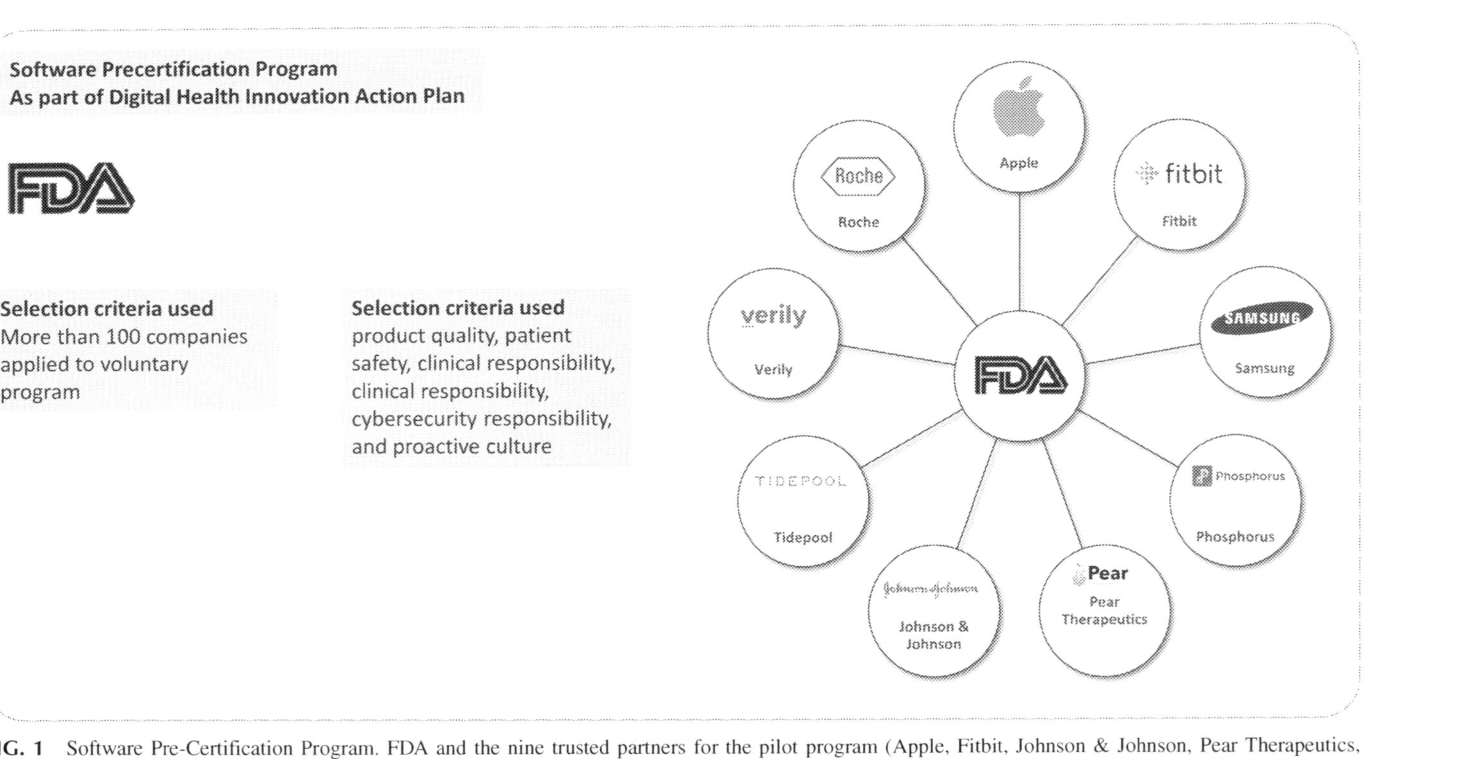

FIG. 1 Software Pre-Certification Program. FDA and the nine trusted partners for the pilot program (Apple, Fitbit, Johnson & Johnson, Pear Therapeutics, Phosphorus, Roche, Samsung, Tidepool, and Verily).

The Good Machine Learning Practice (GMLP) proposed by the FDA details the critical steps to manage TPLC of AI/ML SaMD throughout the model development, production modeling, and device modifications process (Fig. 3). The critical steps include CQOE, Premarket Assurance of Safety and Effectiveness, Review of SaMD Pre-Specification (SPS) and Algorithm Change Protocol (ACP) as well as Real-World Performance Monitoring (Figs. 2 and 3).

(1) CQOE: FDA intends to understand, appraise, and audit the AI model development process at the organizational level from five principles:
  a. patient safety
  b. product quality
  c. clinical responsibility
  d. cybersecurity responsibility
  e. proactive culture.

For instance, the data selection and management (e.g., annotation, preprocessing, missing data, etc.) is a practice that FDA expect organization to establish maturely due to the criticality not only for the model performance, but

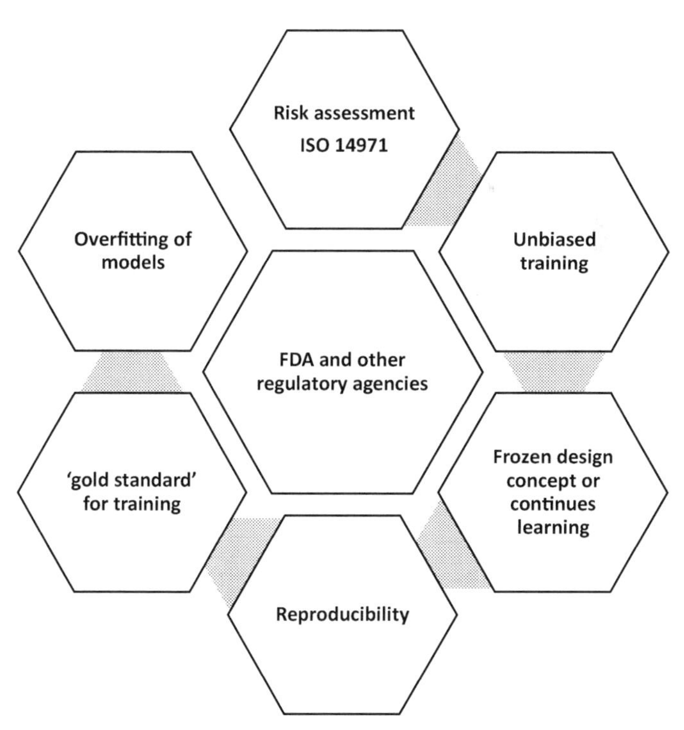

FIG. 2 Key aspects for artificial intelligence (AI) and Machine Learning (ML)-based software in health care. There are many factors that are important in the regulatory framework of AI/ML-based software as medical device (SaMD) that include risk assessment, unbiased training, frozen concept or continues learning, reproducibility a 'golden standard' for training, and overfitting of models.

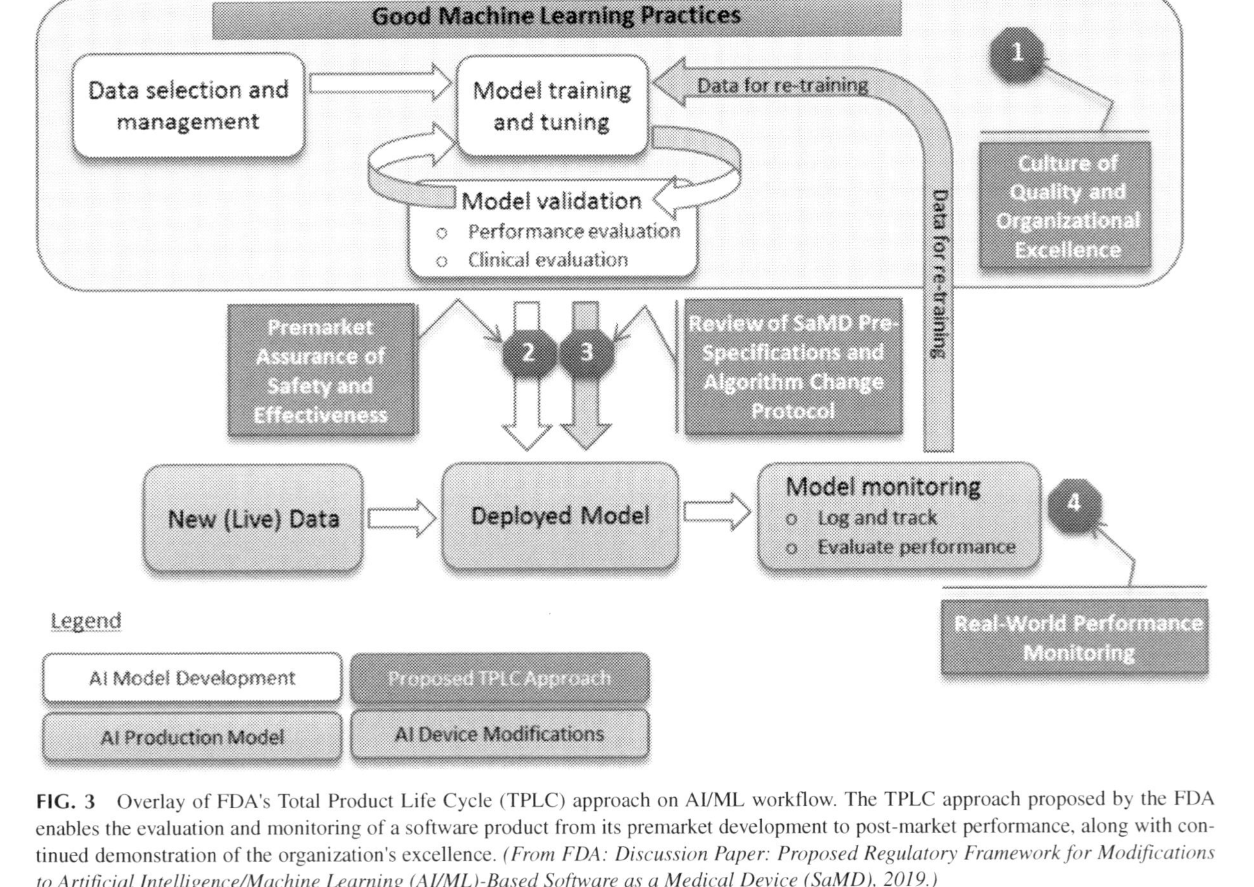

FIG. 3 Overlay of FDA's Total Product Life Cycle (TPLC) approach on AI/ML workflow. The TPLC approach proposed by the FDA enables the evaluation and monitoring of a software product from its premarket development to post-market performance, along with continued demonstration of the organization's excellence. *(From FDA: Discussion Paper: Proposed Regulatory Framework for Modifications to Artificial Intelligence/Machine Learning (AI/ML)-Based Software as a Medical Device (SaMD), 2019.)*

also for many of the principles mentioned above. FDA intends to move from episodic oversight to continuous oversight that enables trust in the organization and a pragmatic check-in with organization and product performance data [26].

(2) Premarket Assurance of Safety and Effectiveness: The traits of AI/ML SaMD such as the iterative, frequent modifications nature, high availability and access to rich real-world data challenge the traditional premarket review process by the FDA. The FDA must exert adequate oversight on SaMDs subjected to frequent updates as a result of ingesting large volume of real world dataset. The manufacturer must establish a pragmatic approach to software updates and demonstrate to the regulatory authority during premarket review to ensure the safety and effectiveness of the intended use after the updates, and to ensure that the updates are interpretable and explainable to the end-user. The scope of premarket assessment suddenly broadens to a whole new level.

(3) Review of SPS and ACP: The "predetermined change control plan" in premarket submissions would include the types of anticipated modifications—referred to as the "SaMD Pre-Specifications"—and the associated methodology being used to implement those changes in a controlled manner that mitigates risks to patients (referred to as the Algorithm Change Protocol or ACP). The concept of "Region of potential changes" around the initial specifications and labeling (e.g., retraining for performance improvement, new data acquisition system, change related to intended use) is particular critical when ACP is being established to explain applied procedures (e.g., data management, re-training, performance evaluation, and update procedures). This concept aims to provide flexibility for frequent iterations to ensure that the performance of products improves over time while avoiding their degradation. One aspect to highlight is that FDA specifically explains that if adaptation is pre-specified, and the methods for determining an appropriate adaptation is clearly delineated, a decision-making framework may be similarly applied for both locked and adaptive algorithms.

(4) Real-World Performance Monitoring: In this approach, the FDA would expect a commitment from manufacturers on transparency and real-world performance monitoring for AI and ML-based SaMD, as well as periodic updates to the FDA on what changes were implemented as part of the approved pre-specifications and the ACP.

The objectives for a tailored regulatory framework from the regulators perspective include **r**easonable **a**ssurance of **s**afety and **e**ffectiveness (RASE), improved time for patients to access high-quality SaMD first in the world, improved submission experience in terms of clarify, predictability, efficiency of review process, and commit to the least burdensome approach.

2.5 Status of those proposed program

The US currently lacks a well-defined AI/ML strategy to address the rapid rise of big data and its impact on digital health technologies. As a result, full adoption of the Software Pre-Cert Program (SPP) appears to be years off, with only traditional FDA approval methods currently available for AI/ML-based health technologies. Other FDA AI/ML-based qualification programs, such as the Medical Device Development Tools program, remain in the nascent stages and have not been widely utilized by AI and ML developers. Thus, today's AI/ML-based SaMD continue to be subjected to lengthy regulatory timelines and risk being outdated by the time they are approved for commercialization. As stated throughout various chapters of the book, AI/ML are poised to revolutionize the field of healthcare. In response, the FDA aimed to refine the Pre-Cert Program with the goal to assess the effectiveness of the program by comparing parallel submissions made to the *de novo* and the traditional route. The elements of a previously-reviewed complete submission will be reviewed retrospectively to iteratively refine the excellence appraisal. Initially, FDA intends to test program components of the model using a mock standard review package. Sponsors who opt-in to the prospective approach will submit traditional and excellence appraisal submission requirements. The FDA's Center for Devices and Radiological Health (CDRH) will review a mock streamlined submission to determine if sufficient information exists to assure reasonable safety and effectiveness. However, an official regulatory decision would be based on traditional regulatory submission. A finding that excellence at the organizational level is correlated with excellence in design, developing and testing a SaMD product would support the use of the excellence appraisal approach as part of the product premarket authorization. Lastly, FDA launched an updated working model that describes the proposed implementation approach and future vision for the program. Despite positive intentions and ambitious plans to accelerate approvals without jeopardizing safety, questions remain. Well-known global companies with substantial product portfolios and a wealth of data will have a significant advantage over small companies and startups. In addition, it is unclear whether FDA will have authority to force a recall on companies and/or their products in the program. Lastly, management of recalls in case of failure to the SaMD might require further refinement. As this chapter was finalized, the FDA published an Action plan in Jan 2021 that provides the background of AI-based medical devices and outlines next steps. Significant focus is put on Good Machine Leanring Practice (GMLP), patient-centered approach, validation of (un)bias, robustness and resilience of algorithms as well as on real-world performance.

2.6 Examples of FDA cleared products using AI in healthcare

It is somewhat controversial when the first AI-based medical device received FDA clearance and which device was the first. It strongly depends on the definition of AI. In 2016 Arterys became the first AI company to receive FDA clearance to use cloud-based deep learning (DL) in a clinical setting (Table 1). The company now has 6 FDA cleared devices, two of which are leveraging AI in clinical applications. This has paved the way for a growing number of examples of FDA cleared products that utilize AI in different ways. One of those examples is the Critical Care Suite, a collection of AI algorithms embedded on a mobile X-ray device for which GE Healthcare received FDA's *510(k)* clearance. Built-in collaboration with UC San Francisco (UCSF), St. Luke's University Health Network, Humber River Hospital, and CARING—Mahajan Imaging—India, using GE Healthcare's Edison platform, the AI algorithms help to reduce the turn-around time it can take for radiologists to review a suspected pneumothorax, a type of collapsed lung. As of January 2021, a total of 71 SaMD leveraging AI have been cleared by the FDA (according to https://medicalfuturist.com/fda-approved-ai-based-algorithms/) (Fig. 4).

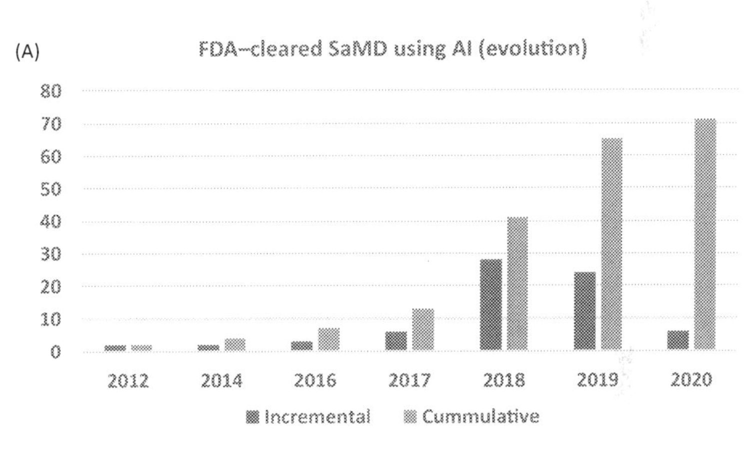

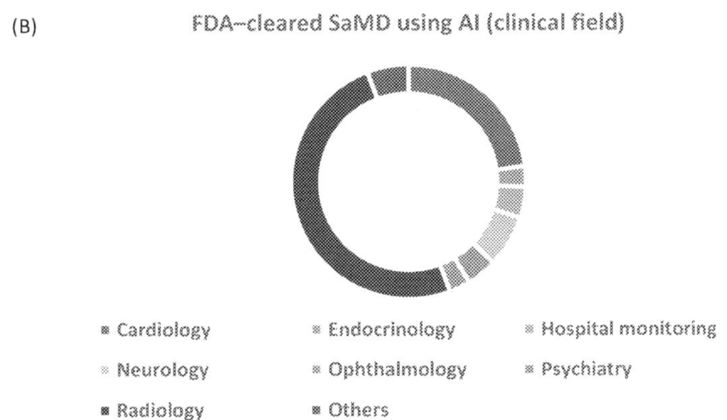

FIG. 4 Figure 4 Food and Drug Administration (FDA) clearance of Software as medical device (SaMD) using artificial intelligence (AI). Panel A. shows the incremental and cumulative FDA approvals of SaMD containing AI. In panel B.) The clinical areas of the devices are indicated.

Continued

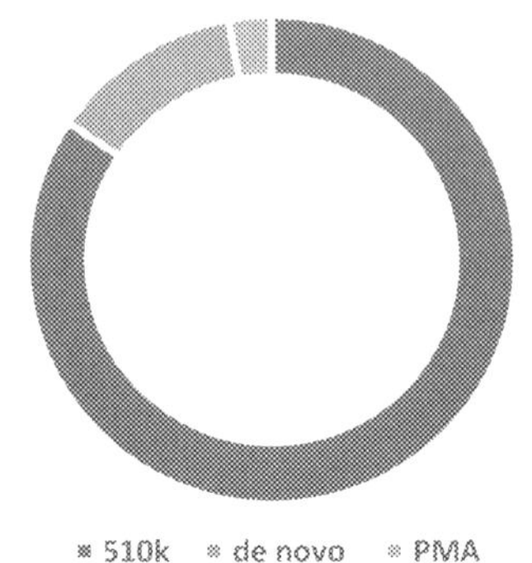

FIG. 4, CONT'D Panel C. illustrates the FDA clearance pathway. Abbreviations: PMA=premarket approval. Data extracted from https://medicalfuturist.com/fda-approved-ai-based-algorithms/.

3 CE mark and in-vitro diagnostic regulation

Until today medical devices are regulated through the Directive for in vitro diagnostic (IVD) which is based on risk assessment of the product and, depending on the product risk, on a self-declaration by the manufacturer. However, this process will change substantially with the new in vitro diagnostic regulation (IVDR) becoming effective May 2022 across all member EU states. The transition from complying with the current IVD Directive to meeting the new IVDR requirements involves significant preparation and data generation to be ready for additional notified body review.

3.1 Software qualification and classification under medical device regulation (MDR)

We are witnessing a drastic regulatory shift as MDR (2017/745/EU) replaces MDD (93/42/EEC) and AIMDD (90/385/EEC), effective May 26th, 2021. As a result, some SaMD companies who already have introduced devices on the market might be subject to more stringent regulatory oversight on their device going forward. Additionally, new startup SaMD companies will have to plan for the increased regulatory and quality burden in the EU.

SaMD that meets the definition of a medical device shall now be considered an active medical device [Article 2, Definition (4), of 2017/745/EU]. Under Rule 12 in MDD (called the "fall back rule" by industry), most SaMDs used to be classified as Class I and it was less complicated to demonstrate compliance for this category. However, due to the added provision of Rule 11 of MDR, manufacturers will have to approach device classification process differently and some devices are now subjected to CE Mark certification issued by a Notified Body (NB).

3.2 In-vitro diagnostic regulation (IVDR)

The European Union's new Medical Devices Regulation (MDR) and IVDR, which introduces new legislation, forces device manufacturers to modify product development, data reporting and quality assurance leading to significantly increased costs and timelines for developing new products, as well as costly new clinical monitoring and evidence generation to recertify many existing devices. The regulations will profoundly affect business models of all medical device companies within the world's second largest device market. The main challenges can be summarized as follows:

1. Reclassified/up-classified devices and scope expansion

   Under IVDR, new devices as well as products already on the market will be reclassified. In addition, the focus will shift from product approval to the entire product lifecycle, requiring greater clinical evaluation before approval.

2. Increased clinical testing requirements

   Another challenge is an increase in clinical testing requirements. Due to reclassification of IVDs, manufacturers that have not previously been required to perform clinical testing will have to develop the ability to run comprehensive clinical trials (either internally or via collaborations or third-party vendors). Furthermore, the clinical evidence required for IVDs is more complex than that of many other medical devices. For medical devices, IVDR requires reassessment of clinical data for devices already on the market.

3. Increased demand on notified bodies

   Until now, NBs served as consultants to support manufacturers to meet CE mark requirements. Under the new MDR and IVDR, NBs will enforce regulations, evaluating all medical devices (excluding IVDs) above Class I and IVDs above Class A. This accounts for approximately 90% of all IVDs, up from about the previous 1%, which will likely lead to NB availability limitations, which in turn will delay product approvals and launches. Additionally, with NBs now being required to review a greater volume of data, timelines will be lengthened, increasing overall costs in the device pipeline.

4. Maintaining product claims

   For legacy products that were developed many years ago under older guidelines for evaluations (e.g., CLSI guidelines), it might not be possible to verify the claims in the labeling. Therefore, amendment of claims might be required which likely will trigger resubmission for the FDA if the same part numbers are used for US and CE mark markets.

5. Emphasis on post-market surveillance

   Finally, under the new IVDR and MDR there will be an increased emphasis on post-market surveillance. This includes proactively monitoring device performance for recertification, annual safety updates for higher-risk class devices, and rapid reporting of safety incidents. While the increased pace and frequency of safety and performance reporting may require significant additional resources for manufacturers, these requirements may provide a feedback mechanism during the TPLC wherein potential issues are identified sooner. Addressing them can protect patients and reduce manufacturer liability.

4 National Medical Products Administration—China

In June 2019, the National Medical Products Administration (NMPA, also known as cFDA) issued a technical guideline on AI-aided software. Given the large Chinese market and eagerness from the government to address the shortage of clinicians, the new initiative is believed to position China as the world leader of AI applications. The Guideline is based on the characteristics of DL technology, combined with the intended use, usage scenarios and core functions, focusing on software data quality control, algorithm generalization ability and clinical risk. In terms of software updates, the Guideline clarified that the software version naming rules should cover algorithm-driven and data-driven software updates, and should list all typical scenarios for major software updates. Minor data-driven software updates can be controlled through a quality management system without applying for Modification Registration. In addition, the Guideline highlights the requirements of non-assisted decision-making software, traditional AI/ML software, third-party databases, and mobile and cloud computing.

Since April 2018 when the FDA approved a novel AI/ML-based SaMD to detect certain diabetes-related eye problems, NMPA has been very busy in preparing the regulatory establishments for AI/ML software devices. Medical Device Classification Catalog, effective on August 1, 2018, included AI/ML diagnostics for the first time. NMPA issued the Notices of Soliciting Manufacturers' Opinion on November 19, 2018 for the Registration Guidelines for AI Medical Devices. Opinions are to be collected in the areas of "Workflow Optimization, Data Processing and Assisted Diagnosis."

NMPA issued the Technical Review Guideline on AI/ML-Aided Software (Draft) on February 1, 2019. The draft guideline lists four main considerations for registration

1. Need analysis
2. Data collection
3. Algorithm design
4. Verification and validation

NMPA established the AI/ML Medical Device Standardization Unit on March 14, 2019. The Unit is to be responsible for standardization of terminology, classification, data set quality management, basic common technology, quality management system, product evaluation process, special evaluation methods for AI/ML-based medical devices.

Considering that many manufacturers will go through the Innovation Pathway, NMPA is to publish the Technical Review Guideline on AI/ML-Aided Software for Innovation Approval this year.

Due to shortages of qualified medical staff and the Chinese focus on AI/ML technology, the NMPA has made a number of recommendations regarding the registration process for AI/ML software in digital health landscape, most recently (March 2020) to provide an assessment framework for product registration of COVID-19-related diagnostic software.

Software that assists with curative or diagnostic processes in China generally must be registered as a medical device with the NMPA. China has two kinds of license for AI/ML medical devices:

1. Assistive products, being those which aid healthcare professionals in their provision of care to patients.
2. Diagnostic products, being those that assist with determining the cause of an ailment.

On 05.03.2020, the Center for Medical Device Evaluation of NMPA released (No.8-2020) the assessment framework for the product registration of pneumonia diagnostic software in China. The assessment applies to SaMDs adopting DL technology to evaluate chest CT scans, to assist triage and to diagnose clinical cases affected by the COVID-19 pandemic (Class III device, software security level B).

5 Data privacy considerations around AI/ML-enabled software

SaMDs tend to be more susceptible to cybersecurity vulnerabilities due to their unique technological characteristics (e.g., process large amount of patient data, interface with other medical devices, facilitate data transfer, etc.). The FDA established the Cybersecurity Working Group within CDRH back in 2013 and has published several guidance documents addressing cybersecurity in the premarket and post-market contexts. Manufacturers are encouraged to leverage National Institute for Standard and Technology's (NIST) well-known cybersecurity framework to assess and improve their ability to present, detect, and respond to address post-market cyber risks, particular for the data privacy considerations. Data that is being queried, created, transferred, leveraged in healthcare system is by nature highly confidential, so the security of the end products and the practices used during production, maintenance and deployment are of paramount concern.

5.1 Health Insurance Portability and Accountability Act

The Health Insurance Portability and Accountability Act (HIPAA) Privacy Rule aims to safeguard all protected health information (PHI), or all personally identifiable information (PII) from being used and disclosed without the patient's consent [28,30]. This rule applies to "covered entities," which are health plans, health care clearing houses, and any health care provider who transmits health information in electronic form in connection with transactions [28]. Development and testing of SaMD using data from patients that have consented to its use is permitted, and HIPAA requires that patients be given the right to direct any covered entity to transmit a copy of their medical records to a designated person or entity of the individual's choice [29]. Even though HIPAA allows sharing of de-identified data, this could result in compromises on some of the AI/ML's effectiveness by making it difficult for the program to match data from different databases [28], and potentially removing important information (e.g., demographic, geographic, etc.) that might be relevant to effective diagnosis [30]. The HIPAA Security Rule established national standards around the protection of patient information on storage, accessibility, and transmission by a covered entity [31]. Encryption and HIPAA-compliant databases and cloud services can be used to mitigate risk of a breach, and the risk analysis to PHI and complete risk management processes are required. It is worth noting that Section 2034 of the 21st Century Cures Act provided FDA with the authority to permit consideration in research which can be exempted from the HIPAA Security rule in their regulatory decision making [32].

5.2 European Union's General Data Protection Regulation

The European Union's General Data Protection Regulation (GDPR) implements new protections to increase the control that individuals have over their data, enabling them to demand access to or request deletion of their information. The GDPR has direct privacy implications on how manufacturers use large amount of medical data for medical purposes such as informing the development of device for treating or monitoring a medical condition. The information required to inform AI/ML-driven medical devices will almost inevitably be classified as special category data under GDPR [33] and hence data controller must comply with data processing regulations (see Article 6 and Article 9). Manufacturers have to pay close attention to the user's rights and obligations around the use of automated decision-making and grants individuals the right to object to any decision made about them if that decision is purely based on automated processing. Manufacturers also have to think about how to process pre-GDPR data as re-applying to the data subject's consent is required, which be a heavy burden on manufacturers.

6 Challenges to the manufacturers when AI/ML SaMD evolve

Manufacturers might find it quite challenging to commercial their ML/AI-based SaMD due to lack of guidance from the regulators and limited literature available from previously cleared SaMD. Such manufacturers have the responsibility to convince the regulatory reviewers that the benefit of their SaMD outweigh the risk [34]. In many scenarios this justification can become very complex and challenging to develop. SaMD often runs on general purpose, non-medical purpose computing platforms (e.g., remote server, workstation, mobile platform). Therefore, the data security requirements and implementations to address conformance to regulations (e.g., HIPAA, cybersecurity risks, GDPR) varies from different computing platforms. Hence, the review process might be subject to delay due to much more resource and expertise needed from the regulatory bodies [35]. Similarly, SaMD often interfaces with other medical devices (e.g., receive DICOM images from CT scanner or other application entities) and it is frequently used in combination with other products (e.g., acquisition of cardiac ultrasound image to provide real-time guidance). The scope of review can expand quite rapidly.

Change control associated with different modifications (e.g., performance, inputs, intended use) require more effort to assess and implement. In addition to the internal change control SOPs that have to be in place, the consolidation of audit trail records of software tools involved in change control can be very time-consuming. The product-oriented risk impact analysis should be developed and practiced repeatedly by cross-functional team throughout the development cycle.

The practice of change on training dataset for AI/ML-enabled SaMD can be particularly complicated. When an AI/ML-based algorithm is shown with limited generalizability, the underlying reasons (e.g., small cohorts, non-representative, ground-truth subjectivity) should be addressed by modifying training datasets or model training practices (e.g., multi-site validation cohort; continuous fine-tuning; federated ML; distributed DL). However, the ability of the manufacturer explaining to end user the impact on algorithm performance (e.g., explainability) or the availability of sufficient information for end user to interpret (e.g., interpretability) can be very challenging [35].

In scenarios when metrics output from the algorithm is not sufficient to allow clinical decision (e.g., confidence of the AI/ML in the output that is produced, human-interpretable reasoning for the output), it is important to consider risk controls which allow better interpretation on device performance. Additional metrics to indicate whether the algorithm is making prediction for the right reason (e.g., pixel saliency maps, text descriptions, feature significant ranking) might be able to help in some clinical context but not others [35]. How to prevent unintended consequences due to erroneous clinical information presentation is the key and most regulatory agencies have discussed about different observer type studies to quantify the explainability and interpretability to the end user.

7 Summary

AI/ML SaMDs have significant potential to improve the healthcare system in many facets which regular software in medical devices is unable to achieve [36]. Significant milestones have been achieved providing hope for future expansion of AI/ML in digital health landscape (see Fig. 5). However, some of the challenges which industry and regulators are facing could generate burden

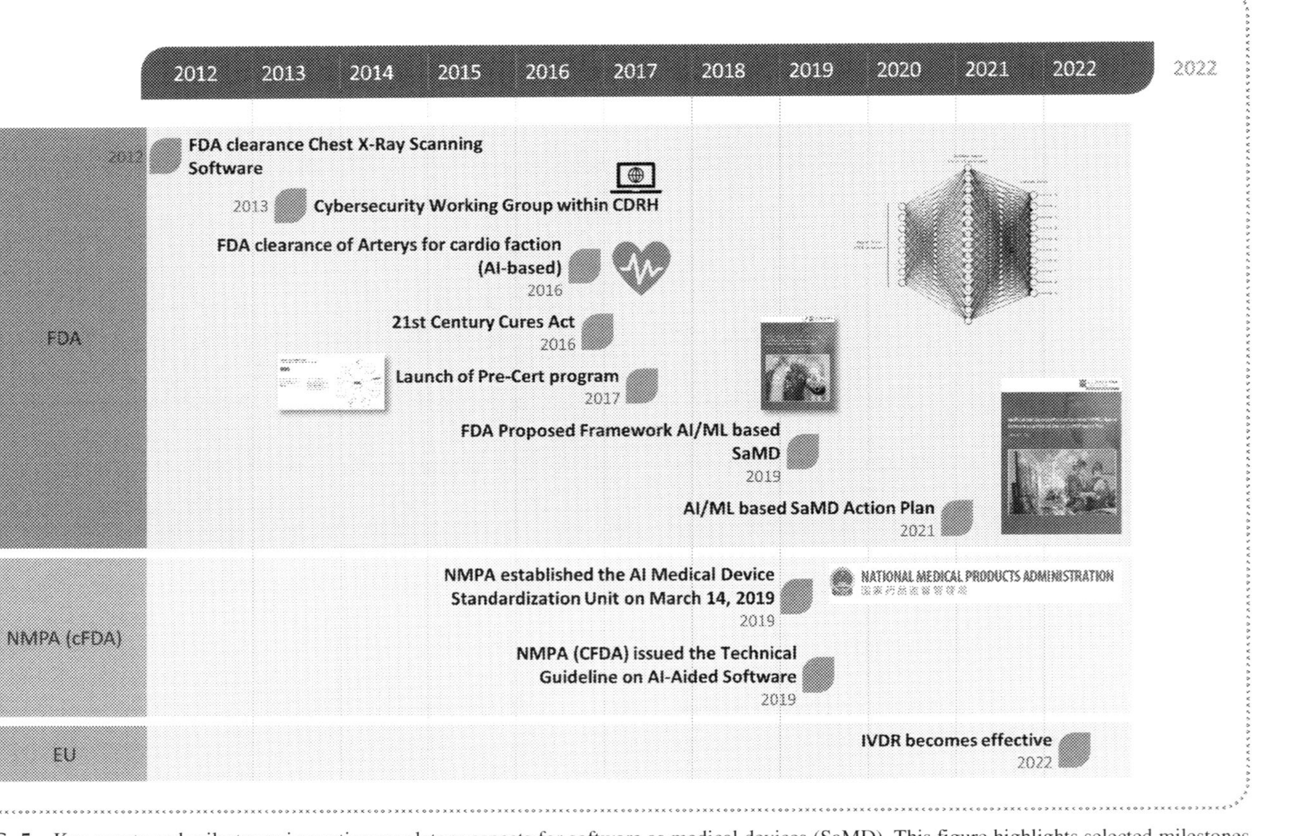

FIG. 5 Key events and milestones impacting regulatory aspects for software as medical devices (SaMD). This figure highlights selected milestones pertaining to regulatory aspects of artificial intelligence in digital health devices. *CDRH*, Center for Devices and Radiological Health.

on both sides [19, 20, 37, 38]. Industry must continue to work with regulators and IMDRF to align regulatory processes and optimize policies with respect to SaMD so that SaMD manufacturers know the regulatory requirements they must meet to commercialize and deploy their product to healthcare system faster [39,40].

References

[1] C.C. Bennett, K. Hauser, Artificial intelligence framework for simulating clinical decision-making: a Markov decision process approach, Artif. Intell. Med. 57 (1) (2013) 9–19.

[2] J. Escudero, E. Ifeachor, J.P. Zajicek, C. Green, J. Shearer, S. Pearson, Machine learning-based method for personalized and cost-effective detection of Alzheimer's disease, IEEE Trans. Biomed. Eng. 60 (1) (2013) 164–168.

[3] A. Torkamani, K.G. Andersen, S.R. Steinhubl, E.J. Topol, High-definition medicine, Cell 170 (5) (2017) 828–843.

[4] M.J. Fritzler, L. Martinez-Prat, M.Y. Choi, M. Mahler, The utilization of autoantibodies in approaches to precision health, Front. Immunol. 9 (2018) 2682.

[5] D. Dana, S.V. Gadhiya, L.G. St Surin, D. Li, F. Naaz, Q. Ali, L. Paka, M.A. Yamin, M. Narayan, I.D. Goldberg, et al., Deep learning in drug discovery and medicine; scratching the surface, Molecules 23 (9) (2018) 2384.

[6] D. Grapov, J. Fahrmann, K. Wanichthanarak, S. Khoomrung, Rise of deep learning for genomic, proteomic, and metabolomic data integration in precision medicine, OMICS 22 (10) (2018) 630–636.

[7] C. Wang, P. Xu, L. Zhang, J. Huang, K. Zhu, C. Luo, Current strategies and applications for precision drug design, Front. Pharmacol. 9 (2018) 787.

[8] K.W. Johnson, S.J. Torres, B.S. Glicksberg, K. Shameer, R. Miotto, M. Ali, E. Ashley, J.T. Dudley, Artificial intelligence in cardiology, J. Am. Coll. Cardiol. 71 (23) (2018) 2668–2679.

[9] M.L. Giger, Machine learning in medical imaging, J. Am. Coll. Radiol. 15 (3 Pt B) (2018) 512–520.

[10] A.M. Williams, Y. Liu, K.R. Regner, F. Jotterand, P. Liu, M. Liang, Artificial intelligence, physiological genomics, and precision medicine, Physiol. Genomics 50 (4) (2018) 237–243.

[11] D. Gruson, T. Helleputte, P. Rousseau, D. Gruson, Data science, artificial intelligence, and machine learning: opportunities for laboratory medicine and the value of positive regulation, Clin. Biochem. 69 (2019) 1–7.

[12] A.A. Seyhan, C. Carini, Are innovation and new technologies in precision medicine paving a new era in patients centric care? J. Transl. Med. 17 (1) (2019) 114.

[13] H. Hampel, A. Vergallo, G. Perry, S. Lista, The Alzheimer precision medicine initiative, J. Alzheimers Dis. 68 (1) (2019) 1–24.

[14] N. Noorbakhsh-Sabet, R. Zand, Y. Zhang, V. Abedi, Artificial intelligence transforms the future of health care, Am. J. Med. 132 (7) (2019) 795–801.

[15] D. Nie, J. Lu, H. Zhang, E. Adeli, J. Wang, Z. Yu, L. Liu, Q. Wang, J. Wu, D. Shen, Multi-channel 3D deep feature learning for survival time prediction of brain tumor patients using multi-modal neuroimages, Sci. Rep. 9 (1) (2019) 1103.

[16] M.B.M.A. Rashid, E.K. Chow, Artificial intelligence-driven designer drug combinations: from drug development to personalized medicine, SLAS Technol. 24 (1) (2019) 124–125.

[17] J. He, S.L. Baxter, J. Xu, J. Xu, X. Zhou, K. Zhang, The practical implementation of artificial intelligence technologies in medicine, Nat. Med. 25 (1) (2019) 30–36.

[18] E.J. Topol, High-performance medicine: the convergence of human and artificial intelligence, Nat. Med. 25 (1) (2019) 44–56.
[19] C.J. Kelly, A. Karthikesalingam, M. Suleyman, G. Corrado, D. King, Key challenges for delivering clinical impact with artificial intelligence, BMC Med. 17 (1) (2019) 195.
[20] D.S. Bitterman, H. Aerts, R.H. Mak, Approaching autonomy in medical artificial intelligence, Lancet Digit. Health 2 (9) (2020) E447–E449.
[21] FDA, Public Workshop: Evolving Role of Artificial Intelligence in Radiological Imaging, 2019.
[22] FDA, Clinical Decision Support Software, Draft Guidance for Industry and Food and Drug Administration Staff, 2019. https://www.fda.gov/media/109618/download. (FDA-2017-D-6569).
[23] Deciding When to Submit a 510(k) for a Software Change to an Existing Device. Guidance for Industry and Food and Drug Administration Staff Document issued on October 25, 2017, 2017. The draft of this document was issued on August 8, 2016 https://www.fda.gov/regulatory-information/search-fda-guidance-documents/deciding-when-submit-510k-software-change-existing-device. (FDA-2016-D-2021).
[24] International Medical Device Regulators Forum: Software as Medical Device (SaMD) Guidances, 2016.
[25] Forum IMDR: SaMD N12: Software as a Medical Device (SaMD): Possible Framework for Risk Categorization and Corresponding Considerations—2014, IMDRF/SaMD WG/N12Final, 2014. http://www.imdrf.org/docs/imdrf/final/technical/imdrf-tech-140918-samd-framework-riskcategorization-141013.pdf.
[26] FDA, Discussion Paper: Proposed Regulatory Framework for Modifications to Artificial Intelligence/Machine Learning (AI/ML)-Based Software as a Medical Device (SaMD), 2019.
[27] FDA, Software Precertification Program: Regulatory Framework, 2019. https://www.fda.gov/media/119724/download.
[28] "Summary of the HIPAA privacy rule." OCR Privacy Brief, HHS, 2003. https://www.hhs.gov/sites/default/files/privacysummary.pdf?language=es.
[29] Individuals' right under HIPAA to access their health information 45 CFR § 164.524, Health Information Privacy. HHS, 2016. https://www.hhs.gov/hipaa/for-professionals/privacy/guidance/access/index.html.
[30] S. Alder, De-identification of protected health information: how to anonymize PHI, HIPAA J. (2017). https://www.hipaajournal.com/de-identification-protected-health-information/.
[31] The Security Rule. Health Information Privacy, HHS, 2017. https://www.hhs.gov/hipaa/for-professionals/security/index.html.
[32] FDA, IRB Waiver or Alteration of Informed Consent for Clinical Investigations Involving No More Than Minimal Risk to Human Subjects, U.S. Department of Health and Human Services, 2017. https://www.fda.gov/downloads/RegulatoryInformation/Guidances/UCM566948.pdf.
[33] T. Reuters, Glossary, Special Category Data. https://uk.practicallaw.thomsonreuters.com/2-200-3468?transitionType=Default&contextData=(sc.Default)&firstPage=true&bhcp=1(2-200-3468.
[34] FDA, Factors to Consider When Making Benefit-Risk Determinations in Medical Device Premarket Approval and De Novo Classifications, 2019. https://www.fda.gov/regulatory-information/search-fda-guidance-documents/factors-consider-when-making-benefit-risk-determinations-medical-device-premarket-approval-and-de(FDA-2011-D-0577.
[35] Duke: Current State and Near-Term Priorities for AI-Enabled Diagnostic Support Software in Health Care, Margolis Center for Health Policy, 2019. https://healthpolicy.duke.edu/sites/default/files/2019-11/dukemargolisaienableddxss.pdf.
[36] I. El Naqa, M.A. Haider, M.L. Giger, R.K. Ten Haken, Artificial intelligence: reshaping the practice of radiological sciences in the 21st century, Br. J. Radiol. 93 (1106) (2020) 20190855.

[37] Q. Zeng-Treitler, S.J. Nelson, Will artificial intelligence translate big data into improved medical care or be a source of confusing intrusion? A discussion between a (cautious) physician Informatician and an (optimistic) medical informatics researcher, J. Med. Internet Res. 21 (11) (2019) e16272.
[38] X. Frank, Is Watson for oncology per se unreasonably dangerous?: making a case for how to prove products liability based on a flawed artificial intelligence design, Am. J. Law Med. 45 (2–3) (2019) 273–294.
[39] H.B. Harvey, V. Gowda, How the FDA regulates AI, Acad. Radiol. 27 (1) (2020) 58–61.
[40] F. Pesapane, C. Volonte, M. Codari, F. Sardanelli, Artificial intelligence as a medical device in radiology: ethical and regulatory issues in Europe and the United States, Insights Imaging 9 (5) (2018) 745–753.

Chapter 12

Precision medicine from the patient's perspective: More opportunities and increasing responsibilities

Kim MacMartin-Moglia and Michael Mahler
Inova Diagnostics, Inc., San Diego, CA, United States

Abbreviations

AI artificial intelligence
IBD inflammatory bowel disease
IoT Internet of things
PCP primary care physician
PM precision medicine
POCT point of care testing
RA rheumatoid arthritis
SLE systemic lupus erythematosus

1. Introduction

Precision medicine (PM) requires several stakeholders to function and to excel, including patients / consumers, health care providers, biopharmaceutical and, diagnostic companies, academic researchers, IT/informatic companies, patient advocacy groups and payers [1]. Although PM opens new opportunities for patients with rare conditions (such as autoimmune diseases), it might also put more burden on the individual patient to become familiarized with the novel approaches that PM can offer [1]. This chapter provides a high level perspective of some of the key aspects for patients in light of PM with special focus on autoimmune diseases.

2. Challenges of autoimmunity

Autoimmune diseases are heterogeneous conditions that (in most cases) slowly evolve from the preclinical (also referred to as the natural history of autoimmunity) into the clinical phase [2–5]. Due to the non-specific symptoms during the early

Precision Medicine and Artificial Intelligence. https://doi.org/10.1016/B978-0-12-820239-5.00001-2

clinical phase, a diagnosis can take many years [6]. During this time, opportunities to stall the disease and prevent damage are missed. This is where PM can utilize patient collected data paired with algorithms to potentially suggest an earlier diagnosis to clinicians, or simply triage patients to the right care giver.

3. Technology and telemedicine

One important component of PM is telemedicine which holds promise to streamline processes around chronic conditions [7–18]. Although human face to face interactions and in-person clinician assessment are difficult to replace, telemedicine using Skype and more advanced systems can often complement or enhance patient management [19]. In July 2020, ACR published a telemedicine position statement reiterating the importance of in-person visits, but also advocating that the value of telerheumatology lies in its ability to reach patients with mobility issues and/or lack of local specialists. ACR intends to study the effects of telemedicine on the outcomes of less frequent in person visits and to determine what the best use of in-person visits are and how best to remotely monitor disease activity. Telerheumatology is believed to provide several benefits to individuals suffering from autoimmune rheumatic diseases [20–33].

Since many novel PM approaches are leveraging technologies [e.g., Internet of things (IoT), smart devices, web-based applications) to collect new and more frequent data on patients, those with higher affinity to embrace novel technologies will have a significant advantage. Especially technologies such as IoT [34–37], smart watches or rings [35, 38–40], potentially smart implants and also voice recognition systems like Amazon Alexa could provide new insights and guidance [41–43]. Unfortunately, this might provide a challenge for elderly or less affluent individuals that are not able to keep up with technological advancements. In this regard, the support of family members and patient advocacy groups might play an important role. Another, although lessening, concern regarding this promising approach is the access to technology such as smartphones (81% of all US adult consumers own smart phones and 21% use a smart watch or fitness tracker regularly) which can be driven by financial factors (only 71% of consumers who make less than $30K/year own a smartphone and 12% use a smart watch or fitness tracker regularly) and age (only 54% of Consumers over the age of 65 own smartphones and 17% use a smart watch or fitness tracker regularly) [44–49]. In some instances [50], health care providers are providing smart watches to their patients who commit to improving their health. As one example, Aetna Healthcare recently teamed up with Apple to launch a new app and wellness program for their patients. As an incentive for participating in the program which tracks their daily activity levels and recommends other measures they can do to stay healthy, Aetna members can earn an Apple Watch or gift cards by meeting their activity goals or other health challenges. Currently, we are already experiencing the use of mobile applications deployed on mobile devices to capture disease activity in rheumatoid arthritis (RA) [17, 51], systemic lupus erythematosus

(SLE) [43], and other conditions. Due to the fragmentation of the market with an overabundance (there are over 300,000 health apps available today) of many, non-regulated, less well-validated applications, adaptation has been limited [52]. However, this will likely change quickly driven by large capital investments by digital health companies (over $9 billion in 2018) including several companies new to healthcare, but not new to technology or consumer behavior, including: Amazon, Google, Microsoft, and Apple [53].

4. Direct to consumer information and marketing

Direct-to-consumer medicine [54, 55] is believed to increase efficiency and decrease health care costs. Although not much data is available as to the benefit of insurance coverage for biometric devices, current knowledge seems controversial [56] and should be objectively researched. Direct-to-consumer medicine is a very broad field that includes information offered to patients, diagnostic tests, and even treatment options. The quality of information and services that are offered to the patient is of key importance regarding the impact on patient care [57, 58]. Several clinical trials are ongoing to study the effectiveness of mobile phone apps to assess the joint function [59].

Point of care testing (POCT) is not novel and has made major impact in several medical fields. However, due to the complexity of autoimmune diseases, simple POCT approaches might not be applicable to autoimmunity. Although POCT have been developed for some autoimmune diseases such as RA [60, 61] or inflammatory bowel disease (IBD) [62], they have not made much inroads. Rather than home testing devices, home collection [63] and central testing might represent a future trend in terms of direct to consumer diagnostics. According to this concept, marketing is directed toward the patient who orders online a collection device that is delivered to the patient. After sample collection, the device is shipped to a central laboratory for analysis. In addition, this concept could be applied for monitoring of disease (activity).

5. The role of the general practitioner/primary care

Depending on the health care system, primary care physicians (PCP) or general practitioners might play an increasingly relevant role in management of patients with chronic conditions and especially in the area of disease prevention. Therefore, educational programs for PCPs might help to keep their knowledge current about the evolution of PM. In addition, tools such as mobile apps might help to keep PCPs engaged in the care of autoimmune patients [64]. This can be difficult as PCPs see such a wide range of conditions on a daily basis and are often already overworked. To the extent that PM applications can help them be more efficient with their time, they will be more likely to adopt these into their clinical practice. This in turn might indicate that the selection of a PCP becomes more and more important for patients (Figs. 1 and 2).

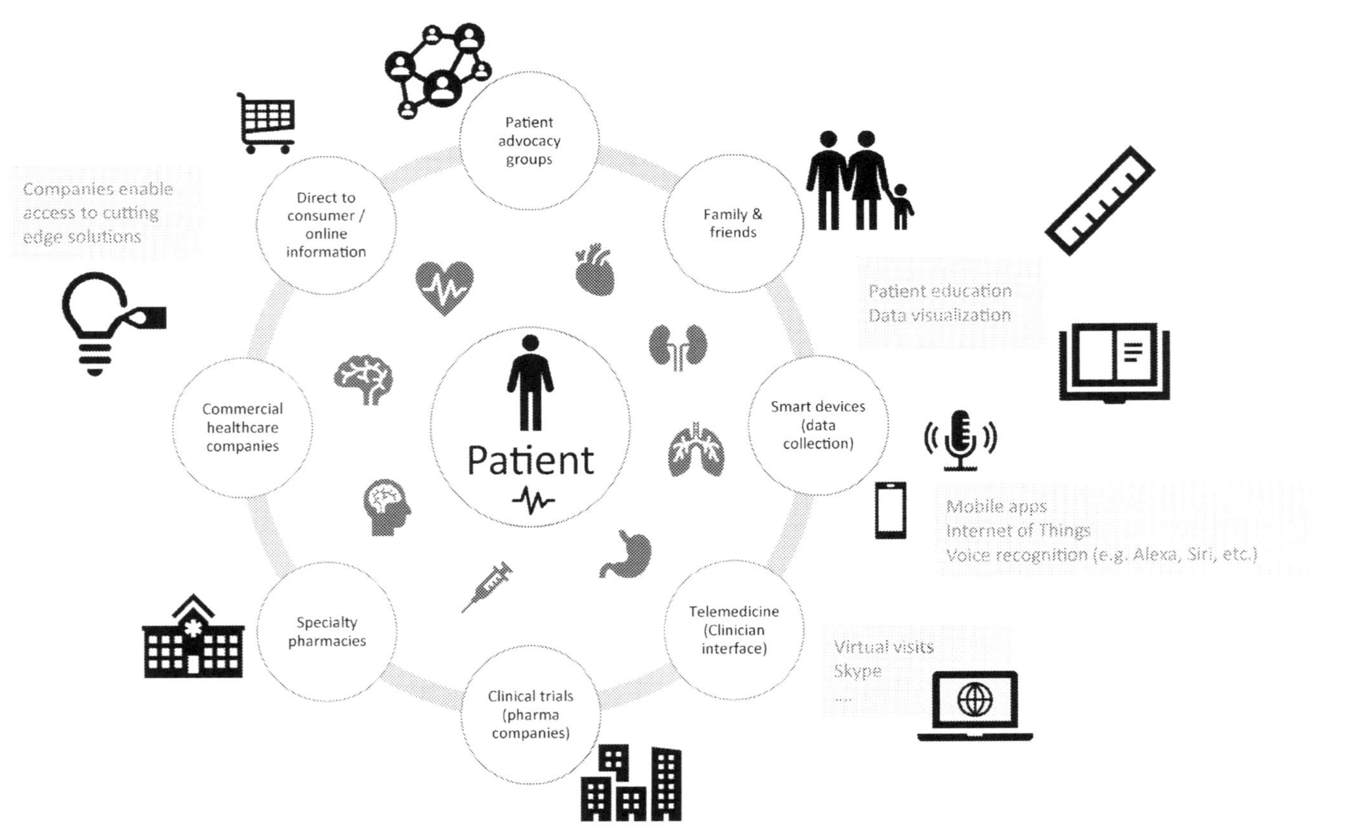

FIG. 1 Schematic overview of the role of the patient in precision medicine (PM). The patient will likely take a central position in future PM concepts. Other components of such initiatives include specialty pharmacies, pharmaceutical companies, telemedicine, smart devices, family and friends, patient advocacy groups, direct to consumer channels, and online information as well as commercial healthcare companies.

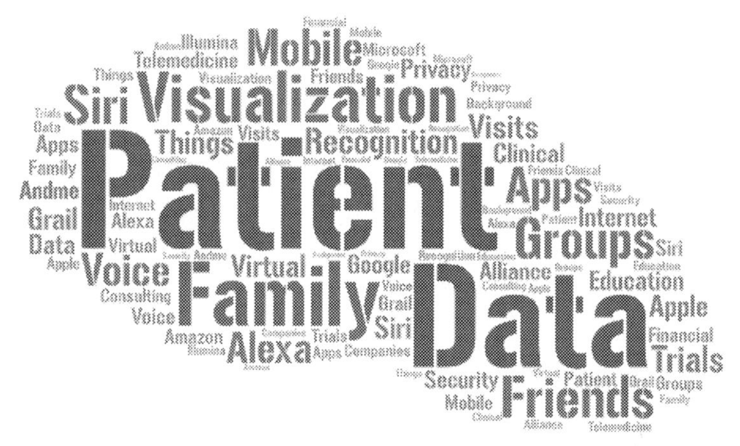

FIG. 2 Word cloud to visualize concepts that impact patients in precision medicine approaches (PM). The graphic was created using https://wordart.com/create.

6. Education and communication

Understanding complex data and information related to PM (e.g., genetic information) can be challenging, even overwhelming, and will require focus on tailored communication for patients at all education levels. This will likely include carefully vetted verbiage as well as visualization of complex data [65]. One example of PM in autoimmunity includes the prediction and prevention in individuals at risk to develop future disease (e.g., family members) [2, 4, 5, 66–70]. Explaining to individuals at risk for future development of an autoimmune disease (e.g., family members of patients) about the risk level as well as outlining treatment options has been shown to be critical for willingness to participate in disease prevention [5, 71, 72]. Details are described in a different chapter of this book. Another example in which patients will require tailored communication for PM will be through the specialty pharmacy [73]. Specialty pharmacists will be utilized as a check point to ensure that patients receive the proper selection and dosing of medications (Table 1).

7. Patient consulting organizations

Several organizations have been created and are emerging to accelerate access of patients to PM which includes patient registries [74], patient advocacy groups as well as commercial entities. Patient advocacy groups typically receive information about new opportunities to manage chronic conditions in autoimmune diseases. Therefore, having close contact with patient advocacy groups might provide valuable insights into opportunities for new treatment options/modalities. Patient advocacy groups can be disease focussed (e.g., My RA Team) or broad (e.g., The American Autoimmune Related Diseases Association, AARDA). Even social media sites are utilized to connect patients

TABLE 1 Selected mobile health applications for autoimmune diseases.

	Goals of application	Disease area
MyVectra	Help to track patient's disease activity levels and symptoms	RA
AthritisPower	Help to track patient's disease and symptoms	RA
TRACK + REACT3	Help to track patient's disease, disease activity, and symptoms	RA
Cliexa-RA	Help to track patient's disease, medication adherence, and symptoms	RA
RheumaHelper	Disease activity calculator	RA
MyRAteam	Social network and support for patients	RA
Lupie Diary	Help to track patient's disease, medication adherence and symptoms, plan and keep medical appointments, prescriptions, contact information	SLE
LupusMinder	The app enables a patient to easily access and review their history with their provider during valuable appointment time. Users can take pictures of their physical symptoms to show their physician at an upcoming appointment, as well as print and save their daily logs as PDF files	SLE
VALUE	VALUE (VAlidation in LUpus of an Electronic Patient Reported Outcomes Tool); Apple ResearchKit platform Monitoring and reporting symptoms https://www.lupusresearch.org/new-smartphone-app-may-enable-lupus-patients-report-symptoms-easily-accurately/	SLE
Flaredown	Flaredown has the ability to share patient's data with scientists and researchers (anonymously) to fuel research	Several

RA, rheumatoid arthritis; *SLE*, systemic lupus erythematosus.

(e.g., Facebook, MeetUp). A summary of some selected examples is presented in Table 2.

The business models of the commercial entities are diverse but are mostly based on generating value by bridging the gap between patients with chronic conditions and pharma companies. One of the largest commercial PM companies, 23andMe, a direct to consumer genetic testing company, which for a nominal cost provides patients (over 10 million people have utilized this service) with information regarding their genetic risk factors for heritable diseases.

[26] T.U. Nguyen-Oghalai, K. Hunter, M. Lyon, Telerheumatology: the VA experience, South. Med. J. 111 (6) (2018) 359–362.
[27] R.A. Matsumoto, et al., Rheumatology clinicians' perceptions of telerheumatology within the veterans health administration: a National Survey Study, Mil. Med. (2020), https://doi.org/10.1093/milmed/usaa203.
[28] A. Akpabio, R.O. Akintayo, U. Effiong, Can telerheumatology improve rheumatic and musculoskeletal disease service delivery in sub-Saharan Africa? Ann. Rheum. Dis. (2020), https://doi.org/10.1136/annrheumdis-2020-218449.
[29] F. Caso, et al., Improving telemedicine and in-person management of rheumatic and autoimmune diseases, during and after COVID-19 pandemic outbreak. Definite need for more Rheumatologists. Response to: 'Can telerheumatology improve rheumatic and musculoskeletal disease service delivery in sub-Saharan Africa?' by Akpabio et al, Ann. Rheum. Dis. 0 (2020) 1–2.
[30] G. Figueroa-Parra, C.M. Gamboa-Alonso, D.A. Galarza-Delgado, Challenges and opportunities in telerheumatology in the COVID-19 era. Response to: 'Online management of rheumatoid arthritis during COVID-19 pandemic' by Zhang et al, Ann. Rheum. Dis. 0 (2020) 1, https://doi.org/10.1136/annrheumdis-2020-217631.
[31] L. Costa, et al., Telerheumatology in COVID-19 era: a study from a psoriatic arthritis cohort, Ann. Rheum. Dis. (2020), https://doi.org/10.1136/annrheumdis-2020-217806.
[32] L. Gupta, et al., Response to: 'Telerheumatology in COVID-19 era: a study from a psoriatic arthritis cohort' by Costa et al, Ann. Rheum. Dis. 0 (2020) 1–2, https://doi.org/10.1136/annrheumdis-2020-217953.
[33] J. Geuens, et al., Mobile health features supporting self-management behavior in patients with chronic arthritis: mixed-methods approach on patient preferences, JMIR Mhealth Uhealth 7 (3) (2019) e12535.
[34] Y. Luo, W. Li, S. Qiu, Anomaly detection based latency-aware energy consumption optimization for IoT data-flow services, Sensors (Basel) 20 (1) (2019) 122.
[35] E. Jovanov, Wearables meet IoT: synergistic personal area networks (SPANs), Sensors (Basel) 19 (19) (2019) 4295.
[36] G. Kyriakopoulos, et al., Internet of things (IoT)-enabled elderly fall verification, exploiting temporal inference models in smart homes, Int. J. Environ. Res. Public Health 17 (2) (2020) 408.
[37] S. Gupta, et al., Radiology, mobile devices, and internet of things (IoT), J. Digit. Imaging 33 (3) (2020) 735–746.
[38] G.T. Sharrer, Personalized medicine: ethics for clinical trials, Methods Mol. Biol. 823 (2012) 35–48.
[39] D. Witt, et al., Windows into human health through wearables data analytics, Curr. Opin. Biomed. Eng. 9 (2019) 28–46.
[40] W.N. Price 2nd, I.G. Cohen, Privacy in the age of medical big data, Nat. Med. 25 (1) (2019) 37–43.
[41] J.N. Weinstein, Artificial intelligence: have you met your new friends; Siri, Cortona, Alexa, Dot, Spot, and Puck, Spine (Phila Pa 1976) 44 (1) (2019) 1–4.
[42] H. Tang, J.H. Ng, Googling for a diagnosis—use of Google as a diagnostic aid: internet based study, BMJ 333 (7579) (2006) 1143–1145.
[43] L.O. Dantas, et al., Mobile health technologies for the management of systemic lupus erythematosus: a systematic review, Lupus 29 (2) (2020) 144–156.
[44] N. Vangeepuram, et al., Smartphone Ownership and Perspectives on Health Apps Among a Vulnerable Population in East Harlem, vol. 4, Mhealth, New York, 2018, p. 31.

[45] A. Singh, S. Wilkinson, S. Braganza, Smartphones and pediatric apps to mobilize the medical home, J. Pediatr. 165 (3) (2014) 606–610.
[46] J.D. Piette, et al., Patient-centered pain care using artificial intelligence and mobile health tools: protocol for a randomized study funded by the US Department of Veterans Affairs Health Services Research and Development Program, JMIR Res. Protoc. 5 (2) (2016) e53.
[47] K. Singh, et al., Patient-facing mobile apps to treat high-need, high-cost populations: a scoping review, JMIR Mhealth Uhealth 4 (4) (2016) e136.
[48] K. Singh, et al., Many mobile health apps target high-need, high-cost populations, but gaps remain, Health Aff. (Millwood) 35 (12) (2016) 2310–2318.
[49] Mobile Fact Sheet, 2019. https://www.pewresearch.org/internet/fact-sheet/mobile/.
[50] B. King, Two Major Health Insurance Companies Now Offer Wellness Programs Featuring Free Apple Watches, Phillyvoice, 2019. https://www.phillyvoice.com/aetna-united-health-care-health-insurance-free-apple-watch-wellness-programs-fitness-activity-trackers/.
[51] D. Luo, et al., Mobile apps for individuals with rheumatoid arthritis: a systematic review, J. Clin. Rheumatol. 25 (3) (2019) 133–141.
[52] W.J. Gordon, et al., Beyond validation: getting health apps into clinical practice, NPJ Digit. Med. 3 (1) (2020) 14.
[53] We Visualize Where Leading Tech Companies Are Placing Strategic Bets on Healthcare Startups Across Clinical Research, Genetics, Drug Delivery, and More, CB Insights, 2019. https://www.cbinsights.com/research/tech-giants-digital-healthcare-investments/#:~:text=The%20most%20active%20tech%20giants,investment%20focuses%20and%20strategies%20vary.
[54] L. Walsh, et al., Harnessing and supporting consumer involvement in the development and implementation of models of care for musculoskeletal health, Best Pract. Res. Clin. Rheumatol. 30 (3) (2016) 420–444.
[55] K. Walsh, Artificial intelligence and healthcare professional education: superhuman resources for health? Postgrad. Med. J. 96 (1133) (2020) 121–122.
[56] A. Jo, et al., Is there a benefit to patients using wearable devices such as fitbit or health apps on mobiles? A systematic review, Am. J. Med. 132 (12) (2019) P1394–1400.E1.
[57] M. Reynolds, A. Hoi, R.R.C. Buchanan, Assessing the quality, reliability and readability of online health information regarding systemic lupus erythematosus, Lupus 27 (12) (2018) 1911–1917.
[58] C. Cutler, et al., A North American perspective of content and quality of websites in the English language on childhood-onset lupus erythematosus, Lupus 27 (5) (2018) 762–770.
[59] V. Hamy, et al., Developing smartphone-based objective assessments of physical function in rheumatoid arthritis patients: the PARADE study, Digit. Biomark. 4 (1) (2020) 26–43.
[60] F. Renger, et al., Immediate determination of ACPA and rheumatoid factor—a novel point of care test for detection of anti-MCV antibodies and rheumatoid factor using a lateral-flow immunoassay, Arthritis Res. Ther. 12 (3) (2010) R120.
[61] G. Zandman-Goddard, et al., A novel bedside test for ACPA: the CCPoint test is moving the laboratory to the rheumatologist's office, Immunol. Res. 65 (1) (2017) 363–368.
[62] A. Heida, et al., Agreement between home-based measurement of stool calprotectin and elisa results for monitoring Inflammatory Bowel Disease Activity, Clin. Gastroenterol. Hepatol. 15 (11) (2017) 1742–1749.e2.
[63] T.G. Liou, et al., Prospective multicenter randomized patient recruitment and sample collection to enable future measurements of sputum biomarkers of inflammation in an observational study of cystic fibrosis, BMC Med. Res. Methodol. 19 (1) (2019) 88.
[64] J. Ayre, et al., Factors for supporting primary care physician engagement with patient apps for type 2 diabetes self-management that link to primary care: interview study, JMIR Mhealth Uhealth 7 (1) (2019) e11885.

[65] I. Ko, H. Chang, Interactive visualization of healthcare data using tableau, Healthc. Inform. Res. 23 (4) (2017) 349–354.
[66] D. Alpizar-Rodriguez, A. Finckh, Is the prevention of rheumatoid arthritis possible? Clin. Rheumatol. 39 (5) (2020) 1383–1389.
[67] A. Di Matteo, et al., Third-generation anti-cyclic citrullinated peptide antibodies improve prediction of clinical arthritis in individuals at risk of rheumatoid arthritis, Arthritis Rheumatol. 72 (11) (2020) 1820–1828.
[68] L.B. Kelmenson, et al., Timing of elevations of autoantibody isotypes prior to diagnosis of rheumatoid arthritis, Arthritis Rheumatol. 72 (2) (2020) 251–261.
[69] M.K. Demoruelle, Improving the prediction of rheumatoid arthritis using multiple anti-cyclic citrullinated peptide assays, Arthritis Rheumatol. 72 (11) (2020) 1789–1790.
[70] E.A. Bemis, et al., Factors associated with progression to inflammatory arthritis in first-degree relatives of individuals with RA following autoantibody positive screening in a non-clinical setting, Ann. Rheum. Dis. (2020), https://doi.org/10.1136/annrheumdis-2020-217066.
[71] J.A. Sparks, et al., Personalized risk estimator for rheumatoid arthritis (PRE-RA) family study: rationale and design for a randomized controlled trial evaluating rheumatoid arthritis risk education to first-degree relatives, Contemp. Clin. Trials 39 (1) (2014) 145–157.
[72] J.A. Sparks, et al., Disclosure of personalized rheumatoid arthritis risk using genetics, biomarkers, and lifestyle factors to motivate health behavior improvements: a randomized controlled trial, Arthritis Care Res. 70 (6) (2018) 823–833.
[73] S. Schwartz, et al., Precision Medicine and the Pharmacist's Role as a Trusted Counselor in Specialty Pharmacy, 2020. https://www.pharmacytimes.com/news/precision-medicine-and-the-pharmacists-role-as-a-trusted-counselor-in-specialty-pharmacy.
[74] G. Lionetti, et al., Using registries to identify adverse events in rheumatic diseases, Pediatrics 132 (5) (2013) e1384–e1394.

Index

Note: Page numbers followed by *f* indicate figures and *t* indicate tables.

Printed in the United States
by Baker & Taylor Publisher Services